Advances in organic dairy cattle farming

It is widely recognised that agriculture is a significant contributor to global warming and climate change. Agriculture needs to reduce its environmental impact and adapt to current climate change whilst still feeding a growing population, i.e. become more 'climate-smart'. Burleigh Dodds Science Publishing is playing its part in achieving this by bringing together key research on making the production of the world's most important crops and livestock products more sustainable.

Based on extensive research, our publications specifically target the challenge of climate-smart agriculture. In this way we are using 'smart publishing' to help achieve climate-smart agriculture.

Burleigh Dodds Science Publishing is an independent and innovative publisher delivering high quality customer-focused agricultural science content in both print and online formats for the academic and research communities. Our aim is to build a foundation of knowledge on which researchers can build to meet the challenge of climate-smart agriculture.

For more information about Burleigh Dodds Science Publishing simply call us on +44 (0) 1223 839365, email info@bdspublishing.com or alternatively please visit our website at www.bdspublishing.com.

Related titles:

Achieving sustainable production of milk Volume 1
Print (ISBN 978-1-78676-044-9); Online (ISBN 978-1-78676-046-3, 978-1-78676-047-0)

Achieving sustainable production of milk Volume 2
Print (ISBN 978-1-78676-048-7); Online (ISBN 978-1-78676-242-9, 978-1-78676-243-6)

Achieving sustainable production of milk Volume 3
Print (ISBN 978-1-78676-977-0); Online (ISBN 978-1-78676-050-0, 978-1-78676-051-7)

Advances in sustainable dairy cattle nutrition
Print (ISBN 978-1-78676-052-4); Online (ISBN 978-1-78676-054-8, 978-1-78676-055-5)

Improving dairy herd health
Print (ISBN 978-1-78676-467-6); Online (ISBN 978-1-78676-469-0, 978-1-78676-470-6)

Improving grassland and pasture management in temperate agriculture
Print (ISBN 978-1-78676-200-9); Online (ISBN 978-1-78676-202-3, 978-1-78676-203-0)

Improving organic animal farming
Print (ISBN 978-1-78676-180-4); Online (ISBN 978-1-78676-182-8, 978-1-78676-183-5)

Understanding the behaviour and improving the welfare of dairy cattle
Print (ISBN 978-1-78676-459-1); Online (ISBN 978-1-78676-461-4, 978-1-78676-462-1)

Chapters are available individually from our online bookshop:
https://shop.bdspublishing.com

climate•SMART•publishing
IN AGRICULTURAL SCIENCE

BURLEIGH DODDS SERIES IN AGRICULTURAL SCIENCE
NUMBER 153

Advances in organic dairy cattle farming

Edited by Dr Mette Vaarst, Aarhus University, Denmark, Dr Stephen Roderick, Duchy College, UK and Dr Lindsay Whistance, Organic Research Centre, UK

burleigh dodds
SCIENCE PUBLISHING

Published by Burleigh Dodds Science Publishing Limited
82 High Street, Sawston, Cambridge CB22 3HJ, UK
www.bdspublishing.com

Burleigh Dodds Science Publishing, 1518 Walnut Street, Suite 900, Philadelphia, PA 19102-3406, USA

First published 2025 by Burleigh Dodds Science Publishing Limited

Library of Congress Control Number: 2024940646

British Library Cataloguing in Publication Data
A catalogue record for this book is available from the British Library

ISBN 978-1-80146-538-0 (print)
ISBN 978-1-80146-539-7 (ePub)
ISBN 978-1-80146-540-3 (PDF)
ISSN 2059-6936 (print)
ISSN 2059-6944 (online)

DOI: 10.19103/AS.2024.0138

Typeset by Deanta Global Publishing Services, Dublin, Ireland

Contents

Series list

Title	Series number
Achieving sustainable cultivation of maize - Vol 1 From improved varieties to local applications Edited by: Dr Dave Watson, CGIAR Maize Research Program Manager, CIMMYT, Mexico	001
Achieving sustainable cultivation of maize - Vol 2 Cultivation techniques, pest and disease control Edited by: Dr Dave Watson, CGIAR Maize Research Program Manager, CIMMYT, Mexico	002
Achieving sustainable cultivation of rice - Vol 1 Breeding for higher yield and quality Edited by: Prof. Takuji Sasaki, Tokyo University of Agriculture, Japan	003
Achieving sustainable cultivation of rice - Vol 2 Cultivation, pest and disease management Edited by: Prof. Takuji Sasaki, Tokyo University of Agriculture, Japan	004
Achieving sustainable cultivation of wheat - Vol 1 Breeding, quality traits, pests and diseases Edited by: Prof. Peter Langridge, The University of Adelaide, Australia	005
Achieving sustainable cultivation of wheat - Vol 2 Cultivation techniques Edited by: Prof. Peter Langridge, The University of Adelaide, Australia	006
Achieving sustainable cultivation of tomatoes Edited by: Dr Autar Mattoo, USDA-ARS, USA and Prof. Avtar Handa, Purdue University, USA	007
Achieving sustainable production of milk - Vol 1 Milk composition, genetics and breeding Edited by: Dr Nico van Belzen, International Dairy Federation (IDF), Belgium	008
Achieving sustainable production of milk - Vol 2 Safety, quality and sustainability Edited by: Dr Nico van Belzen, International Dairy Federation (IDF), Belgium	009
Achieving sustainable production of milk - Vol 3 Dairy herd management and welfare Edited by: Prof. John Webster, University of Bristol, UK	010
Ensuring safety and quality in the production of beef - Vol 1 Safety Edited by: Prof. Gary Acuff, Texas A&M University, USA and Prof. James Dickson, Iowa State University, USA	011
Ensuring safety and quality in the production of beef - Vol 2 Quality Edited by: Prof. Michael Dikeman, Kansas State University, USA	012
Achieving sustainable production of poultry meat - Vol 1 Safety, quality and sustainability Edited by: Prof. Steven C. Ricke, University of Arkansas, USA	013
Achieving sustainable production of poultry meat - Vol 2 Breeding and nutrition Edited by: Prof. Todd Applegate, University of Georgia, USA	014

Chapter 1

Organic dairy farming in Europe: building on the past to develop the future

Mette Vaarst, Aarhus University, Denmark; Lindsay K. Whistance, Organic Research Centre, UK; and Steve Roderick, Duchy College, UK

1 Introduction: advances in organic farming

Organic farming was first formally defined and established in different context-relevant forms during the past century and is one of the leading alternatives to contemporary industrialised farming and food systems. It has developed into a farming system defined through its own set of standards, which are now embedded within European Union (EU) regulations. EU countries are committed to 25% of their national farmland being cultivated under organic certification by 2030. The global organisation IFOAM - Organics International (also known as the International Federation of Organic Agriculture Movements) defines a basis for organic farming through the four organic principles formulated in 2005:

- Health.
- Ecology.

http://dx.doi.org/10.19103/AS.2024.0138.01

- Fairness.
- Care.

These principles provide robust guiding principles for future development across a wide range of farming and food systems. Whilst standards and certification systems provide the legal framework for implementing these principles within organic farming, it is the founding principles which guide the overall ethos of organic farming in an ever-changing environment. These organic principles also serve as a reference throughout this book, helping to encourage thoughtful interpretations and analyses of current and future advances from a range of perspectives across different topics.

Dairy cattle farming has played a key role in the development of organic principles, regulations and practices. From a farm perspective, cattle, as ruminants, have been important beneficial components on organic farms, eating nitrogen-fixing legumes and providing the farm with manure, contributing to soil life and fertility. Organic milk is an important dietary component for many families, especially children, and is a key product within organic markets. However, unlike vegetables and fruit, that provide good opportunities for close connections between farmers and consumers through short supply chains, milk typically has a different route to consumption through more centralised processing before milk products reach consumers.

Organic farms compete with other farming systems for access to markets. As consumers have now become fully accustomed to accessing products all year round, despite many of these products, including milk, being more seasonal in nature (e.g. given changes in pasture availability), the challenge of increasing availability stretches farm resources and, with this, principles of circularity and nearness when feed is purchased. Organic dairy farmers receive limited support from agriculture-associated industries and research bodies which continue to focus predominantly on aspects of intensive, high-production systems including breeding, feeding and housing systems. The resulting practices and knowledge then have to be made sense of, adapted or discarded in an organic context. This is very much the real world that organic dairy farmers are living in. This means the principles of Health, Ecology, Fairness and Care become increasingly important as a guide for farmers facing the challenges and gaps between the 'real' and the 'ideal' where the aims of organic food production are realised. These principles, and the way they are defined, act as four clear ethical pillars for reviewing and rethinking challenges and existing practices.

The dairy sector generally relies on increasingly large dairy companies and processing sites (processing milk from many farms) with the route from farm to consumer not easily traced. Although there are good examples of on-farm bottling, processing and direct selling, including box schemes and self-service vending, organic milk and meat products in Europe are generally part of either national or globalised food systems. Similarly, organic animal feeds are often traded and imported across continents. This means that the manner in which milk and milk products are marketed, distributed and consumed provides a set of additional challenges over and above those directly concerned with organic farm practices.

2 Future challenges facing organic dairy farming

We envisage the future development of the sector as contributing to solving some of the significant current problems inherent in contemporary dairy farming. In many ways, organic dairy farming has been successful, becoming a well-established and important component of the growing organic farming sector. The market for organic milk has also been growing, and the production of local/regional organic cheeses and milk products has been welcomed by consumers throughout Europe and beyond. Cattle may be considered as the 'engine' for organic development, because of their ability to transform grass into human food and to provide manure as a key source of fertilisation in many organic farming systems.

However, during the past few years, other debates have also influenced dairy farming and the perception of milk and meat from dairy cattle. This has partly been due to the increasing impacts of climate change on our everyday lives, and the growing awareness of the extent of greenhouse gas (GHG) emissions associated with cattle.

The contribution that organic farming plays within the wider social landscape is as important as its role in producing food. The context of European organic dairy farming is a combination of the social, ecological, and agricultural landscapes connected to our food systems. These have changed dramatically over the past 60–70 years, including the growth of the global human population by four billion in the last half of the twentieth century (Bongaarts, 2009), which means that it has increased by approximately one billion people every 12 years since the mid-1900s. An increase in urbanisation in most countries (Kundu and Pandey, 2020) has removed many people away from contact with, and knowledge of, food production. At the same time, consumption patterns in many regions have changed towards higher consumption of more processed and industrialised food, more animal products, resulting in increased sugar, fat and starch levels. This also includes a greater reliance on environmentally damaging products such as palm oil used in a wide range of manufactured products.

Land, nature and resources are being exploited in ways that have increasingly visible impacts (Steffen et al., 2004; Richardson et al., 2023), such as the extremes of weather associated with climate change, biodiversity and species loss, water and soil pollution and the related occurrence of hunger, inequality and conflict (Cook, 2018). Current food systems also create high levels of losses and waste at all stages, from production to consumption, increasing the burden on associated natural systems. When measured at the level of global land use for agriculture, an estimated 28% of food produced is wasted annually (FAO, 2013). These dramatic changes in consumption, the inefficiency of current agri-food systems, and its resulting negative ecological impacts, demand a rethink in the way we produce, distribute and consume our food and fibre as well as the manner in which we connect to our physical and social landscapes.

Organic dairy production has to be seen in this context. Although the current unsustainable model of producing animal-derived products to meet increasing human demand calls for a shift in focus from animal-based to more plant-based food, animal farming – including dairy farming – can still play an important role within our food production systems, albeit with a sharper prioritisation of the needs of individual animals, the ecological health of the landscapes to which dairy cattle can contribute substantially and to satisfy societal expectations concerning food cultures and norms.

Over the past decades, in some parts of the world (including Europe), an increasing interest in local and seasonal food and a more conscious and careful use of resources can be observed. Local food, and access to healthy fresh food, clean water and land, are in the interest of everybody, alongside local food cultures and traditional crops and varieties. The organic movement includes a diverse range of peoples and types of agriculture, and the principle of fairness emphasises equal access and more balanced stewardship of resources such as seeds, breeds, water and ancestral land.

Although perhaps not easily identifiable within European dairy farming, these issues are relevant to dairy farming in the twenty-first century. For example, imported dietary ingredients such as soya are grown on farmland that could otherwise have an alternative agricultural purpose of feeding and sustaining local populations and cultures. This is one of the areas of responsibility where organic dairy farming builds on principles that strongly emphasise that other and more fair, caring and system-oriented solutions must be found.

In addition to addressing some of the negative environmental impacts of dairy farming, particularly by reducing GHG emissions associated with inputs and redressing the feed versus food dilemma (where land is used to grow crops for animal feed rather than for human consumption), the development of organic farming also needs to deal with concerns about animal welfare and the treatment of disease and associated problems of antimicrobial resistance (AMR). Whilst organic principles provide a framework for addressing these significant societal issues, particularly those linked to intensive dairy farming, success is not guaranteed. The development of the sector again needs to ensure that it focuses on key principles, and is not dictated by the systemic issues inherent in our current, and failing, food and farming systems.

3 Organic dairy cattle as robust and resilient animals

The animals themselves, the cows, calves and cattle of all age groups that live on a dairy farm, are living sentient beings. The organic principle of fairness very strongly emphasises our obligations towards animals: *This principle insists that animals should be provided with the conditions and opportunities of life that accord with their physiology, natural behaviour and well-being.* In addition to being individual animals with senses and feelings, they are an important part of farms and the entire ecological, social and cultural context in which our organic dairy farming exists. This also affects the vocabulary we use and develop around these animals, which reflects our views about them. We, therefore, try to insist that we talk about farmed animals rather than 'livestock' and 'calves from dairy farms' rather than 'surplus calves'. This latter term does not seem to give respect to the calves as living sentient beings and may invite a negative attitude to an animal as a commodity, for which we also need to care and take responsibility, just as much as any other living animal born in our care. In this way, we consider the way in which we write and talk about animals and encourage other authors to consider this too.

A key aim of organic farming is to provide systems that enable animals to be robust and resilient, and, in achieving this, the choice of breeds is an important focus and challenge. Whilst organic dairy cattle are kept in various environments, not all of the conditions we find across these diverse situations are suitable for all breeds to thrive. Furthermore, the self-imposed limits on organic farming are such that only certain breeds and types are permitted.

High-yielding animals, bred for conventional systems that rely on year-round housing and precise feeding regimes based on high-concentrate usage, are unlikely to be sufficiently robust to thrive in more extensive, low-input systems, where energy gained from grazing must be balanced with the energy required to forage and return to the parlour for milking. Yet many of these same high-yielding dairy breeds are also found on organic farms. The reasons for this situation can include dairy farmers converting to organic with existing herds which they continue to use after conversion; the economic reality where organic farmers are competing for market share and where costs are high, meaning higher returns per cow remain important; and a lack of appropriate breeds in sufficient numbers, further exacerbated by the decline in breed diversity across most regions of Europe.

To live up to the principles, choosing the right breed of animal for the conditions in which it will live, produce and behave naturally, is a crucial determining factor when considering the type of dairy farms we want for the future, and therefore the type of animals we should include within these systems. During the development of our current milk production systems, the breeding of a specialised animal may be seen as taking us in a direction where animals are no longer viewed as systems partners within an organic farm, but are 'designed' with the single purpose of producing milk. Yet contrary to this development, there is a growing debate as to how we get better quality meat from dairy animals as well as addressing the moral issue of unwanted male calves.

Seen from the point of view of organic principles, diversity is an important part of creating robustness and resilience. In some European countries, such as e.g. Switzerland, Hungary and Portugal, there is a stronger focus on breed diversity among some farmer groups and government programmes (e.g. FOAG (2021) in Switzerland), while the Dutch Cooperative for Organic Cattle Farming (https://bioki.org/) works to increase the breeding of animals specifically suited to organic farming.

In Chapter 2, Kargo and co-authors introduce breeding concepts and perspectives, including breed diversity and how to consider the right animal for the right conditions, including discussing dual-purpose breeds in contrast to specialised breeds. Over decades, there have been discussions about 'special organic lines', and attempts have been made to develop organic breeding programmes for cattle to be more robust and better suited to organic conditions involving more grazing, outdoor living and walking. The chapter also examines the role that genome breeding values may play in the development of such breeding programs, as well as its limitations.

A priority in organic farming is outdoor living and access to species-specific feed and natural elements within the landscape, contributing to animal longevity – because they are healthy and robust – and natural behaviour. Grazing and foraging behaviour is natural for ruminating animals, and local conditions should play a major role in developing and selecting breeds that best fit the climate, weather, available feed-stuffs and the landscapes where cattle live and which they become part of.

4 The health and welfare of animals and the humans who take care of them in organic farming

The concept of health, one of the cornerstones in the organic principles, was unfolded by Vaarst (2019) as addressing physical, mental, emotional and social health, i.e. so much more than 'absence of physical signs of disease', as it is often assumed. The One Health concept is fundamentally embraced within organic farming given that that concept applies across many levels, from soil, plants, animals and humans to the planet. The concept of animal welfare is closely connected to this understanding of One Health, and also includes the way in which we as humans interact and live with our animals, which is an important part of the overall concept of 'One Welfare'. Welfare of all groups of cattle, addressed by Whistance (ibid., Chapter 5), acknowledges the intertwined lives of animals and humans who take care of them, by providing them with surroundings which allow natural needs to be met, and how this mutuality and shared lives is embraced by the organic principles of care, health and ecology.

The responsibilities of humans to care for animals, including good living conditions which support their health at all levels, are also addressed by Morgans and co-authors in Chapter 7 (ibid.). They focus on health-supporting and disease-preventative approaches in farm situations with less reliance on antibiotics, or even working towards a complete phasing out of antibiotics. Acknowledging the complexity of the health concept, they also acknowledge the importance of communication with the humans around the animals, as well as introducing different forms of animal health advisory services applied to dairy herds across Europe.

The rapid development of a wide spectrum of technological innovations for use in dairy herds has occurred over recent years, and organic farming requires an ethical framework to evaluate its potential role in organic farming. Some forms of genomic and genetic techniques are avoided, whilst other technological developments, such as milking robots and various forms of automated, sensor-based surveillance and daily data collection for use in decision-making, are embraced and widely implemented in organic dairy herds. Rose and co-authors (Chapter 10, ibid.) discuss this complex issue of acceptability of technology by directly addressing the organic IFOAM principles and consider an inclusive, structured decision-making process to judge how individual technologies may match organic values and standards.

5 Disease prevention and management strategies

Disease prevention in organic farming is based on the principles that an animal that is allowed to exhibit natural behaviour is not subject to stress, is fed optimal (organic) feed, and will have a higher ability to cope with infections than animals reared in a conventional way (Kijlstra and Eijck, 2006). This makes them more resilient, which is a core of the organic health principle (IFOAM, 2005). Animal disease, in some form, is probably inevitable, no matter how strong and resilient animals are, and how supportive their environment may be.

In organic systems, beyond focusing on system designs in which health is optimised, the aim must be to seek a reduction in both incidence and severity of disease, for example, by early recognition and responsible treatments or changes in management. In Chapter 8 (ibid.), Groot and co-authors present and discuss different approaches to treat diseased animals with more natural methods, such as phytotherapy, minerals and homoeopathy, with a strong emphasis on the scientific evidence identifying their strengths and limitations. Since AMR is of increasing environmental concern in both human and veterinary medicine, it seems obvious to explore responsible ways of handling animal diseases, ensuring appropriate healthcare but without risking the loss of important and powerful drugs for the future.

The management of parasite burdens is a continuous challenge for many organic farmers. Parasite behaviour, impacts and control are influenced by changing conditions, including environmental factors such as climate and biodiversity change. Based on a wealth of established and current recent research, Thamsborg and co-authors (Chapter 9, ibid.) provide an overview of the life cycle and epidemiology of key parasites under dairy farming conditions and outline prevention and treatment options and challenges within an organic farming context.

6 Strengthening the connection between organic dairy cattle and the wider ecosystem

In acknowledgement of the fact that the industrialisation of dairy farming is considered as a significant contributor to GHG emissions, organic dairy farming can be considered as part of the solution to this existential challenge. Whereas the conventional farming model has sought to specialize and maximize production within increasingly controlled environments, organic farming is concerned with a more holistic approach, looking at entire systems and aiming at finding solutions which encompass both the welfare of animals and humans, conserve biodiversity and balance the sharing of landscapes between farmed, semi-natural and natural elements.

In this regard, innovative solutions and strategies have been proposed to enhance the resilience of organic milk producing systems typically found in the temperate regions of Europe and the United States (Leiber et al. 2019). A challenge is to acknowledge the importance of grazing animals as integral and natural to ecosystems and not merely as an external 'tool' for humans to utilize when managing landscape biodiversity.

Feeding organic dairy cattle requires a balance between meeting the nutritional requirements of the animals at various stages of their development, meeting the behavioural and physiological needs of the animal and ensuring that the feed production operates within closed nutrient cycles. The concept of 'naturalness' includes feeding behaviours, and allowing animals to access feed through foraging and grazing, at least during the growing season. Such pasture-based dairy farming is discussed by Roderick and Jones in Chapter 3 (ibid.), including the benefits that such systems can accrue with regard to building soil quality and enhancing carbon sequestration.

Further to this, Simonson in Chapter 12 (ibid.) examines the impact of organic grazing practices on pasture biodiversity, including at the landscape level, emphasizing the importance of farmer engagement and understanding. The integration of dairy cattle into the whole farming system, so as to add resilience through complementarity, is explored by Roderick and Whistance in Chapter 11, again with opportunities linked to the development of circular economies at the farm and landscape level.

7 Valuing products from organic dairy farms

The are many extrinsic factors such as animal welfare, biodiversity and the treatment of labour along the supply chain that can be used to judge the success of organic dairy farming, with the nutritional quality of milk providing the most important indicator of its value to the human diet. In Chapter 4 (ibid.), Butler and Davis review the recent research and development on the quality of organic milk, including the impacts of organic practices around pasture-based feeding and breed choice as well as innovations such as cow–calf contact (CCC) systems that allow calves to remain with their mother during part of the lactation.

This points to an issue which is less explored in this book. We address milk quality, but not the quality of meat from animals born on or having lived their lives on a dairy farm. The importance of not viewing any animals born into organic farming as 'surplus animals' shifts the focus to consideration of male calves born in dairy farms. In Chapter 6 (ibid.) Barth and co-authors explore both sides of the frequently imbalanced view of the different animals born and living in dairy herds, including the development of different CCC-systems and the possible strategies we may adopt to re-value and re-purpose calves not intended for milk production.

8 Organic dairy farming as part of markets and food systems

Organic dairy products were a driver of much of European production during the latter decade of the twentieth century and, as such, dairy farming has been a very important part of the overall growth of European organic farming. More recently, there has been a growth in products labelled as having similar societal benefits, such as high welfare and pasture-based products that are not organic. There has also been a very recent and growing marketing of plant-based products, including some intended to substitute cow's milk. The manner in which the population now buy and consume food has also changed, driven by multi-national companies and the influence of the internet. In light of these changes, Schmid and co-authors (Chapter 13) provide an insight into consumer preferences and the organic milk market over recent years.

Many organic dairy farms operate under significant economic pressures, and are part of the same food system as industrial farming, with the aim of producing large quantities of cheap food. Looking into the future, organic farming generally needs to develop in response to the increasingly visible impacts of climate change and environmental degradation. For organic dairy production, in particular, there is an urgent need to determine how it can fit comfortably and sustainably within natural and semi-natural environments. It is also necessary that consumption patterns change and that meat and dairy products, including organic, become a smaller part of a nutritionally balanced diet at the individual level. How can organic dairy farming further develop holistic approaches to farming that enable the transition to more sustainable food systems?

The last chapter in the book attempts to view organic dairy farming from a systems perspective, with a particular emphasis on social and food systems. The basic premise is that, while dairy farming can conform well to organic regulations and principles, the surrounding social and food systems also need to represent the values of organic food and farming, by being circular in nature and being fair to all of those involved. The challenges and opportunities for a more holistic organic dairy farming that encompasses these ambitions are confronted and developed within this final chapter.

9 Conclusion

Organic farming was established during the past century as a leading alternative to increasingly industrialised farming and food. It was based on its own set of standards, which are now embedded within EU regulation. The founding principles, formulated by the many actors in the movement, serve as a reference throughout this book. Organic dairy cattle farming advanced into a successful and well-established 'engine' in the broader development of organic practices, and as ruminants, cattle offer many important benefits to the organic farming systems. However, there are still challenges and gaps between the 'ideal' and the 'real', and we envisage continued future development of the sector, contributing to solutions for some of the significant current issues raised in present day dairy farming, such as environmental crises linked to climate change and biodiversity loss.

10 References

Bongaarts, J. (2009). *Human population growth and the demographic transition.* The Royal Society Publishing, Special Issue, Published: 27 October 2009. https://doi.org/10.1098/rstb.2009.0137

Cook, S. (2018). *The spice of life: The fundamental role of diversity on the farm and on the plate.* Discussion Paper. IIED and Hivos, London and The Hague. http://pubs.iied.org/G04305/

FAO. (2013). Food wastage footprint, impacts on natural resources. Online: https://www.fao.org/3/i3347e/i3347e.pdf [accessed 20 December 2023] ISBN 978-92-5-107752-8

FOAG. (2021). *Farm animal genetic resources in Switzerland*. Federal Office for Agriculture FOAG, CH-3003 Bern, pp. 102. https://issuu.com/blw_ofag_ufag/docs/brosch_re_tiergenetische_ressourcen_en_mid/1 (accessed 11th December 2024)

IFOAM. (2005). https://www.ifoam.bio/principles-organic-agriculture-brochure

Kijlstra, A. and Eijck, I. A. J. M. (2006). Animal health in organic livestock production systems: A review. *NJAS* 54: 77–94.

Kundu, D. and Pandey, A. K. (2020). World urbanisation: Trends and patterns. In Kundu, D., Sietchiping, R. and Kinyanjui, M. (eds.), *Developing National urban policies*. Springer, Singapore. https://doi.org/10.1007/978-981-15-3738-7_2

Leiber, F., Müller, A., Maurer, V., Schader, C. and Bieber, A. (2019). Organic farming towards sustainability. In Vaarst, M. and Roderick, S. (eds.) 2019. *Improving organic animal farming*. Burleigh Dodds Series in Agricultural Sciences. 46. chapter 11, 225–243.

Richardson, K., Steffen, W., Lucht, W., Bendtsen, J., Cornell, S. E., Donges, J. F., Drüke, M., Fetzer, I., Bala, G., von Bloh, W., Feulner, G., Fiedler, S., Gerten, D., Gleeson, T., Hofmann, M., Huiskamp, W., Kummu, M., Mohan, C., Nogués-Bravo, D., Petri, S., Porkka, M., Rahmstorf, S., Schaphoff, S., Thonicke, K., Tobian, A., Virkki, V., Wang-Erlandsson, L., Weber, L. and Röckström, J. (2023). Earth beyond six of nine planetary boundaries. *Science Advances*. 9: eadh2458. https://doi.org/10.1126/sciadv.adh2458

Steffen, W., Sanderson, R. A., Tyson, P. D., Jäger, J., Matson, P. A., Moore III, B., Oldfield, F., Richardson, K., Schellnhuber, H.-J., Turner, B. L. and Wasson, R. J. 2004. *Global change and the earth system: a planet under pressure*. Series: Global Change – The IGBP Series (closed). Springer, Germany. pp. 332. http://www.igbp.net/publications/igbpbookseries/igbpbookseries/globalchangeandtheearthsystem2004.5.1b8ae20512db692f2a680007462.html (accessed 17 January 2024)

Vaarst, M. (2019). Rethinking and engaging with animal health in organic farming. In Vaarst, M. and Roderick, S. (eds.) *Improving organic animal farming*. Burleigh Dodds Series in Agricultural Science. 46, chapter 4, 59–78.

Part 1

Improving breeding and nutrition

Chapter 2

Organic dairy cattle breeding: possibilities and challenges

Morten Kargo, Julie Brastrup Clasen and Margot Slagboom, Aarhus University, Denmark

1 Introduction

Organic farming is expected to be an important component of future cattle production. However, until now, improvements in the efficiency of organic dairy production have mainly occurred through better management, with improved feeding, health and production strategies while, in most cases, using the same breeding stock as in conventional production systems.

The principles behind organic production, as well as the restrictions set by both national and international regulations for organic production, may result in:

1 different breeding goals for organic dairy cattle (as described by Slagboom et al. (2018));
2 genotype by environment interactions (G by E) between conventional and organic production systems, meaning that cows of the same

http://dx.do .org/10.19103/AS.2024.0138.02

genotype (bred to achieve the same breeding goal) perform differently in the two environments;
3. the need to include new traits in organic breeding goals, which may have little or no value in conventional production systems; and
4. different attitudes to and usage of new breeding technologies.

Depending on the differences between breeding goal traits and the extent of G by E interactions, the optimal solution will either be customised indices for organic farmers or establishing a separate organic breeding programme with specialied breeding goals.

2 The right animal for the right production system

Overall, organic breeding is about finding the right animal for the right production system, i.e. for organic production systems. It is important to keep in mind that breeding of any kind follows the same biological laws. Organic production systems are, however, different within different geographical areas and within different types of organic production systems, meaning that many types (breeds) of cows are needed for different organic production systems. The performance of a cow – the phenotype (P) – depends on several factors: feeding, housing, hygiene level and many other elements related to the production system. All these elements together form the environment (E). Describing and understanding these environmental effects of production systems are the main subject of this book. In addition, there is an effect of genetics (G), which is the topic of this chapter.

One of the basic equations in animal breeding is P = G + E. An important drawback within this basic formula is the missing interaction between G and E. This interaction is one of the reasons why specific breeds or lines of breeds are needed for optimal organic dairy production. There is a fundamental difference between the effects of changes in environmental and genetic effects. When changing environmental effects, the phenotypic effect will occur rapidly in many cases, e.g. when increasing the feeding level. On the other hand, these effects will disappear as soon as the environment is changed back to its previous state. In contrast, genetic changes caused by selective breeding will influence the phenotype over a longer period. Furthermore, the effects of genetic changes are cumulative and continue even when selective breeding is discontinued.

The performance of an organic dairy cow is therefore dependent on its genetic profile in interaction with the production environment. Genetics can be changed through selective breeding and need to be adjusted to the environment, which, to some degree, can be changed by management practices. However, the production environment is heavily dependent on factors such as climate and market conditions. The intensification of dairy production in European countries since the Second World War has led to a more homogeneous dairy cattle population in Europe. Resources have been focused on fitting the environment to the cows, such as feeding animals with feed that is not grown in the local environment, keeping them in controlled indoor settings (that do not reflect local climatic conditions), etc.

However, the alternative approach of fitting the cow to the environment may require fewer resources and be more sustainable in the long run. This can be done by genetic improvements – or by utilizing the genetic variation of breeds. Changing the environment is most often a short-term solution, and performance falls back to its previous level if the change is reversed. On the other hand, changing the genetics is a long-term, permanent and cumulative solution, in which the genetic gain achieved within one generation will be kept over the following generations.

The most prevalent dairy breed in Europe and the rest of the world is the Holstein (Augère-Granier, 2018), which has been bred to produce high amounts of milk. The historic focus on increasing milk yield has suppressed other traits and negatively impacted genetic traits affecting fertility, health and longevity (Stefani et al., 2018; Ma et al., 2019). For the same reason, Holstein cows require high-energy diets to reach their genetic potential and often struggle in low-input environments, such as mountainous and pasture-based farming, compared to cows of other breeds (Horan et al., 2006; Kiessé et al., 2022; Johansen et al., 2021). Feeding high-energy diets to dairy cattle often requires the supply of human-edible feed, such as wheat, maize and soybean, which competes with food production and accelerates the kinds of changes in land use that fuel climate change.

Cattle can eat forage that is inedible to humans and can be used strategically to produce beef and dairy products in areas that are not suitable for crop production (Wilkinson and Lee, 2018). However, this requires cattle breeds that can thrive on pasture and in the range of environmental conditions characteristic of different types of grassland areas. Additionally, rising temperatures and extreme weather conditions due to climate change are negatively affecting dairy and beef production, requiring more resilient animals that are able to cope with and adapt to different environments (Gauly et al., 2013; Bengtsson et al., 2022).

3 Setting up an organic breeding scheme: prerequisites and breeding goals

Before setting up and running an organic breeding scheme, with the goal of moving towards the ideal animal for an organic production system, there are a number of factors to consider. These are very well described in the book *Sustainable Management of Animal Genetic Resources* (Woolliams et al., 2015) where ten important points for running sustainable breeding schemes are set out. They include the following:

1. Ensure that the market for the use of an improved animal is well defined, including the definition of production systems as well as any restrictions on production.
2. Have a well-defined breeding goal taking into account both income and costs of production as well as any consequences related to health and welfare.
3. Be aware of external factors such as fluctuations in the market, risk of diseases reducing or destroying the breeding nucleus, awareness of food safety issues among consumers as well as acceptance of reproduction techniques used in relation to running the breeding plan.
4. Ensure there are enough economic, technical and human resources for running the breeding scheme.
5. Be aware that the population size and selection strategies ensure genetic diversity within the future population.
6. Have enough well-defined registrations for ensuring progress in developing new traits defined in the breeding goal and to avoid undesired correlated responses.
7. Estimate expected genetic progress from the suggested breeding programme.
8. Monitor genetic progress.
9. Evaluate the costs and benefits of the breeding scheme for breeders, producers and consumers.
10. Evaluate the profitability of the breeding scheme.

The breeding goal defines which traits are to be improved and how much weight is given to each trait (Groen et al., 1997). In dairy cattle breeding, the weighting factors are often based on modelling. These weighting factors are called economic values (EVs). EVs can be derived using a profit equation or a bio-economic model.

EVs are derived as partial derivatives of the profit equation, or the marginal effect on monetary output of a one-unit change in the bio-economic model (Groen, 1989). EVs are therefore marginal, which means measuring the effect of one unit improvement of the trait while keeping the remaining traits constant. EVs are therefore functions of production conditions, depending on differences in production systems and differences in economic conditions, e.g. output prices (milk and meat) and input (primarily feed) prices. This creates different EVs for different traits in conventional and organic production systems (Kargo et al., 2015).

These EVs are based on economic models which, for some traits, such as traits related to welfare and robustness, can be difficult to derive. In such cases, one can choose to assign these traits with so-called non-market values (Nielsen et al., 2005). Non-market values are values that can be added on to the breeding goal weight for a trait, that reflect trait improvements that have no measurable economic effect. EVs can also be adjusted based on farmer preferences which can be incorporated into the model to calculate EVs (Slagboom et al., 2016, Slagboom et al., 2017).

Breeding goal weights can also be calculated by means of a desired gains index, where the weight of a breeding goal trait does not directly translate to an economic unit but is rather calibrated based on a desired improvement in certain traits. This method can be used to adjust breeding goals to conform to particular production principles, e.g. the principles of organic production, as shown by Slagboom et al. (2018).

4 Breeding traits or goals of specific importance for organic cattle production

Many traits are important for both conventional and organic dairy cattle production. The weighting of traits may, however, be different. The four principles of organic agriculture, defined by IFOAM (IFOAM, 2023), illustrate that traits related to climate impact, robustness, welfare, feed efficiency related to roughage consumption, and ability to withstand heat (or cold) will be ranked high in organic dairy production. Furthermore, traits related to product quality are expected to have higher importance.

Milk fatty acid composition is an example of a quality trait that can be improved through breeding. Danish analyses have shown different compositions of milk fatty acids in different breeds, with Jersey cows having the highest amount of saturated fatty acids, which, in most situations, is considered undesirable. Fatty acid traits are heritable at the same level as the total amount of fat (Hein et al., 2018), and they are weakly correlated with other traits in the breeding goal. This means that it is possible to select for improved fat quality of milk without negatively impacting the genetic improvement of other traits (Buitenhuis et al., 2023).

If we focus on the core organic principle of Health, organic agriculture should sustain and enhance the health of the soil, plants, animals, humans and the planet as one. This principle suggests that the health of individuals and communities cannot be separated from the health of ecosystems – healthy soils produce healthy crops that foster the health of animals and people. Health is the wholeness and integrity of living systems. It is not simply the absence of illness but the maintenance of physical, mental, social and ecological well-being. Immunity, resilience and regeneration are key characteristics of health.

All this must be considered when setting the breeding goal. The same holds for the other three principles: Ecology, Fairness and Care.

Apart from improvements in breeding traits, another goal of breeding programmes is to maintain ample genetic diversity through an acceptable rate of inbreeding. This both relates to ensuring an acceptably low rate of inbreeding within the populations under selection and ensuring sustainable use of as many breeds as possible, since approximately half of the variation within a species is within populations, and half of the variation is between populations.

A key issue to consider is where to separate or combine dairy and beef production. Historically, increasing demands for beef and dairy products have intensified production, with cattle breeds being highly specialised for either beef or dairy production (Roche et al., 2018). In Europe, two-thirds of European cows are specialised dairy cows, which means that more than half of beef production comes from culled dairy cows and young dairy bulls (Eurostat, 2022a, 2022b). There is both an economic potential (Ettema et al., 2017; Clasen et al., 2021) and an environmental benefit (Knapp et al., 2014; Johansen et al., 2022) to improve beef output from dairy production with a minor impact on overall milk production. This can be achieved by mating bulls from beef breeds with dairy cows for giving birth to beef × dairy crossbred calves, combined with an improved focus on beef production in the breeding goal of dairy cattle breeds (Bazzoli et al., 2014). With increased focus on beef in the dairy cattle breeding goal, a more dual-purpose breed will develop.

The cattle sector is responsible for about 9% of global greenhouse gas (GHG) emissions (Gerber et al., 2013), which can potentially be decreased by optimising production efficiency and farm management (Crosson et al., 2011; Knapp et al., 2014). It is known that enteric methane emissions from cattle can potentially be reduced by up to 30% by increasing beef output from dairy herds (Knapp et al., 2014), producing beef × dairy crossbred slaughter calves from dairy cows (van Selm et al., 2021), or having dual-purpose cattle breeds (Mazzetto et al., 2020). This means that sustainable dairy production should use dual-purpose breeds with the aim of producing as much beef as possible alongside milk production – until the demand for meat becomes less than the production of meat from dairy populations.

To solve the challenges that have emerged from the homogenisation of cattle production, future dairy and beef production needs to be more heterogeneous again (Roche et al., 2018; Brito et al., 2021). This includes re-considering the use of diverse cattle breeds in different European areas. As early as 1992, the Convention on Biological Diversity (1992) recommended conserving local cattle breeds to maintain genetic variety as insurance for future cattle farming – and that future is now. There are fortunately still small European populations of local dairy and dual-purpose breeds (producing both meat and milk) that perform well in local environments (Hiemstra et al., 2010).

There is an unexploited genetic potential from locally conserved cattle breeds that can be used to optimize adaptability and resource utilisation in future beef and dairy production. Additionally, crossbreeding between different cattle breeds creates more flexibility in the utilisation of genetic resources and has been shown to create more robust, healthier and efficient animals (Shonka-Martin et al., 2019; Clasen, 2021; Clasen et al., 2022).

A key question is whether dual-purpose breeds are economically feasible. In a study investigating the economic performance of Holstein and German Angler breeds with German Red and White dual-purpose cattle, in a typical production system in Schleswig-Holstein in Germany, it was shown that both Holstein and German Angler cattle achieved a higher profit per cow compared to German Red and White dual-purpose cattle (Schmidtmann et al., 2021). On the other hand, the income per kg of milk was highest in German Red and White dual-purpose cattle.

5 When to divide a breed into different lines

The main reason for the absence of distinct organic breeding lines, i.e. animals selected based on an organic breeding goal in a certain group of the cattle population, is most likely because breeding goals were too similar. The similarity between breeding goals can be measured as the correlation between the breeding goals calculated with the equation used in Buch et al. (2009). If the correlation between the breeding goals is low, it is beneficial to divide a population into separate lines. On the other hand, if the correlation is high, it makes no sense to keep separate breeding lines. A correlation higher than approximately 0.8 has been shown to cause a loss of genetic gain when dividing populations with less than a million cows in lines (Smith and Banos, 1991). When populations are small or of different sizes, the 'break-even' correlation will be lower. However, this assumption was developed when genetic gain in cattle breeding schemes was primarily driven by the selection of bulls based on their progeny performance (Kargo et al., 2018). In creating an organic breeding line, it will be possible to increase the overall genetic gain measured in economic terms based on EVs from organic production systems, as illustrated in Figure 1.

In case the difference between the organic – and the conventional breeding goal is too small – corresponding to a correlation between breeding goals being near 1, or the cost of running an extra breeding scheme is too high compared to the number of animals in the cluster ('asking' for animals with a profile defined by that specific organic breeding goal), then it will be possible to select breeding animals – often bulls – based on a customised index.

A customised index (a herd index) is a total merit index with EVs fitted for a specific herd or group of herds. Such an index offers the possibility for selecting breeding animals fitted for an organic production system among the

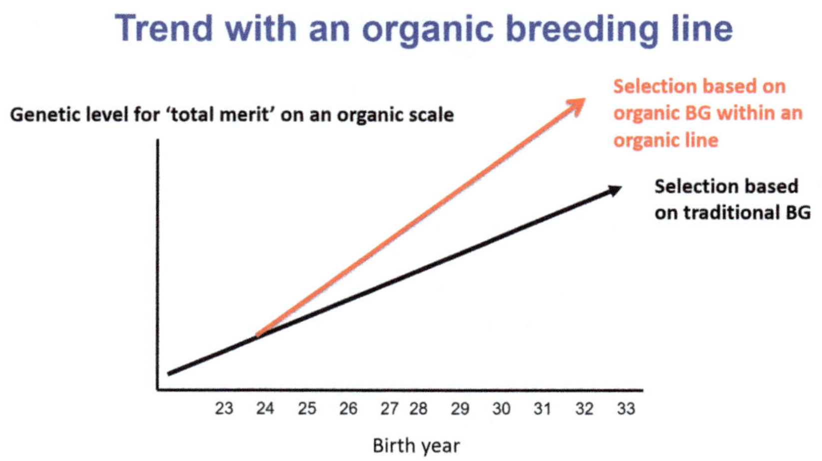

Figure 1 Genetic trend over years in an organic breeding line using either a breeding goal (BG) using EVs based on organic production systems (red line) or a breeding goal (BG) using EVs based on conventional production systems (black line).

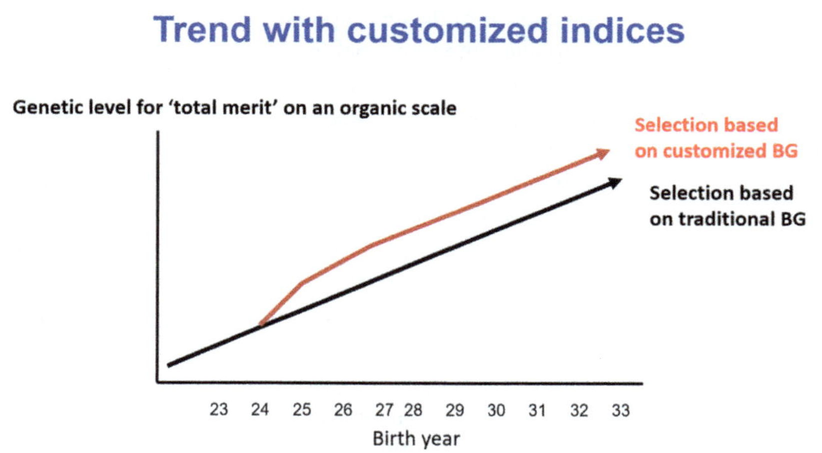

Figure 2 Genetic trend over years in a population selecting breeding animals based on customised total merit using EVs originating from an organic breeding goal (BG; red line) compared to a genetic trend using a conventional breeding goal (black line).

group of breeding animals within the conventional breeding scheme. By using a customised index, it is possible to select organic breeding animals with an organic profile among conventional bulls. However, it is not possible to follow an organic trend in the long run, as the overall genetic trend is driven by the breeding scheme from which you select – in this case, the conventional scheme, as illustrated in Figure 2.

6 The role of phenotypic data and genomic breeding values

For the purposes of this book, a general overview and understanding of estimated breeding values is sufficient. The estimated breeding value of an animal is a proxy of the true genetic breeding value of the animal. The first and most important issue in relation to traditional breeding value estimation is having available phenotypes. This was the situation before genomic information was used in breeding value estimation. It is still the situation when using genomic information in breeding value estimation, resulting in estimated breeding values for young animals. Phenotypes are also of great importance for the optimisation of management in cattle herds and for monitoring developments at the herd and population levels.

There are many phenotypes relevant for breeding, including those related to:

1 Production of milk and meat both in terms of quantity and quality.
2 All aspects of health for all groups of animals: calves, young stock and cows.
3 Welfare and behavior.
4 Fertility.
5 Calving ease.
6 General feed efficiency and efficiency related to the utilisation of specific nutrients, e.g. protein.
7 Grazing ability and production efficiency based on roughage consumption.
8 Milkability and temperament.

As well as these, there may be other types of relevant phenotypes for specific production systems, e.g. skin quality in case a population is used for production of leather goods.

Based on these phenotypes and pedigree information, traditional breeding values can be calculated. Breeding values are an estimation of the true breeding value for a specific trait or the true breeding value for a combination of traits – a so-called total merit index (Kargo et al., 2014). The quality of the breeding value is scored on either the accuracy or the reliability of the breeding value. Accuracy is the genetic correlation between the estimated breeding value and the true breeding value. Reliability is the squared accuracy. Reliability can be interpreted as the amount of total genetic variation explained by the estimated breeding value.

Due to the Mendelian sampling variance, the highest possible accuracy of an EBV based on family information is 0.5. This is because the genetic variation in a group of full siblings is one-quarter of the genetic variation within the sire group, one-quarter of the variation within the dam group, plus the Mendelian sampling term, which is half of the genetic variation. The Mendelian sampling term is due to the fact that which of the two alleles at a loci is inherited by the offspring is random. This means that increasing the accuracy of breeding values above 0.5 requires information on performance, information from offspring or genomic information.

The information used for predicting genomic breeding values is based on a large number of single-nucleotide polymorphism (SNP) markers along the chromosome. This information can be derived from blood, tissue or semen samples. Based on a group of animals having both SNP information and information on phenotypes, a correlation between SNPs and phenotypes is established. The group of animals having both genomic and phenotypic information is called the reference group. Through this correlation, it is possible to estimate breeding values for animals based on information from a genomic test. This means that young animals can have genomic breeding values, including information about the Mendelian sampling term, without having expressed this phenotypically.

The accuracy of the genomic breeding values relies on the level of linkage disequilibrium (LD) between SNPs and quantitative trait loci and the size of the reference group. It is also necessary that the animals for which we estimate genomic breeding values are genetically related to the reference population (De Roos et al., 2009; Goddard, 2009; Clark et al., 2012; Vandenplas et al., 2016).

It can be concluded from this discussion that the efficiency of organic breeding schemes can be improved by good recording of phenotypic data and, at the same time, ensuring enough genomic tests are carried out in organic cattle production systems.

7 Breeding scheme optimisation

Breeding schemes can be organised either at population level or at herd level. At population level, the overall goal is to ensure the largest possible genetic gain (ΔG) for the overall breeding goal while, at the same time, ensuring an acceptable rate of future inbreeding. Control of inbreeding is achieved via the control of the average relationship among breeding animals in the next generation. Optimal contribution selection can be used for that purpose (Meuwissen, 1997; Grundy et al., 2000; Woolliams et al., 2015). The overall breeding goal at population level is the result of a compromise among groups of stakeholders (cattle farmers) making use of that population.

Breeding at herd level involves finding the right breed or combinations of breeds, when applying systematic crossbreeding, based on an assessment of the traits of importance within the herd. Ideally, this means setting up a customised total merit herd index for the selection of bulls and females to give birth to replacement animals. In pure-breed herds, it is also necessary to consider inbreeding in the next generation.

At population level, the amount of genetic response per year (ΔG) is calculated using the breeder's equation:

$$\Delta G = \frac{i \cdot r_{IA} \cdot \sigma_A}{L}$$

where i is the selection intensity corresponding to the frequency of animals being selected as parents for the next generation, r_{IA} is the accuracy of the estimated breeding values used when selecting parents for next generation, S_a is the genetic variation, and L is the average generation interval between offspring and parents.

This equation is to be used in each selection step, with the selection criterion being the overall total merit index. In an ideal situation, all these selection steps should use optimal contribution selection. This involves finding the right balance between the average level of total merit of the next generation's offspring and the average relationship among breeding animals in the next generation. The average relationship among breeding animals in the next generation defines future inbreeding and is therefore of great importance. The higher the average relationship, the higher the future inbreeding.

8 Issues in developing specific organic breeding schemes

So far, there have been very few organic breeding programmes for cattle, and only on a small scale. In the pre-genomic era, there were good arguments for this, as genetic progress was strongly dependent on a substantial number of tested young bulls per year to provide information about the Mendelian sampling term for each bull, thus ensuring high accuracy in breeding values. This required at least 50 daughters per bull. Relatively large populations were required to ensure enough genetic progress. In the genomic era, progeny testing is no longer essential to provide information on Mendelian sampling. A high ΔG can thus be achieved in quite small populations. This simply requires the selection of bulls and heifers for the breeding programme among many genomically tested calves. The larger the group of genomically tested calves, the higher the selection intensity will be, and thus, a larger ΔG. It is, therefore, possible to ensure significant genetic progress in smaller populations or lines of populations. This has significantly enhanced the opportunity to establish dedicated organic breeding lines.

Options for organic breeding programmes are discussed in detail in Slagboom (2019). This shows how breeds or lines of breeds can be changed to meet organic objectives. However, it also shows that changes in ΔG will be insufficient unless newer reproduction technologies, like multiple ovulation and embryo transfer (MOET), are used. Simulations of breeding programmes by Slagboom et al. (2020) suggest that selection can be carried out in two ways:

- Using a conventional total merit index as the selection criterion; and
- Using an organic total merit index as a selection criterion.

To reduce the complexity of the simulations undertaken in the study, the total merit index only contained three traits: milk production, fertility and mastitis resistance. The EVs given to the traits in each of the two breeding goals were based on model EVs for conventional and organic Nordic dairy cattle production (Sørensen et al., 2018). This meant that no non-market values were included. The correlation between the conventional breeding goal and the organic breeding goal was 0.94. The simulated breeding schemes were based on the Nordic Holstein breeding scheme. Both a conventional and an organic breeding scheme were simulated. In one scenario, breeding animals for the organic breeding scheme were selected across these two breeding schemes (organic BG). In another scenario, breeding animals were only selected within the organic breeding scheme (Within). Both schemes were simulated with and without the use of embryo transfer (MOET). Embryo transfer is at present prohibited within organic production systems. However, the method was included in the simulations because it is assumed that a higher genetic gain in an organic production system can be achieved using MOET.

The overall organic total merit genetic gain, using organic EVs as defined by Sørensen et al. (2018), was similar between breeding schemes. The Organic BG breeding scheme delivered 1% more gain than the conventional scheme, while the Within breeding scheme gave a 4% lower gain. The types of gain within the different schemes were, however, different (as seen in Figure 3). Milk yield gain was reduced by approximately 20% in the Organic BG breeding scheme. Approximately 40% less genetic gain was achieved by the Within breeding scheme. The costs of this reduction in genetic gain for yield are offset by a much higher genetic gain for fertility and mastitis resistance within the organic breeding schemes – especially in the situations where breeding animals were only selected from the organic breeding scheme and not from the conventional breeding scheme.

The conclusions from Slagboom (2019) is that running an organic breeding scheme results in approximately the same economic outcome as running a conventional breeding scheme. This means that buying semen from a conventional breeding scheme results in the same economic outcome for

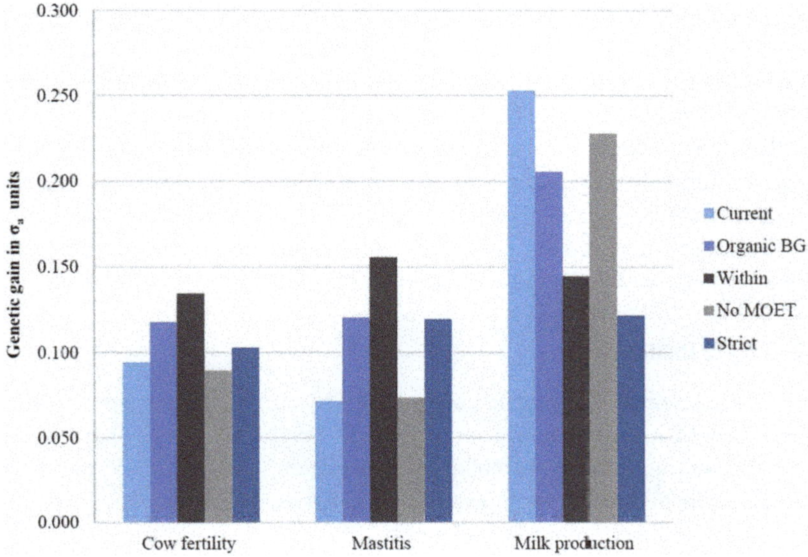

Figure 3 Genetic gain per year in genetic standard deviations units (σ_a) for each trait and scenario, for the Holstein breed. Scenarios: 'Current' = a conventional breeding goal, MOET, and the possibility to select conventional bulls in the organic breeding programme. 'Organic BG' = an organic breeding goal, MOET, and the possibility to select conventional bulls in the organic breeding programme. 'Within' = an organic breeding goal, MOET, and no possibility to select conventional bulls in the organic breeding programme. 'No MOET' = an organic breeding goal, no MOET, and the possibility to select conventional bulls in the organic breeding programme. 'Strict' = an organic breeding goal, no MOET, no possibility to select conventional bulls in the organic breeding programme. (Source: Slagboom et al., 2020.

an organic dairy producer as buying from an organic breeding scheme. The profile of specific trait improvements is, however, very different. This may be of great importance for future organic production, allowing a combination of both organic management and distinct organic genetics.

8.1 Use of advanced reproduction technologies

Advanced reproduction technologies include the use of MOET. The theoretical reasoning for considering this technology is that the selection intensity on the female side can be greatly improved. This increased selection intensity results in a large increase in genetic gain. The overall genetic gain in a breeding scheme without the use of MOET in the organic part of the breeding scheme, selecting across the organic and the conventional production system (No MOET), reduced the overall economic gain by 7% (Slagboom et al., 2020). If the organic breeding scheme is run without MOET, and without being able to use breeding animals from the conventional scheme (Strict), then the reduction in economic gain is as much as 24%. The large difference between the two scenarios is that the No MOET organic breeding programme can select breeding animals from the conventional programme which benefits from the use of MOET. Excluding this in the Strict scenario results in lower overall economic gain and less genetic gain for traits included in the organic total merit index. Rejecting the use of advanced reproduction technologies on a limited number of animals greatly reduces the ability to develop a distinctive organic genetic profile. It also makes it impossible to buy breeding and production animals from conventional herds as nearly all animals in these herds have pedigree bulls born using MOET.

8.2 G by E interaction

G by E interactions for a specific trait between two environments are measured as the genetic correlation between the traits in the two environments. The larger this interaction is, the smaller the correlation, corresponding to a larger reranking of estimated breeding values. This makes the use of genetic material from the other environment less efficient.

Looking at correlations between countries confirms that G by E interactions exist (Interbull, 2023). Even for easily measurable traits like production traits, there are correlations below 0.8 between mainly intensive, indoor production systems in West European countries and extensive, grass-based production systems in New Zealand with minimal feeding of concentrates. This suggests that there may also be G by E interaction between organic, grass-based production systems and more intensive, indoor-based systems using a high amount of concentrate and high-quality silage in feeding.

However, for G by E interactions vary within the literature. In Sweden, G by E interaction between organic and conventional dairy farms was found for fertility traits in second lactation only, and no G by E interaction was found for production traits (Sundberg et al., 2010). In the Netherlands, moderate G by E interactions were found for milk production traits (Nauta et al., 2006). The reason for these either negligible or inconsistent estimates may be due to a low number of registrations in these analyses and, in some cases difficulties, in assigning different herds to the right production classes. However, the expectation is that dairy cattle breeding in the future will be faced with more diversified production systems, some very intensive and some more extensive systems, and therefore will also be faced with the existence of G by E interactions.

8.3 Herd-level decisions

At the herd level, one needs to find the right breed or combinations of breeds in applying systematic crossbreeding. Breeds need to be evaluated based on current specific traits of interest for each organic herd and on the expected overall economic gain to be achieved from using that breed. Optimally, this evaluation needs to be decided among the relevant group of advisors (with expertise in nutrition, breeding health, herd management and economics) connected to a particular herd. In most cases, the breed(s) will be chosen from mainstream breeds or semi-mainstream breeds.

9 Leveraging genetic diversity: use of conservation breeds and crossbreeding

One of the core values of organic production is diversity. Traditional original cattle breeds are therefore important in organic cattle production. The reason is that, as a rule of thumb, half of the genetic variation within animal species is within breeds and the remaining variation is between breeds. In addition, as well as being part of the cultural heritage of particular regions, many of these breeds may contain specific genes which may become important in future management systems. As a result, most countries have signed the Interlaken Declaration on Animal Genetic Resources (FAO, 2023) to protect endangered breeds. In the case of dairy cattle, many of these breeds are described in the publication 'Local cattle breeds in Europe. Development of policies and strategies for self-sustaining breeds' (Heimstra et al., 2010). The requirements set out in the Interlaken Declaration align with the organic value of diversity. These breeds can, for example, be considered as production animals within some low-input organic systems to which they are particularly well adapted. Another option is to include a few cows from endangered breeds within conventionally bred herds.

Crossbreeding between different cattle breeds creates more flexibility in the utilisation of genetic resources and has been shown to create more robust, healthier and more efficient animals (Sørensen et al., 2008; Shonka-Martin et al., 2019; Clasen, 2021; Clasen et al., 2022). Using systematic crossbreeding enables the producers to combine complementary strengths of different breeds. When crossing, heterosis is also built on top of the additive genetic potential of the breeds included in the crossbreeding programmes.

The most important systematic crossbreeding systems are rotational crossbreeding and terminal crossbreeding systems (Østergård et al., 2018). For various reasons, systematic crossbreeding has been less applied within dairy production than in pig and poultry production, though it is now increasing. In New Zealand, e.g. approximately 50% of the dairy population is crossbred. In Denmark, 20–21% of the approximately 70 000 organic dairy cows are crossbred.

It is clear that organic dairy producers have many choices related to the use of breeds or breed combinations, in choosing different mainstream breeds and, in some cases, endangered conservation breeds for use in low-input systems or for niche production. More generally, crossbreeding has been found to improve the overall health and robustness of a dairy herd.

10 References

Augère-Granier, M.-L. 2018. The EU dairy sector: main features, challenges and prospects. https://policycommons.net/artifacts/1335407/the-eu-dairy-sector/1941893/

Bazzoli, I., De Marchi, M., Cecchinato, A., Berry, D. P. and Bittante, G. 2014. Factors associated with age at slaughter and carcass weight, price, and value of dairy cull cows. *J Dairy Sci* 97(2):1082–1091.

Bengtsson, C., Thomasen, J. R., Kargo, M., Bouquet, A. and Slagboom, M. 2022. Emphasis on resilience in dairy cattle breeding: possibilities and consequences. *J Dairy Sci* 105(9):7588–7599.

Brito, L. F., Bedere, N., Douhard, F., Oliveira, H. R., Arnal, M., Penagaricano, F., Schinckel, A. P., Baes, C. F. and Miglior, F. 2021. Review: genetic selection of high-yielding dairy cattle toward sustainable farming systems in a rapidly changing world. *Animal* 15(Suppl 1):100292.

Buch, L. H., Sørensen, A. C., Lassen, J., Berg, P., Christensen, L. G. and Sørensen, M. K. 2009. Factors affecting the exchange of genetic material between Nordic and US Holstein populations. *J Dairy Sci* 92:4023–4034. https://doi.org/10.3168/jds.2008-1541

Buitenhuis, A. J., Hein, L., Sørensen, L. P. and Kargo, M. 2023. Correlation between breeding values for milk fatty acids and the Nordic Total Merit index traits for Danish Holstein and Danish Jersey. *J Dairy Sci* 106:5554–5561.

Clark, S. A., Hickey, J. M., Daetwyler, H. D. and van der Werf, J. H. J. 2012. The importance of information on relatives for the prediction of genomic breeding values and the implications for the makeup of reference data sets in livestock breeding schemes. *GSE* 44(1):4. https://doi.org/10.1186/1297-9686-44-4

Clasen, J. 2021. *Crossbreeding as a strategy in dairy cattle herds*. Department of Animal Breeding and Genetics. Doctoral Thesis. Swedish University of Agricultural Sciences.

Clasen, J. B., Kargo, M., Østergaard, S., Fikse, W. F., Rydhmer, L. and Strandberg, L. 2021. Genetic consequences of terminal crossbreeding, genomic test, sexed semen, and beef semen in dairy herds. *J Dairy Sci* 104(7):8062–8075.

Clasen, J. B., Nielsen, H. M., Hein, L., Johansen, K., Vestergaard, M. and Kargo, M. 2022. *Performance of first parity Holstein, Danish Red and their crosses in an experimental setting*. In 73rd Annual Meeting of the European Federation of Animal Science, Porto, Portugal.

Crosson, P., Shalloo, L., O'Brien, D., Lanigan, G. J., Foley, P. A., Boland, T. M. and Kenny, D. A. 2011. A review of whole farm systems models of greenhouse gas emissions from beef and dairy cattle production systems. *Anim Feed Sci Technol* 166–167:29–45.

De Roos, A. P. W., Hayes, B. J. and Goddard, M. E. 2009. Reliability of genomic predictions across multiple populations. *Genetics* 183(4):1545–1553. https://doi.org/10.1534/genetics.109.104935

Ettema, J. F., Thomasen, J. R., Hjortø, L., Kargo, M., Østergaard, S. and Sørensen, A. C. 2017. Economic opportunities for using sexed semen and semen of beef bulls in dairy herds. *J Dairy Sci* 100(5):4161–4171.

Eurostat. 2022a. Bovine population – annual data. Vol. 2022. https://ec.europa.eu/eurostat/web/agriculture/data/database

Eurostat. 2022b. Slaughtering in slaughterhouses – annual data. Vol. 2022. https://ec.europa.eu/eurostat/web/agriculture/data/database

FAO. 2007. Global plan of action for animal genetic resources and the Interlaken declaration. https://www.fao.org/3/a1404e/a1404e.pdf (Accessed November 30 2023)

Gauly, M., Bollwein, H., Breves, G., Brugemann, K., Danicke, S., Das, G., Demeler, J., Hansen, H., Isselstein, J., Konig, S., Loholter, M., Martinsohn, M., Meyer, U., Potthoff, M., Sanker, C., Schroder, B., Wrage, N., Meibaum, B., von Samson-Himmelstjerna, G., Stinshoff, H. and Wrenzycki, C. 2013. Future consequences and challenges for dairy cow production systems arising from climate change in Central Europe – a review. *Animal* 7(5):843–859.

Gerber, P. J., Steinfeld, H., Henderson, B., Mottet, A., Opio, C., Dijkman, J., Falcucci, A. and Tempio, G. 2013. *Tackling climate change through livestock – A global assessment of emissions and mitigation opportunities*. Food and Agriculture Organization of the United Nations (FAO), Rome.

Goddard, M. 2009. Genomic selection: prediction of accuracy and maximisation of long term response. *Genetica* 136(2):245–257. https://doi.org/10.1007/s10709-008-9308-0

Groen, A. F. 1989. *Cattle breeding goals and production circumstances*. PhD Thesis, Department of Animal Breeding, Wageningen Agricultural University, the Netherlands.

Groen, A. F., Steine, T., Colleau, J. J., Pedersen, J., Pribyl, J. and Reinsch, N. 1997. Economic values in dairy cattle breeding, with special reference to functional traits. Report of an EAAP-working group. *Livest Prod Sci* 49:1–21.

Grundy, B., Villanueva, B. and Woolliams, J. A. 2000. Dynamic selection for maximixing response with constrained inbreeding in schemes with overlapping generations. *Anim Sci* 70:373–382.

Hein, L., Sørensen, L. P., Kargo, M. and Buitenhuis, A. J. 2018. Genetic analysis of predicted fatty acid profiles of milk from Danish Holstein and Danish Jersey cattle populations. *J Dairy Sci* 101:2148–2157. https://doi.org/10.3168/jds.2017-13225

Hiemstra, S. J., De Haas, Y., Mäki-Tanila, A. and Gandini, G. 2010. *Local cattle breeds in Europe. Development of policies and strategies for self-sustaining breeds*. Wageningen Academic Publ., Wageningen, The Netherlands.

Horan, B., Faverdin, P., Delaby, L., Rath, M. and Dillon, P. 2006. The effect of strain of Holstein-Friesian dairy cow and pasture-based system on grass intake and milk production. *Anim Sci* 82(4):435–444.

IFOAM. 2023. The four principles of organic agriculture. https://www.ifoam.bio/why-organic/shaping-agriculture/four-principles-organic (Accessed November 30 2023)

Interbull. 2023. http://www.interbull.org/ib/maceev_archive (Accesed November 25 2023)

Johansen, K., Hein, L., Spleth, P., Vestergaard, M., Nielsen, H. M. and Kargo, M. 2021. *Comparing growth and feed intake in pure- and crossbred heifers fed different energy concentrations*. In 72nd Annual Meeting of the European Federation of Animal Science, Davos, Switzerland.

Johansen, K., Kargo, M., Bjerring, M., Løvendahl, P. and Buitenhuis, A. 2022. *Phenotypic differences and genetic parameters for methane concentration in BeefxDairy crossbred slaughter calves*. In 12th World Congress on Genetics Applied to Livestock Production, Rotterdam, The Netherlands.

Kargo, M., Ettema, J. F., Hjortø, L., Pedersen, J. and Østergaard, S. 2015. *Derivation of economic values for breeding goal traits in conventional and organic dairy production*. Pages 403–403 in Book of Abstracts of the 66th Annual Meeting of the European Association for Animal Production, Wageningen Academic Publishers, The Netherlands.

Kargo, M., Hjortø, L., Toivonen, M., Eriksson, J. A., Aamand, G. P. and Pedersen, J. 2014. Economic basis for the nordic total merit index. *JDS* 97(12):7879–7888.

Kargo, M., Thomasen, J. R., Wallenbeck, A., Hjortø, L. and Slagboom, M. 2018. *Organic dairy breeding lines? – Possibilities and Requirements*. Proceedings of the World Congress on Genetics Applied to Livestock Production, 11.536, Auckland, Nerw Zeeland.

Kiessé, T. S., Corson, M. S. and Wilfart, A. 2022. Analysis of milk production and greenhouse gas emissions as a function of extreme variations in forage production among French dairy farms. *J Environ Manage* 307:114537.

Knapp, J. R., Laur, G. L., Vadas, P. A., Weiss, W. P. and Tricarico, J. M. 2014. Invited review: enteric methane in dairy cattle production: quantifying the opportunities and impact of reducing emissions. *J Dairy Sci* 97(6):3231–3261.

Ma, L., Sonstegard, T. S., Cole, J. B., VanTassell, C. P., Wiggans, G. R., Crooker, B. A., Tan, C., Prakapenka, D., Liu, G. E. and Da, Y. 2019. Genome changes due to artificial selection in US Holstein cattle. *BMC Genomics* 20(1):1–14.

Mazzetto, A. M., Bishop, G., Styles, D., Arndt, C., Brook, R. and Chadwick, D. 2020. Comparing the environmental efficiency of milk and beef production through life cycle assessment of interconnected cattle systems. *J Clean Prod* 277:124108.

Meuwissen, T. H. E. 1997. Maximizing the response of selection with predefined rate of inbreeding. *J Anim Sci* 75:934–940.

Nauta, W. J., Veerkamp, R. F., Brascamp, E. W. and Bovenhuis, H. 2006. Genotype by environment interaction for milk production traits between organic and conventional dairy cattle production in the Netherlands. *J Dairy Sci* 89:2729–2737.

Nielsen, H. M., Christensen, L. G. and Groen, A. F. 2005. Derivation of sustainable breeding goals for dairy cattle using selection index theory. *J Dairy Sci* 88:1882–1890.

Østergaard, S., Ettema, J. F., Clasen, J. B. and Kargo, M. 2018. SimHerd *Croosbred for estimating the economic effects of systematic dairy crossbreeding*. Proceedings of the World Congress on Genetics Applied to Livestock Production, 11.927.

Roche, J. R., Berry, D. P., Delaby, L., Dillon, P. G., Horan, B., Macdonald, K. A. and Neal, M. 2018. Review: new considerations to refine breeding objectives of dairy cows for increasing robustness and sustainability of grass-based milk production systems. *Animal* 12(s2):s350–s362.

Shonka-Martin, B. N., Hazel, A. R., Heins, B. J. and Hansen, L. B. 2019. Three-breed rotational crossbreds of Montbeliarde, Viking Red, and Holstein compared with Holstein cows for dry matter intake, body traits, and production. *J Dairy Sci* 102(1):871–882.

Slagboom, M., Hjortø, L., Sørensen, A. C., Mulder, H. A., Thomasen, J. R. and Kargo, M. 2020. Possibilities for a specific breeding program for organic dairy production. *J Dairy Sci* 103: 6332–6345.

Slagboom, M., Kargo, M., Edwards, D., Sørensen, A. C., Thomasen, J. R. and Hjortø, L. 2016. Organic dairy farmers put more emphasis on production traits than conventional farmers. *J Dairy Sci* 99:9845–9856.

Slagboom, M., Kargo, M., Edwards, D., Sørensen, A. C., Thomasen, J. R. and Hjortø, L. 2017. Herd characteristics influence farmers' preferences for trait improvements in danish red and danish jersey cows. *Acta Agric Scand Sect A Anim Sci* 66:177–182.

Slagboom, M., Wallenbeck, A., Hjortø, L., Sørensen, A. C., Rydhmer, L., Thomasen, J. and Kargo, M. 2018. Simulating consequences of choosing a breeding goal for organic dairy production. *JDS* 101:11086–11096.

Schmidtmann, C., Thaller, G., Kargo, M., Hinrichs, D. and Ettema, J. 2021. Derivation of economic values for German dairy breeds by means of a bio-economic model – with special emphasis on functional traits. *J Dairy Sci* 104:3144–3157.

Smith, C. and Banos, G. 1991. Selection within and across populations in livestock improvement. *J Anim Sci* 69:2387–2394.

Stefani, G., El Faro, L., Santana, M. L. and Tonhati, H. 2018. Association of longevity with type traits, milk yield and udder health in Holstein cows. *Livest Sci* 218:1–7.

Sundberg, T., Rydhmer, L., Fikse, W. F., Berglund, B. and Strandberg, E. 2010. Genotype by environment interaction of Swedish dairy cows in organic and conventional production systems. *Acta Agric Scand Sect A Anim Sci* 60:65–73.

Sørensen, M. K., Norberg, E., Pedersen, J. and Christensen, L. G. 2008. Crossbreeding in dairy cattle: a Danish perspective. *JDS* 91:4116–4128.

Sørensen, L. P., Pedersen, J., Kargo, M., Nielsen, U. S., Fikse, F., Eriksson, J.-Å., Pösö, J., Stephansen, R. S. and Aamand, G. P. 2018. Review of nordic total merit index full report. November 2018. https://www.nordicebv.info/wp-ontent/uploads/2018/11/2018.11.06-NTM-2018-report-Full.pdf

Vandenplas, J., Calus, M. P. L., Sevillano, C. A., Windig, J. J. and Bastiaansen, J. W. M. 2016. Assigning breed origin to alleles in crossbred animals. *Genet Sel Evol*, 48(1):61. https://doi.org/10.1186/s12711-016-0240-y

Van Selm, B., de Boer, I. J., Ledgard, S. F. and van Middelaar, C. E. 2021. Reducing greenhouse gas emissions of New Zealand beef through better integration of dairy and beef production. *Agric Syst* 186:102936.

Wilkinson, J. M. and Lee, M. R. F. 2018. Review: use of human-edible animal feeds by ruminant livestock. *Animal* 12(8):1735–1743.

Wooliams, J., Berg, P., Mäki-Tanila, A., Meuwissen, T. and Fimland, E. 2005. *Sustainable Management of Animal Genetic ressources*. Nordic Gene Bank Farm Animals. ISBN 92-893-1089-8

Woolliams, J. A., Berg, P., Dagnachew, B. S. and Meuwissen, T. H. E. 2015. Genetic contributions and their optimization. *J Anim Breed Genet* 132(2):89–99.

Chapter 3

Optimizing soils, pasture and grassland management for organic dairy farming

Stephen Roderick, Duchy College, UK; and Hannah Jones, Farm Carbon Toolkit, Hill View Farm, UK

1 Introduction

This chapter explores the opportunities for pasture-fed milk production, and in particular enhanced grassland management, to meet the health and welfare needs of organic dairy cattle and impact climate change mitigation and adaptation, while also satisfying other social and economic objectives. In particular, we will focus on describing and exploring the relative attributes and otherwise of rotational grazing systems and the use of diverse species pastures within an organic dairy farming context and how these influence the resilience of these farming systems. Soil health is key to successful organic farming, including organic dairy farming. As grassland management can be an effective means to sequester carbon and thus potentially contribute to mitigating greenhouse gas emissions (Soussana et al., 2010; Jordon et al., 2024), we will focus on how pasture-based systems offer viable solutions for organic milk production while enhancing soil health, and soil carbon storage. We will discuss the importance of the animal, and how successful organic grassland

http://dx.doi.org/10.19103/AS.2024.0138.03

farming relies on selecting appropriate breeds and types so as to achieve the highest level of animal welfare while remaining economically viable.

2 Pasture-fed dairy systems

Organic dairy cows should rely on homegrown forages, particularly pasture, and a minimum of 60% of an animal's dry matter ration should be from forages. The animals should have access to available pasture, and should be kept at stocking rates defined by the organic regulations, which are generally lower than those found in comparable non-organic dairy farms.

Restrictions on the use of artificial fertilizers mean that most farms rely heavily on legumes, particularly clovers, to provide a source of nitrogen for pasture growth (Watson et al., 2002). The proportion of clovers within a sward, including their consistency and persistency, will have considerable influence on the nutrient content, and in particular protein, supplied to grazing animals within an organic system. While there is increasing practical knowledge and scientific evidence to show the value of very diverse species within pastures, clovers remain a mainstay of organic dairy farms.

Most commercial dairy farms rely on varying proportions of non-forage feeds to increase the nutrient density and hence intake of animals whose milk production potential is limited by their dry matter intake capacity. While restricted feeding of concentrated feeds to organic dairy animals is permitted, such an approach has the potential to compete with feedstuffs that could otherwise be utilized as human food and can therefore be viewed as unsustainable (van Zanten, 2018 and van Zanten, 2019). Furthermore, the feeding of concentrated feeds does not fit well with the concept of naturalness associated with organic farming (Verhoog et al., 2003), in that the rumen has developed as an organ suited to the digestion of fibrous material.

It has been proposed that the best way to achieve profitable and sustainable organic dairy farming is to maximize milk production from forage and to minimize the use of purchased concentrates and grains (Nicholas et al., 2004), although forage-only organic herds are not very common due to the limited dry matter intake and the resulting low milk yield (Ertl et al., 2017). Achieving this requires animals to have high levels of voluntary feed intake from forage, with the associated grassland management practices that optimize forage digestibility. Compared with confined systems, although producing less milk, pasture-based systems can be economically efficient, with lower operating costs and higher net incomes (White et al., 2002). Such systems are best suited to regions where the environmental conditions allow long grazing seasons of grass growth (Butler, 2014).

Pasture-based systems rely heavily on seasonal breeding, whereby cows reproduce at the start of the main growing season (spring) and are thus able to meet their increasing energy and protein requirements when the nutrients from pasture are most abundant and available, i.e. the nutritional requirements of the cow are matched to the seasonal availability of nutrients. A critical management aim is to ensure cows have the optimum body condition at calving and to keep the inevitable lactational loss of condition to a minimum (Butler, 2014). It is important to maintain production and to ensure ongoing reproductive success in the subsequent year, as nutritional status impacts fertility.

3 Key considerations on organic dairy farms

Organic dairy farms should be designed to be reliant as much as possible on natural cycles, including ecological, nutrient, seasonal and biological cycles, while also ensuring the highest standards of health and welfare of individuals within a herd. Many of the legal restrictions that are applied to organic farms are designed to ensure the achievement of these aims. Some of the key limitations embedded in the regulations include the following requirements:

- Lower stocking rates, partly consequential of maximum permissible rates of nitrogen application, including manures, to 170 kg N/ha;
- Restrictions on the proportion of concentrates in the diets of organic animals (% of DMI);
- A requirement to feed animals, as far as possible, from feed produced on the farm; and
- Animals on organic farms should always be of a breed or crossbreed that is well adapted to the climatic and other natural conditions, as well as being well-suited to the organic production system.

The lower stocking rates observed on organic farms are in part due to the limits on concentrate use applied by organic farming standards and the need to feed animals, wherever possible, with feed produced on the farm (Nicholas et al., 2004). Producing milk predominantly from grass requires the breeding to be matched to the seasonal availability of pasture so that the nutritional requirements of lactating cows are synchronized to the seasonal fluctuations in nutrients supplied by the pasture (Horn et al., 2013).

4 Grazing management systems

The productivity of an organic dairy system is reliant on forage production and preservation, and the manner in which these are utilized. Inevitably grassland productivity is dominated by the factors which affect the photosynthetic capacity of the grass, namely sunlight, rainfall, and nutrient supply. The management of the available forage is the key factor influencing not only the efficiency, but the wider ecological and ethological objectives. Assessment of grass growth and then consequent adaptation of stocking rates can be a significant component of pasture management.

There are various methods of grazing management employed by dairy farmers that can be defined by different factors and parameters, including the grazing intensity (number of animals per unit area), the period of grazing, the condition of the pasture at the time of starting and cessation of grazing and the rest period between subsequent grazing. A range of terms has been used to describe the extent of some of these approaches, including set stocking, continuous grazing, rotational grazing, paddock grazing, cell grazing, mob grazing, and holistic grazing. While there is a degree of contradiction in the literature regarding precise definitions, Allen et al. (2011) provide a consensus.

4.1 Rotational or paddock grazing

Rotational grazing involves frequently rotating animals between small, confined paddocks, often created by the sub-division of larger fields. Each paddock is rested between grazing to allow grass to regrow. This is a different principle from set stocking, which involves leaving animals within larger fields for longer periods with no rest periods. Rotational grazing has a number of potential advantages (Jordon et al., 2023), including increasing the quantity of forage available to grazing animals over the season as plants grow more rapidly after resting (Jordon et al., 2022). Pasture growth is generally stimulated by grazing, and the extent of the subsequent grass regrowth resulting from increased manuring will be dependent on other conditions, including soil type and climate. Adaptive grazing practices allow the periods of grazing and rest to be adapted according to climate and seasonal growing conditions, and the resulting impacts on forage growth and quality.

A key advantage of paddock grazing relates to the potential improvement in soil health. Plant roots are larger and soil organic matter is increased. Also, there is increased productivity as plant growth is more rapid after resting from grazing. Rotational grazing reduces risks of soil compaction and has a positive impact on soil carbon and greenhouse gas emissions from soil (Ma et al., 2021; Reed and Morrissey, 2022; Baronti et al., 2022; Abagandura et al., 2024). There are also potential improvements in soil fertility, partly as a consequence of more effective spreading of manure as animals are moved between paddocks. The use of paddocks to increase the fertility of crops grown in rotation within an integrated cropping-dairy system is described by Roderick and Whistance in Chapter 12.

There are some disadvantages of rotational grazing, including the associated additional infrastructure and labour requirements (Jordon et al., 2023). Paddock grazing frequently relies on temporary electric fencing to periodically allocate fresh pasture. These paddocks often require both front and back fencing for more efficient pasture management and to reduce soil damage. The labour requirement for this activity can be a significant practical and economic factor. The development of virtual fencing, which relies on collar-mounted GPS devices capable of delivering audio cues or an electric shock, has been used to restrict animals within a specific grazing area. There are welfare issues to consider with the use of these virtual systems (Lomax et al., 2019), including situations such as the increasingly popular option of keeping lactating cows and their calves together (see Barth et al., Chapter 6). Virtual systems could be beneficial to unrestrained calves, allowing them access to superior grazing, but conversely could create stress and anxiety for virtually restrained mothers unable to access calves. As with all situations involving electric shock to animals, there should be clear ethical justification for the use of this form of containment that cannot be practically achieved by other means (Grumett and Butterworth, 2022).

Rotational grazing applied under wet conditions can also cause soil damage, or poaching, potentially causing problems with diffuse water pollution, loss of productivity as well as animal welfare issues although these risks can be minimized by setting effective paddock size and grazing intervals. The effect of trampling under high grazing intensity can alter the soil microbial balance, with a resulting negative impact on soil organic carbon content (Abdalla et al., 2018).

There are other important animal welfare considerations inherent to confined grazing systems such as paddock grazing. The time cows spend walking to paddocks can influence cows' behaviour during their time spent within paddocks. In line with organic farming principles of enabling animals more opportunities to meet their natural behavioural needs (Lund and Weary, 2005), providing animals with shorter walking distances allows more time for lying, resting and ruminating (Neave et al., 2021).

The use of temporary paddocks to facilitate rotational grazing can also raise issues around the provision of drinking water (Jensen and Vestergaard, 2021), shade (Kendall et al., 2006., Schütz et al., 2014), and shelter. Heat stress caused by the absence of shade can have negative impacts on feed intake, milk production, respiration rate, and body temperature (Fisher et al., 2008). The introduction of trees integrated within pastures can offset some of these adverse effects (Karki and Goodman, 2009; Carnevalli et al., 2020).

4.2 Mob grazing

Mob grazing is a form of rotational grazing defined as a 'method of stocking at a high grazing pressure for a short time to remove forage rapidly' (Allen et al., 2011). It is a method of pasture management that aims to partly mimic the natural systems of grazing of large herds of wild herbivores in the rangelands. It involves short periods of grazing animals at very high stocking densities followed by long periods of resting and has been shown to have positive impacts on soil organic matter and soil quality. Short periods of grazing and long periods of recovery are intended to improve pasture regrowth as a consequence of the large quantities of vegetation being left un-grazed, which in turn can result in more photosynthesis occurring and more extensive root complexes developing.

Over time, it has been claimed that this approach encourages a shift within pastures toward greater species diversity with taller grass species, with enhanced overall productivity (Wagner et al., 2023). Long pasture recovery periods may also result in better utilization of mature swards (Leach et al., 2014). However, it should be noted that evidence of the benefits of mob grazing is not universally conclusive and experiences and measurable impacts diverge (Briske et al., 2014; Jordon et al., 2024). Most examples are from extensive beef farms, particularly in the United States and Australia, and management approaches vary with regard to timing, intensity of grazing, and periods of rest. In the United Kingdom, some beef farmers adopting mob grazing also included sheep grazing and forage conservation as part of their management (Wagner et al., 2023). For many farmers practicing this approach, it is seen as an important part of a wider holistic or regenerative approach to farming (Wagner et al., 2023).

5 Grazing management and soil quality

Soil health is key to ensuring effective pasture-based dairy farming, particularly in organic systems that rely on efficient re-cycling of nutrients through natural processes. Our current understanding of soil health extends beyond the potential to sustain crop production to the role as an essential living ecosystem responsible for ensuring multiple ecosystem services, including water retention and carbon sequestration (Lehmann et al., 2020).

Grazing management can have a major effect on soil carbon (Conant et al., 2001; Jones and Donelly, 2004; Sanderman et al., 2017), although impacts can vary significantly (McSherry and Ritchie, 2013) and the extent to which grasslands sequester carbon in the soil will depend on many factors, including grassland grazing intensity, the composition of the pasture, and the species of grazing animal (Abdalla et al., 2018). While in organic farming, the number of animals grazing an area of pasture across a farm is regulated, it is the timing, intensity, duration, and frequency of grazing (the grazing pattern) that has the greatest influence on soil organic matter rather than total numbers (Stanley et al., 2024).

Developing effective grazing management approaches requires an understanding of the likely impacts of grazing on soil carbon accumulation and storage. In grassland situations, soil carbon turnover is strongly associated with changes in root mass, which in turn is influenced by grazing practice (Chen et al., 2015). Therefore, decisions on management practices, including grazing pressure, should reflect the productivity potential of the pasture.

In the short term, plant defoliation through grazing impacts photosynthetic capability, and the extent of the impact will vary between pasture types depending on factors such as scale of defoliation, species composition, and season and stage of plant growth. The impact of grazing patterns on soil organic matter will be cumulative over time. Stanley et al. (2024) postulate that these long term impacts will be determined by ground and canopy cover, productivity, input allocation (i.e. root:shoot ratio), input quality (ratio of plant soluble:structural components), and species diversity. Optimal grazing would enhance all or some of these elements, which in turn impact the beneficial physiological processes that facilitate soil organic carbon sequestration. Within natural pastures, reductions in greenhouse gas emissions, including changes in soil organic carbon, have been shown to be most effective under moderate rather than high stocking rates (Liebig et al., 2010).

Grazing can impact soil microbial communities, which in turn may change the extent of decomposition of soil organic matter. Under certain soil and rainfall conditions, at high levels of grazing intensity, there is a risk of soil structure degradation and a reduction in soil organic carbon (Reed and Morrissey, 2022) with the associated losses in a range of ecosystem benefits, including water retention and nutrient capacity. Conversely, high grazing intensity for short periods can have positive impacts on soil carbon and plant diversity. Such impacts are also partly dependent on the extent of vegetation removal, with complete or severe removal likely to have negative effects. Soussana et al. (2010) identify light grazing rather than heavy grazing and increasing the duration of grass leys as grassland management practices that are most likely to aid in reducing carbon loss and increase carbon sequestration.

Multi-paddock grazing involves grazing small paddocks at high stocking densities for short periods, as described above. This method of grazing management can result in changes in moisture content, nutrient availability, and plant growth, providing potentially more favourable conditions for the soil microbial communities and processes that influence soil carbon dynamics (Khatri-Chhetri et al., 2022). When well-managed, rotational grazing can increase soil carbon although the extent and scale are not consistent (Jordon et al., 2024). Multi-paddock or adaptive grazing has been shown to have a positive influence on greenhouse gas processes and fluxes (Shrestha et al., 2020), sequestering carbon in the soil as well as retaining soil nitrogen (Mosier et al., 2021) and improving soil water infiltration (Döbert et al., 2021; Baronti et al., 2022).

It has been noted above that grazing density has been shown to be particularly influential, although the nature and direction of this relationship can differ under different climatic, soil, and vegetation types. Higher fungal:bacterial ratios are normally associated with higher soil carbon levels (Grayston et al., 2004; Jastrow et al., 2007), but not always (Strickland and Rousk, 2010; Chen et al., 2015). High grazing densities can result in a negative change in this fungus-to-bacteria relationship (Khatri-Chhetri et al., 2022).

In relation to the impacts on soil organic carbon, stocking rate is not the sole determinant of grazing intensity. Other factors such as timing, frequency, and duration are also important influential factors. The phenological growth stage of a plant can influence the response to being defoliated, the period between grazing impacts the plant's ability to recover following grazing and the length of time that a pasture is grazed can affect the extent of defoliation (Stanley et al., 2024). Rotational systems with short periods of rest reduce the risk of plants senescing, providing animals with more palatable and digestible forage, and thus enabling potentially higher levels of animal productivity. Conversely, longer periods of rest between grazing results allow plants to reach maturity, with greater forage production and more root growth. The impact on, and variations in, emissions between these various approaches are not easily quantifiable and are highly context-specific (Jordon et al., 2024).

Given the seasonal and unpredictable impact of climate on primary production of grassland, it is not advisable to base grazing management decisions solely on standard levels of grazing intensity. Evidence-based grassland management approaches need to incorporate flexible decisions in relation to animal numbers, timing, duration, and frequency of grazing. Adaptive grazing practices, based on monitoring and evaluation of the productive elements that influence the soil organic carbon pathways, allows flexibility in response to inherent climatic fluctuations. Grazing strategies aimed at reducing greenhouse gas emissions also require consideration of environmental and soil conditions, including soil moisture content and bulk density (Ma et al., 2021).

6 Diverse and multi-species pastures

From a sustainability perspective, the desired attributes of an organic pasture include:

- The ability to provide sufficient nutrition to enable the production of milk without dependence on other forms of feed;
- Resilience to variable and acute fluctuations in climatic conditions;
- The ability to flourish without the addition of artificial forms of fertilizer; and
- Provision of ecological conditions that enable biodiversity to flourish.

Increasingly, multiple species pastures, also referred to as herbal or diverse leys, are being recognized as a means by which these sustainability objectives can be achieved, as well as being a potential source of carbon sequestration. Increasing plant biodiversity and including more nitrogen-fixing legumes, not only improves animal nutrition but boosts biodiversity in the wider ecosystem, helping to increase soil nitrogen levels and carbon sequestration (Teague and Kreuter, 2020).

A summary of the main mechanisms and factors that may result in increased pasture productivity of multi-species swards is provided by Jordon et al. (2022) as

- enhanced niche complementarity resulting from species diversity, including:
  - more root variation and root depth giving more consistent forage quality throughout the season as a result of greater drought tolerance; and
  - greater access to a wider soil profile improving nutrient uptake;
- the ability of legumes to fix atmospheric nitrogen and increase nitrogen supply to other plants;
- more diverse species increasing the ability to thrive across a wide range of growing conditions;
- a higher nutritive value of individual species giving a higher overall protein content;
- better palatability; and
- a higher mineral content.

A multi-species pasture will commonly include a combination of one or more grass, legume and forb (herb) species. Figure 1 shows an image of a mixed-species pasture in Cornwall, England, containing a number of different legume species, including white clover (*Trifolium repens*), red clover (*Trifolium pretense*), Lucerne (*Medicago sativa*), and bird's-foot trefoil (*Lotus corniculatus*), as well as forb species such as plantain (*Plantago lanceolata*) and chicory (*Cichorium intybus*), alongside common grass species including perennial ryegrass (*Lolium perenne*) and cocksfoot (*Dactylis glomerata*). Generally, there is a positive relationship between species diversity and pasture biomass production, and the diversity of functional traits exhibited within a multi-species sward is probably more important than the number of species present (Huyghe et al., 2012). The relative abundance of species and functional groups is a key factor in their impact and persistence within the sward. Species that are complementary and not competitive are most likely to survive and flourish (Huyghe et al., 2012; Jing et al., 2017). Functional diversity is particularly relevant to organic systems, where nutritional resources available for plant growth may be limiting at certain

Figure 1 A diverse species pasture in Cornwall, UK containing mixtures of grasses, legumes, and forbs.

periods, and contrasting functional attributes with regard to, for example, rooting depth, enable greater complementarity between species with the resulting positive impacts on productivity.

6.1 Nutritional benefits

A combination of plant species ensures a variety of growth patterns, providing consistency of forage supply throughout the year and the flattening of the peak of forage production and an extension to the grazing season (Cooledge et al., 2022). Although the nutritive value of multi-species pasture can differ from the values predicted by their single-species constituents (Moloney et al., 2020), their value is largely influenced by the dominant species rather than the complexity of the species mixture (Deak et al., 2007). As a result of the cumulative effect of improved pasture nutritive value and increased pasture dry matter intake, the milk production performance of cows grazing on multi-species mixtures has been shown to be greater than that of animals grazed on simpler ryegrass and clover mixtures (Roca-Fernández et al., 2016; McCarthy et al., 2020). Lower milk fat percentages may occur when animals graze multi-species pastures, possibly a consequence of the lower fibre content of forbs resulting in reduced rumination and a lower rumen pH (Bryant et al., 2017).

6.2 Species selection

Knowledge of plant growth habits, seasonality, and resource utilization capabilities under local environmental conditions are important considerations when choosing species for inclusion in diverse pastures. Including species with complementary traits to improve the species-richness of plant communities allows greater access to resources, enabling a rapid adaptation to climatic changes and more consistent productivity.

Introducing a wider range of species within a pasture, so as to increase the functional diversity, can result in greater flexibility in response to changing growing conditions, and increased resilience to adverse weather conditions in particular. Deep-rooting species such as cocksfoot (*Dactylis glomerata*), timothy (*Phleum pratense*), and chicory (*Cichorium intybus*) improve drought resistance, and allows for deep soil carbon sequestration, which helps improve microbial activity (Kemp et al., 2010). Chicory also has nutritional benefits, including high mineral concentration and high organic matter digestibility. However, at high inclusion rates in pastures, the low dry matter content of chicory may have a lowering effect on overall pasture intake (Li and Kemp, 2005). Nitrogen losses from dairy cattle grazing on multispecies swards containing forbs such as chicory and plantain have been shown to be lower than that with grass/clover pastures, with no effect on milk production (Totty et al., 2013; Bryant et al., 2017).

6.3 Pasture persistency

The persistency of the grazed sward is an important economic and environmental consideration. Grazing management can have the most important impact on the presence and persistence of species diversity. The structure and the composition of a sward are influenced by the frequency and height of grazing, the extent of overgrazing, soiling (by faeces and urine) on the sward heterogeneity, and the extent to which grass seeds are disseminated by grazing animals (Huyghe et al., 2012). A key to achieving yield persistency and species diversity is the selection of appropriate and compatible species, coupled with compatible management. Some of the advantages of diversity are dependent on the cutting or grazing strategy, as forbs and legumes are impacted by continuous grazing (Jing et al., 2017). Evidence shows that, in general, high grazing intensities reduce plant species diversity. However, if well managed, rotational grazing systems can maintain or even increase plant diversity over time (Reed and Morrissey, 2022).

6.4 Carbon sequestration

Plant species density is positively associated with higher soil organic matter (Jordon et al., 2022), although the age of a multi-species pasture is a critical factor influencing the extent of carbon sequestration that can be achieved. Evidence suggests that, depending on other factors such as soil type and management, functional diversity has to continue over many years before reliable increases in soil carbon concentration occur (Skinner and Dell, 2016; Reed and Morrissey, 2022; Thomson and Albornoz, 2023). Compared with simpler pastures, rooting complexity and the associated below-ground biomass appear to be a major contributor to improved soil quality, including soil carbon content (McNally et al., 2015). It may be the case that the increased productivity advantage of long-term, multi-species pastures could result from increased soil carbon, even when species composition has started to decline with time (Skinner and Dell, 2016).

6.5 Barriers to uptake

For some, a lack of knowledge on the suitability of lesser-known grasses and herbs for farm-specific conditions, particularly soil and climatic conditions, may be a barrier to the use of multiple species. Factors relating to mixture composition (choice, cost, and grazing tolerance) were identified as key barriers to diverse ley uptake amongst farmers (organic and non-organic) in the United Kingdom (Stagg et al., 2021).

The majority of organic farmers in Denmark who professed to grow herb species in pasture did so primarily as a source of minerals, and while chicory is generally considered the most widely grown pasture herb, interestingly the most popular grown within the Danish organic and biodynamic farmer cohort were caraway and parsley (Smidt and Brimer, 2005).

There are a number of perceived and real disadvantages of multi-species swards, including the risk of loss of palatability if certain species such as chicory are allowed to become too mature (Jordon et al., 2023). While the relatively high cost of seed may be a discouragement for some farmers, the variable success of germination and establishment and the loss of desirable species is of widespread concern to farmers (Jordon et al., 2023). Knowledge of local suitability is important as is regular monitoring of species persistence. Over time, the loss of desirable species such as chicory and clovers may require re-introduction (Sanderson et al., 2005).

7 Grassland management and the farm carbon balance

Grassland soils provide an immense reservoir of soil carbon that can have a significant impact on the farm's carbon balance or footprint. Consequently, grassland management, particularly effective rotational grazing and the use of diverse species pastures, has been mooted as a tool that could contribute to the reduction of greenhouse gas emissions from farming (Jordon et al., 2022; Vogel and Beber, 2022).

Although there are different approaches to the quantification of the carbon footprint of a farm, the basic premise of carbon auditing involves adding together all of the emissions from the different farm components and balancing these against an estimate of the carbon sequestered or stored on the farm.

Generally, there is a lack of clear empirical evidence showing the comparative carbon footprint superiority of organic dairy farming, and comparisons that exist may have suffered from methodological shortcomings. More recently, a large-scale study of French dairy farms, incorporating land-use-related emissions, has shown the carbon footprint of organic milk to be 11–19% lower than that measured on non-organic dairy farms (Lambotte et al., 2023). Land-use factors relate to the fact that while the area of pasture land required on organic farms is proportionally higher than non-organic, land required to grow concentrated feeds is significantly lower.

At comparable production levels, pasture-based dairy systems can produce milk with a lower carbon footprint than more intensive, confined systems (Lorenz et al., 2019). Considerable variation has been noted in the carbon footprint of organic dairy farms (Hietala et al., 2015). The carbon footprint per kg of milk on organic farms is heavily influenced by the proportion of pasture, a result of the greater capacity for carbon sequestration in grasslands (Guerci et al., 2013).

It would be expected that the conversion of a farm to organic would result in a reduction in the footprint, largely as a consequence of the cessation of artificial fertilizers and enhanced nutrient re-cycling. However, this may not always translate to a lower footprint per unit of milk production, as yield per cow is likely to be lower post-organic conversion. Gross et al. (2022) showed that the carbon footprint per unit of energy-corrected milk dropped by 9% during conversion to organic certification. Even though methane emissions per kg of milk increased in the first year of conversion, this was offset by large decreases in emissions emanating from both on-farm and purchased feeds.

It has been demonstrated that increases in milk yield, pasture intake, and feed efficiency can significantly decrease the carbon footprint of dairy farming across a range of systems. Of course, without supplementary feeds, one of the issues with pasture-based systems is that they tend to produce lower milk yields per cow than other systems, although the disparity can be narrowed through appropriate genetic selection and good grazing management. Nutritional and genetic attributes are also areas that could enhance the mitigation opportunities on lower-yielding dairy farms, provided the wider organic aspirations associated with increased biodiversity and soil carbon sequestration are accounted for (Hietala et al., 2015).

8 Breed suitability for pasture systems

Breed suitability and selection of genotypes that are adapted to the prevailing conditions are fundamental to organic farming principles and to effective management, as discussed in Chapter 2. Here, we consider this in relation to grassland management and the ambition to produce milk either entirely or largely from pasture and conserved forage.

The importance of adopting selection indices with economic traits that are relevant to the production environment has been demonstrated (McCarthy et al., 2007). In milk production systems that depend largely or entirely on pasture as the primary nutrient source, the aim is to ensure that the nutrient demand of lactating animals coincides with seasonally available forage. Breeds selected for high-input, high yielding production scenario are clearly not suited for pasture-based systems as under these conditions, cows will be underfed relative to their intake potential (Nauta et al., 2006; Horn et al., 2013; Butler, 2014). Animals that are able to maintain a higher body condition score are able to draw on fat reserves during periods when pasture quality or quantity may be seasonally limited, and therefore better able to maintain production and remain fertile (Pryce and Harris, 2006).

A requirement of dairy cows in pasture-based systems is the ability to efficiently convert forage, particularly grazing, into milk (Davis et al., 2020). In grazing systems, Jersey cows tend to utilize feed more efficiently during periods when feed is limited, and they have been shown to require less total feed than Holstein-Friesian cows for the production of a unit of milk fat and protein, as well as consuming comparatively higher dry matter intake per unit body weight (Prendiville et al., 2009, Spaans et al., 2018). Dual-purpose breeds, such as the Normandy, are more flexible and better adapted to the production of milk solely or largely from grassland (Peyraud et al., 2014).

Seasonal breeding to coincide with seasonal pasture growth patterns can be critical in pasture-based systems, and breeding strategies need to either emphasize reproductive efficiency or focus on traits associated with longer calving intervals and extended lactations. There can also be significant differences in survival and reproduction between breeds managed in forage-based systems (Dillon et al., 2003b). Under non-organic, low-input conditions in Ireland, crossbred Jersey × Holstein-Friesian cows have been shown to have similar production levels per cow to Holstein-Friesian cows kept under the same conditions, but far superior fertility and survival, as well as a lower carbon footprint of milk (Peyraud et al., 2014). Although Dutch Holstein-Friesian cows have been shown to produce more milk from pasture-based systems than other breeds as a result of being able to mobilize more body reserves in early lactation and having lower live weight gain in later lactation (Dillon et al., 2003a), their poor reproductivity showed them to be less suited to this type of system (Dillon et al., 2003b).

There are differences between strains of the same breed with regard to their feed efficiency and, hence, their suitability for pasture-based systems (Coleman et al., 2009). While the Holstein Friesian breed has been developed for high input, concentrate-based dairy farming, Horn et al. (2013) showed that some strains of this breed, demonstrating superior feed intake per unit of metabolic body weight and reproductive performance, may be better suited to Alpine, pasture-based, low-input milk production systems than the more traditional strain of the Brown Swiss breed.

The benefits of crossbreeding for pasture-based systems are well established, and there is evidence to indicate the superiority of crossbreeds under these conditions (Buckley et al., 2014; McClearn et al., 2020) with regard to reproductive efficiency, body condition score, and longevity compared with pure-bred animals (Washburn and Mullen, 2014). Crossbreeding allows organic farmers to benefit from some of the desired functional traits of low heritability that are found in locally adapted breeds, and the production potential of higher-yielding breeds (Sørensen et al., 2008; Davis et al., 2020). The Jersey, Montbéliarde, Normande, and Norwegian Red breeds have all been shown to be well-suited to crossbreeding with the Friesian breed in pasture-based systems (McClearn et al., 2020).

Cattle kept solely on pasture-based systems may have superior health status and greater longevity, partly as a consequence of spending less time standing and walking on concrete, which may result in a different emphasis being placed on the selection for fitness traits (Washburn and Mullen, 2014). Selection indices reflecting yield traits are also likely to be less relevant for pasture-based systems, as these would have most probably been estimated from animals produced under high production conditions (Washburn and Mullen, 2014). However, there is considerable information and practical knowledge from the non-organic sectors in countries such as New Zealand and Ireland on the production potential of breeds kept under intensive pasture systems that is relevant for organic farming (Rodríguez-Bermúdez, 2019).

9 Conclusion

Effective grazing and forage management are the cornerstone of organic dairy farming, allowing the nutritional and behavioral needs of herd members to be met largely from the grazing and the conservation of pasture. In addition, reducing dependence on concentrated feed sources and optimizing the production of diverse species pastures can contribute to long-term environmental objectives through reduced dependence on fossil fuels, enhanced soil quality and life, recycling nutrients, and maximizing photosynthetic processes. The sustainability of these systems is further enhanced by selecting breeds and crossbreeds that can thrive, produce, and reproduce under predominantly outdoor conditions on forage-based diets. The impact of lower milk yields on the carbon footprint of organic dairy herds can be balanced by good pasture management.

10 Where to look for further information

The Toolbox of Multi-species Swards (TOMS) https://www.multispeciessward.co.uk/ project researched the uses and benefits of multi-species swards, with a particular emphasis on growing conditions in Cornwall. The website contains summarized findings from a review of published research into the key benefits of multi-species swards and results from a range of field experiments, including forage quality and mineral content. The site also presents results from a survey of farmers perceptions of herbal leys, as well as any barriers to uptake. There are links to the Multi-Species Sward app, which has been developed to provide farmers with easy-to-access information about herbal leys and to help them to identify and monitor different species within their leys.

The Farm Carbon Toolkit https://farmcarbontoolkit.org.uk/ is dedicated to understanding greenhouse gas emissions in agriculture and providing tools and services to measure impact. The website contains a free farm carbon calculator and toolkit as well as learning resources for farmers.

- Vaarst, M. and Roderick, S. (2019). Improving organic animal farming. Burleigh Dodds Science Publishing Limited. Contains relevant chapters on organic dairy farming and on greenhouse gas emissions from organic farming.
- Vaarst, M., et al. eds. (2004). Animal health and welfare in organic agriculture. CABI. Contains relevant chapters on dairy nutrition and animal welfare and ethics.

11 References

Abagandura, G. O., et al. (2024). Soil carbon and nitrogen after eight years of rotational grazing in the Nebraska Sandhills meadows. *Geoderma*, 442, 116776.

Abdalla, M., et al. (2018). Critical review of the impacts of grazing intensity on soil organic carbon storage and other soil quality indicators in extensively managed grasslands. *Agriculture, Ecosystems & Environment*, 253, 62–81.

Allen, V. G., et al. (2011). An international terminology for grazing lands and grazing animals. *Grass and Forage Science*, 66, 2–28.

Baronti, S., et al. (2022). Rotational pasture management to increase the sustainability of mountain livestock farms in the Alpine region. *Regional Environmental Change*, 22(2), 50.

Briske, D. D., et al. (2014). Commentary: a critical assessment of the policy endorsement for holistic management. *Agricultural Systems*, 125, 50–53.

Bryant, R. H., et al. (2017). Milk yield and nitrogen excretion of dairy cows grazing binary and multispecies pastures. *Grass and Forage Science*, 72(4), 806–817.

Buckley, F., et al. (2014). Crossbreeding: implications for dairy cow fertility and survival. *Animal*, 8(s1), 122–133.

Butler, S. T. (2014). Nutritional management to optimize fertility of dairy cows in pasture-based systems. *Animal*, 8(s1), 15–26.

Carnevalli, R. A., et al. (2020). Shade controls the ruminating and idleness times of dairy heifers in tropical integrated systems. *Agroforestry Systems*, 94, 779–790.

Chen, W., et al. (2015). Improved grazing management may increase soil carbon sequestration in temperate steppe. *Scientific Reports*, 5(1), 1–13.

Coleman, J., et al. (2009). The influence of genetic selection and feed system on the reproductive performance of spring-calving dairy cows within future pasture-based production systems. Journal of Dairy Science, 92(10), pp.5258–5269.

Conant, R. T., et al. (2001). Grassland management and conversion into grassland: effects on soil carbon. *Ecological Applications*, 11(2), 343–355.

Cooledge, E. C., et al. (2022). Agronomic and environmental benefits of reintroducing herb and legume rich multispecies leys into arable rotations: a review. *Frontiers of Agricultural Science & Engineering*, 9(2), 245–271.

Davis, H., et al. (2020). Evidence that forage-fed cows can enhance milk quality. Sustainability, 12(9), p.3688.

Deak, A., et al. (2007). Production and nutritive value of grazed simple and complex forage mixtures. *Agronomy Journal*, 99(3), 814–821.

Dillon, P., et al. (2003a). A comparison of different dairy cow breeds on a seasonal grass-based system of milk production: 1. Milk production, live weight, body condition score and DM intake. *Livestock Production Science*, 83(1), 21–33.

Dillon, P., et al. (2003b). A comparison of different dairy cow breeds on a seasonal grass-based system of milk production: 2. Reproduction and survival. *Livestock Production Science*, 83(1), 35–42.

Döbert, T. F., et al. (2021). Adaptive multi-paddock grazing improves water infiltration in Canadian grassland soils. *Geoderma*, 401, 115314.

Ertl, P., et al. (2017). Effects of supplementation of a forage-only diet with wheat bran and sugar beet pulp in organic dairy cows. Renewable Agriculture and Food Systems, 32(5), pp.446–453.

Fisher, A. D., et al. (2008). Effects of shade provision on the behaviour, body temperature and milk production of grazing dairy cows during a New Zealand summer. *New Zealand Journal of Agricultural Research*, 51(2), 99–105.

Guerci, M., et al. (2013). Parameters affecting the environmental impact of a range of dairy farming systems in Denmark, Germany and Italy. *Journal of Cleaner Production*, 54,133–141.

Grayston, S. J., et al. (2004). Assessing shifts in microbial community structure across a range of grasslands of differing management intensity using CLPP, PLFA and community DNA techniques. *Applied Soil Ecology*, 25(1), 63–84.

Gross, A., et al. (2022). The carbon footprint of milk during the conversion from conventional to organic production on a dairy farm in central Germany. *Agronomy for Sustainable Development*, 42(3), 37.

Grumett, D. and Butterworth, A. (2022). Electric shock control of farmed animals: welfare review and ethical critique. *Animal Welfare*, 31(3), 373–385. https://doi.org/10.7120/09627286.31.4.006

Hietala, S., et al. (2015). Carbon footprints of organic dairying in six European countries–real farm data analysis. *Organic Agriculture*, 5, 91–100.

Horn, M., et al. (2013). Suitability of different dairy cow types for an Alpine organic and low-input milk production system. *Livestock Science*, 153(1–3), 135–146.

Huyghe, C., et al. (2012). Agronomic value and provisioning services of multi-species swards. *Grassland Science in Europe*, 17, 35–46.

Jastrow, J. D., et al. (2007). Mechanisms controlling soil carbon turnover and their potential application for enhancing carbon sequestration. *Climate Change*, 80, 5–23.

Jensen, M. B. and Vestergaard, M. (2021). Invited review: freedom from thirst–do dairy cows and calves have sufficient access to drinking water?. *Journal of Dairy Science*, 104(11), 11368–11385.

Jing, J., et al. (2017). Species diversity effects on productivity, persistence and quality of multispecies swards in a four-year experiment. *PLoS One*, 12(1), e0169208.

Jordon, M. W., et al. (2022). Rotational grazing and multispecies herbal leys increase productivity in temperate pastoral systems–a meta-analysis. *Agriculture, Ecosystems & Environment*, 337, 108075.

Jordon, M. W., et al. (2023). Advantages, disadvantages, and reasons for non-adoption of rotational grazing, herbal leys, trees on farms and ley-arable rotations on English livestock farms. *Agroecology and Sustainable Food Systems*, 47(3), 330–354.

Jordon, M. W., et al. (2024). A restatement of the natural science evidence base concerning grassland management, grazing livestock and soil carbon storage. *Proceedings of the Royal Society B*, 291(2015), 20232669.

Jones, M. B. and Donnelly, A. (2004). Carbon sequestration in temperate grassland ecosystems and the influence of management, climate and elevated CO2. *New Phytologist*, 164(3), 423–439.

Karki, U. and Goodman, M. S. (2009). Cattle distribution and behavior in southern-pine silvopasture versus open pasture. *Agroforestry Systems*, 78, 159–168. *Proceedings of the Nutrition Society*, 62, 383–392.

Kemp, P. D., et al. (2010). The use of legume and herb forage species to create high performance pastures for sheep and cattle grazing systems. *Revista Brasileira de Zootecnia*, 39, 169–174.

Kendall, P. E., et al. (2006). The effects of providing shade to lactating dairy cows in a temperate climate. *Livestock Science*, 103, 148–157.

Khatri-Chhetri, U., et al. (2022). Adaptive multi-paddock grazing increases soil nutrient availability and bacteria to fungi ratio in grassland soils. *Applied Soil Ecology*, 179, 104590.

Lambotte, M., et al. (2023). Organic farming offers promising mitigation potential in dairy systems without compromising economic performances. *Journal of Environmental Management*, 334, 117405.

Leach, K., et al. (2014). Diverse swards and mob grazing for dairy farm productivity: a UK case study. In: Rahmann, G. and Aksoy, U. (Eds.) *Building Organic Bridges*, Johann Heinrich von Thünen-Institut, Braunschweig, Germany, 4, Thuenen Report, no. 20, pp. 1155–1158.

Lehmann, J., et al. (2020). The concept and future prospects of soil health. *Nature Reviews Earth & Environment*, 1(10), 544–553.

Li, G. and Kemp, P. D. (2005). Forage chicory (Cichorium intybus L.): a review of its agronomy and animal production. *Advances in Agronomy*, 88, 187–222.

Liebig, M. A., et al. (2010). Grazing management contributions to net global warming potential: a long-term evaluation in the northern great plains. *Journal of Environmental Quality*, 39(3), 799–809.

Lomax, S., et al. (2019). Does virtual fencing work for grazing dairy cattle?. *Animals*, 9(7), 429.

Lorenz, H., et al. (2019). Is low-input dairy farmingmore climate friendly? A meta-analysis of the carbon footprints of different production systems. *Journal of Cleaner Production*, 211, 161–170. https://doi.org/10.1016/j.jclepro.2018.11.113

Lund, V. and Weary, D. (2005). Understanding animal behaviour and its use in organic animal husbandry. In: Vaarst, M., et al. (Eds.) *Animal Health and Welfare in Organic Agriculture*, CABI Publishing, Wallingford. pp. 95–115.

Ma, Z., et al. (2021). Photosynthetic carbon allocation to live roots increases the year following high intensity defoliation across two ecosites in a temperate mixed grassland. *Agriculture, Ecosystems & Environment*, 316, 107450.

McCarthy, S., et al. (2007). Economic comparison of divergent strains of Holstein-Friesian cows in various pasture-based production systems. *Journal of Dairy Science*, 90(3), 1493–1505.

McCarthy, K. M., et al. (2020). Herb species inclusion in grazing swards for dairy cows–a systematic review and meta-analysis. *Journal of Dairy Science*, 103(2), 1416–1430.

McClearn, B., et al. (2020). An assessment of the production, reproduction, and functional traits of Holstein-Friesian, Jersey× Holstein-Friesian, and Norwegian Red×(Jersey× Holstein-Friesian) cows in pasture-based systems. *Journal of Dairy Science*, 103(6), 5200–5214.

McNally, S. R., et al. (2015). Root carbon inputs under moderately diverse sward and conventional ryegrass–clover pasture: implications for soil carbon sequestration. *Plant and Soil*, 392, 289–299. https://doi.org/10.1007/s11104-015-2463-z

McSherry, M. E. and Ritchie, M. E. (2013). Effects of grazing on grassland soil carbon: a global review. *Global Change Biology*, 19(5), 1347–1357.

Moloney, T., et al. (2020). Herbage nutritive value of binary-and multi-species swards relative to single-species swards in intensive silage systems. *Irish Journal of Agricultural and Food Research*, 59(1), 167–184.

Mosier, S., et al. (2021). Adaptive multi-paddock grazing enhances soil carbon and nitrogen stocks and stabilization through mineral association in southeastern US grazing lands. *Journal of Environmental Management*, 288, 112409.

Nauta, W. J., et al. (2006). Genotype by environment interaction for milk production traits between organic and conventional dairy cattle production in the Netherlands. *Journal of Dairy Science*, 89(7), 2729–2737.

Neave, H. W., et al. (2021). Do walking distance and time away from the paddock influence daily behaviour patterns and milk yield of grazing dairy cows? *Animals*, 11, 2903. https://doi.org/10.3390/ani11102903

Nicholas, P. K., et al. (2004). Organic dairy production: a review. *Biological Agriculture and Horticulture*, 22, 217–249.

Peyraud, J. L., et al. (2014). Multi-species swards and multi scale strategies for multifunctional grassland-base ruminant production systems: an overview of the FP7-MultiSward project. In *25th EGF General Meeting on "EGF at 50: The Future of European Grasslands*, pp. 695–715.

Prendiville, R., et al. (2009). An evaluation of production efficiencies among lactating Holstein-Friesian, Jersey, and Jersey× Holstein-Friesian cows at pasture. *Journal of Dairy Science*, 92(12), 6176–6185.

Pryce, J. E. and Harris, B. L., (2006). Genetics of body condition score in New Zealand dairy cows. Journal of dairy science, 89(11), pp.4424–4432.

Reed, K. and Morrissey, E. M. (2022). Bridging ecology and agronomy to foster diverse pastures and healthy soils. *Agronomy*, 12(8), 1893.

Roca-Fernández, A. I., et al. (2016). Pasture intake and milk production of dairy cows rotationally grazing on multi-species swards. *Animal*, 10(9), 1448–1456.

Rodríguez-Bermúdez, R., et al. (2019). Breeding for organic dairy farming: what types of cows are needed?. *Journal of Dairy Research*, 86(1), 3–12.

Sanderman, J., et al. (2017). Soil carbon debt of 12,000 years of human land use. *Proceedings of the National Academy of Sciences*, 114(36), 9575–9580.

Sanderson, M. A., et al. (2005). Forage mixture productivity and botanical composition in pastures grazed by dairy cattle. *Agronomy Journal*, 97(5), 1465–1471.

Schütz, K. E., et al. (2014). A field study of the behavioral and physiological effects of varying amounts of shade for lactating cows at pasture. *Journal of Dairy Science*, 97(6), 3599–3605.

Shrestha, B. M., et al. (2020). Adaptive multi-paddock grazing lowers soil greenhouse gas emission potential by altering extracellular enzyme activity. *Agronomy*, 10(11), 1781.

Skinner, R. H. and Dell, C. J. (2016). Yield and soil carbon sequestration in grazed pastures sown with two or five forage species. *Crop Science*, 56(4), 2035–2044.

Smidt, N. W. and Brimer, L., (2005). The use of herbs in pastures: an interview survey among bio-dynamic and organic farmers with dairy cattle. Agriculture and Human Values, 22, pp.355–363.

Sørensen, M. K., et al. (2008). Invited review: crossbreeding in dairy cattle: a Danish perspective. *Journal of Dairy Science*, 91, 4116–428.

Spaans, O. K., et al. (2018). Dairy cow breed interacts with stocking RATE in temperate pasture-based dairy production systems. *Journal of Dairy Science*, 101(5), 4690–4702.

Stagg, B. C., et al. (2021). Toolbox of multi-species swards: investigating UK farmers' perceptions about costs and benefits of diverse leys. In: Pattinson, S. E. and Goatman, T. P. (Eds.). *Proceedings of the BGS 13th Research Conference Multi-Species Swards*. Online. March 2021.

Stanley, P. L., et al. (2024). Ruminating on soil carbon: applying current understanding to inform grazing management. *Global Change Biology*, 30(3), e17223.

Strickland, M. S. and Rousk, J. (2010). Considering fungal: bacterial dominance in soils - Methods, controls and ecosystem implications. *Soil Biology and Biochemistry*, 42, 1385–1395.

Soussana, J. F., et al. (2010). Mitigating the greenhouse gas balance of ruminant production systems through carbon sequestration in grasslands. *Animal*, 4(3), 334–350.

Teague, R. and Kreuter, U. (2020). Managing grazing to restore soil health, ecosystem function, and ecosystem services. *Frontiers in Sustainable Food Systems*, 4, Article 534187.

Thomson, A. L. and Albornoz, R. I. (2023). Multispecies forages in the Australian dairy feedbase: is there a biological business case? *Animal Production Science*. https://doi.org/10.1071/AN23066

Totty, V. K., et al. (2013). Nitrogen partitioning and milk production of dairy cows grazing simple and diverse pastures. *Journal of Dairy Science*, 96(1), 141–149.

Van Zanten, H. H., et al. (2018). Defining a land boundary for sustainable livestock consumption. Global change biology, 24(9), pp.4185–4194.

Van Zanten, H. H., et al. (2019). The role of farm animals in a circular food system. Global Food Security, 21, pp.18–22.

Verhoog, H., et al. (2003). The role of the concept of the natural (naturalness) in organic farming. *Journal of Agricultural and Environmental Ethics*, 16, 29–49.

Vogel, E. and Beber, C. L. (2022). Carbon footprint and mitigation strategies among heterogeneous dairy farms in Paraná, Brazil. *Journal of Cleaner Production*, 349, 131404.

Wagner, M., et al. (2023). Mob grazing: a Nature-based solution for British farms producing pasture-fed livestock. *Nature-Based Solutions*, 3, 100054.
Washburn, S. P. and Mullen, K. A. E. (2014). Invited review: genetic considerations for various pasture-based dairy systems. *Journal of Dairy Science*, 97(10), 5923–5938.
Watson, C. A., et al. (2002). Managing soil fertility in organic farming systems. Soil use and management, 18, pp.239–247.
White, S. L., et al. (2002). Milk production and economic measures in confinement or pasture systems using seasonally calved Holstein and Jersey cows. *Journal of Dairy Science*, 85(1), 95–104.

Chapter 4

Assessing milk quality from organic production

Gillian Butler and Hannah Davis, Newcastle University, UK

1 Introduction

This chapter discusses quality in organic milk – so a useful starting point might be to define what is meant by 'quality' and why it matters. The quality of organic milk is important for processors converting raw milk into a myriad of dairy products, is relevant for dairy consumers but also wider society, as well as the farmers. The obvious aspects of quality are things we can measure and quantify in milk and dairy products. These include key nutrients such as minerals, protein and fat compositions, possible contaminants like pesticide or drug residues, or maybe microbial activity and associated mycotoxins. However, many consumers opt for organic products based on their extrinsic qualities (Aertsens et al., 2009), many of which might be intangible or lacking evidence at the point of sale, without further investigation. The most obvious aspects in this category are the impact of production systems on animal welfare, biodiversity and climate change but might also include fair treatment of everybody along the supply chain. Another highly relevant and topical aspect might be the influence of the production system on antimicrobial resistance and the evolution of superbugs.

There could be other aspects not recognised here but this is still a significant list of topics relevant for defining the intrinsic and extrinsic quality of organic

http://dx.doi.org/10.19103/AS.2024.0138.05

milk and dairy products – some of which will be discussed later. The state of science on this range of quality attributes varies incredibly; milk fat and protein composition have been widely investigated for 40–50 years with lots of peer-reviewed evidence. In contrast, the concept of fairtrade in dairy production in Western society is relatively under-researched Table 1 shows the outcome of a series of Scopus database searches using each topic, combined with 'organic milk', giving a crude indication of research activity over the last 50 years.

Due to the nature of databases, it is possible that many papers identified in these searches are not primarily focused on these topics linked to organic dairying, but the numbers and dates give a relative indication of the volume and history of relevant research. To review all these aspects of quality is way beyond the scope of this chapter, which largely focuses on the nutritional composition of organic dairy foods. The remainder could be the subject of a whole other book.

2 Comparing organic and conventional milk composition

Milk and dairy products have a unique role to play in our diet for many around the world, supplying key nutrients; this, perhaps, is not surprising since milk has evolved into nature's complete food for young mammals. Although a few years old now, Haug et al. (2007) is an excellent review of its contribution to human nutrition. Since the millennium, there has been considerable research comparing the quality of milk and dairy foods produced under organic

Table 1 Number of peer-reviewed publications and date of the earliest paper on 'Organic Milk' and quality parameters from Scopus.

Search topic (+ organic milk)	Peer-reviewed papers (1973-2023)	Date of earliest paper
Protein composition	2,645	1973
Fat composition	1,886	1977
Minerals	526	1974
Vitamins	336	1973
Antioxidants	327	1985
Microbial activity	268	1982
Drug residues	207	1973
Pesticide residues	175	1973
Animal welfare	150	1995
Biodiversity	91	1998
Antimicrobial resistance	76	2001
Climate change (LCA)	51	2000
Fairtrade	5	2009

management with those from non-organic systems – too many papers to list all individually, although some examples are given. A recent overview has been provided by Linehan et al. (2024). This research has mostly focused on milk composition, particularly:

- fat (Larsen et al., 2010, Butler et al., 2011a, Butler et al., 2011b, Stergiadis et al., 2012, Benbrook et al., 2013)
- antioxidants (Butler et al., 2008; Slots et al., 2009)
- proteins (Brodziak et al., 2021)

This multitude of studies has been reviewed and subjected to meta-analysis – finding both similarities and differences in composition. Palupi et al. (2012) compared the nutritional quality of conventional and organic dairy products based on a review of 13 studies and this was followed by a more comprehensive meta-analysis a few years later based on 170 studies (Średnicka-Tober et al., 2016). As the title of this meta-analysis indicates, there is published evidence of "Higher [in] PUFA and n-3 PUFA, conjugated linoleic acid, α-tocopherol and iron, but lower [in] iodine and selenium concentrations in organic milk". However, these authors identified many constituents, including milk proteins, which were similar in organic and non-organic milk – but interestingly, also considerable variation within 'organic milk'. Perhaps this is not surprising, since there are a wide range of management practices permitted under certification. This variation is a good indication of the scope to improve nutritional quality in organic dairy products.

The proportion of fat and protein in milk is highly relevant for most producers and is often routinely assessed by purchasers, as many payment schemes are based on 'milk solids' rather than volume. The composition of milk protein is important in cheese making, where a high proportion of caseins relative to soluble whey proteins (typically about 75–80% (Walker et al., 2004)) reflects a higher yield of cheese from each litre of milk. Unfortunately, to date, there are few comparable schemes rewarding producers for milk fat composition, despite potential benefits to consumers' health from altering the profile of fatty acids (more later). The mineral content of milk is also rarely assessed (other than under experiments or surveys) and again, does not influence payment.

The scope to alter or improve milk composition for the benefit of producers, processors or consumers is driven by a combination of dairy cow's genetics and environmental factors, usually dominated by feeding management on farms. This is briefly summarised here but if readers are interested in greater detail, the classic review "Effects of nutrition and management on the production and composition of milk fat and protein" (Walker et al., 2004) is recommended.

3 Protein composition

Typically milk has 3.0–4.5% protein (McDonald, 2011) depending on breed, cow nutrition and stage of lactation (Walker et al., 2004). Protein is of high biological value, supplying many essential amino acids with some proteins and peptides having bioactive functions (Haug et al., 2007). There is general agreement that genetics has a greater influence over both the amount and composition of milk protein rather than nutrient intake or other aspects of dairy management (Walker et al., 2004). That said, dietary starch can boost milk protein content – although it is questionable if we ought to feed cereal grains to ruminant animals like our dairy cows (another topic not discussed here!).

Despite (or possibly because of) the apparent volume of research activity on milk protein composition suggested in Table 1, there is conflicting evidence about differences between protein composition in organic and non-organic milk, possibly due to the diversity of dairy management (and hence milk composition) seen in both organic and non-organic production systems. Palupi et al. (2012) report organic milk has a higher protein content than non-organic milk, in a review of five studies. However, a later and more comprehensive meta-analysis (Średnicka-Tober et al., 2016) found "no significant difference in [total fat and] protein content between organic and conventional milk" from 127 research papers reviewed, although neither paper gives details of protein levels or compositions reported.

This conflicting evidence on the impact of organic management on milk protein content and composition is also borne out in unpublished PhD work at Newcastle University (Stergiadis, 2009). The ratio of casein:whey proteins (important for cheese making) was found to be higher in milk from organic cows with high grazing intakes in South Wales compared with comparable non-organic herds, also with high grazing intakes (6.0:1 vs 5.6:1 $p < 0.01$). However, in contrast, another study on winter diets with silage feeding found organic cows produced milk with less total protein (33.0 vs 36.8 g/kg milk), casein (27.1 vs 30.1 g/kg milk), and whey (5.9 vs 6.6 g/kg milk) proteins than cows under 'conventional management' – this time with no difference in the ratio of casein:whey proteins (4.6:1). A retail survey of UK organic and conventional milk failed to identify any differences in total casein or whey proteins in milk purchased in the summer or winter (Stergiadis, 2009).

There is genetic variation in milk protein, both between but also within breeds and milk protein content and yield have long been recognised in breeding indices, especially for dairy herds supplying manufacturing outlets (AHDB, 2023). Several 'alternative' breeds particularly Jersey and Brown Swiss (often used in organic and other low-input dairy systems) produce milk with more total protein than Holstein cows. This explains the difference in the ratios of casein:whey found in the grazing survey relative to the winter-feeding trial mentioned above. The former involved 10 dairy herds, all with a high proportion Jersey, Brown Swiss and Scandinavian Red cows compared with the Holstein/Friesian cows on the silage feeding trial. Stergiadis et al. (2013) identified the higher total protein content of retail Jersey and Guernsey milk to be driven by more casein rather than the proportion of whey proteins, making it suited to cheese production. Readers wishing to explore details on the range of genotypes for the wide range of individual milk proteins and their influence over milk properties and functionality are directed to a review by Gai et al. (2021).

A higher protein content in organic milk is likely to be driven mainly by genetics. Further improvement needs to be directed by breeding choices rather than changes in feeding and other management. It is important to note that, while differences identified for protein content and composition in some cases might be statistically significant, they are relatively small in the case of liquid milk consumption (be it for human or young bovine consumers) and may not make much difference (except for quantity and quality of colostrum for new-born calves – but yet another topic!). On the other hand, for herds with milk destined for processing, especially for cheese or yogurt making, milk protein yield and casein content ought to be considered when choosing breeds, as well as individual cows or bulls for breeding replacements.

4 Fatty acid composition

Milk fat is more variable than protein and can range from 2.5% to 5.6% in raw milk. Due to its relatively simple separation from the top of non-homogenised milk, the designation 'full-fat milk' for sale is often standardised between 3.5% and 4.0%. Whilst milk fat content is a function of both genetics and feeding, the composition of that fat or butter is strongly influenced by feeding, with very minor genetic influences (Walker et al., 2004). As with protein, milk from Jersey cows and other less-used breeds (often used in organic dairying) is widely recognised to be higher in fat than Holstein cows (McDonald, 2011).

Milk fat comprises 93% triglycerides, with as many as 400 different fatty acids (FA) of dietary importance. A typical breakdown is

- 65–80% saturated fatty acids (SFA)
- 20–25% monounsaturated fatty acids (MUFA)
- 2–5% polyunsaturated fatty acids (PUFA)

Butler (2014) and Benbrook et al. (2018) suggest, if we are to enhance the nutritional quality of milk fat, the target is to reduce the proportion of SFA and increase PUFA, especially those categorised as omega-3 (n-3) rather than omega-6 (n-6). However, if we are to instigate such changes, we need to understand fatty acid metabolism and identify how this can be achieved. Predicting the relative proportion of individual FA milk in milk is complex with five potential influencing factors (discussed below).

Milk FA are either (Collomb et al., 2004; Destaillats et al., 2005; Chilliard et al., 2007; Stoop et al., 2009):

- transferred, unchanged, from the diet
- modified through biohydrogenation of dietary FA by microbes in the rumen (removing double bonds along the carbon chains)
- synthesised *de novo* in the mammary gland (from rumen volatile fatty acids (VFA))
- modified by desaturase enzymes in the mammary gland (adding double bonds to carbon chains)
- released from mobilised fat stores (when the cow is losing weight, not uncommon in early lactation)

Dairy diets which are high in forage plants are generally low in lipids or fats, dominated by PUFA (α-linolenic acid, ALA, c9,12,15 C18:3) in cell and organelle membranes. Despite 90–95% of alpha linolenic acid (ALA) and other PUFA being *lost* in the rumen (microbial activity adds hydrogen atoms, removing double bonds in the carbon chain – i.e. hydrogenation), diet and PUFA intake has a major influence over the final milk FA profile (Chilliard et al., 2007).

The ultimate product of rumen microbial lipid hydrogenation is stearic acid (C18:0), although rumen turnover (with digesta progressing along the gut) interrupts this process, leading to many intermediate FA being transferred to the mammary gland (Collomb et al., 2004; Destaillats et al., 2005). Some intermediates end up in milk or may undergo further metabolic changes in the udder. Mammary cells synthesise short-chain saturated fatty acids, with between 4 and 16 carbon atoms, using plasma volatile fatty acids (acetate and butyrate originating in the rumen) as precursors (Chilliard et al., 2007). In addition, enzyme activity in the udder (especially Δ^9-desaturase) removes hydrogen, creating double bonds along the FA chains, thus converting absorbed or synthesised SFA to MUFA and circulating MUFA to PUFA (Griinari et al., 2000; Destaillats et al., 2005).

4.1 Effect of organic management on milk fatty acid profile

The impact of organic management on milk quality has been widely researched (Beaulieu and Palmquist, 1995, Palladino et al., 2010, Butler et al., 2011c, Stergiadis et al., 2019) and reviewed (Palupi et al., 2012; Schwendel et al., 2015; Średnicka-Tober et al., 2016). The findings with the greatest repeatability are that organic milk contains more vaccenic acid (VA, t11 C18:1), conjugated linoleic acid (CLA, c9t11 C18:2), ALA, eicosapentaenoic acid (EPA, c5,8,11,14,17 C20:5), docosapentaenoic acid (DPA, c7,10,13,16,19 C22:5), total n-3 and has a lower n-6:n-3 ratio, compared with conventional milk. All of these are considered positive for consumer health as outlined by Butler (2014) Benbrook et al. (2018) – although this has yet to be clinically tested.

Although many organic dairy herds use 'alternative' breeds (as opposed to high-yielding Holstein cows), most of the superiority in composition have been shown to be driven by feeding management (Średnicka-Tober et al., 2016). One of the main differences between feeding in organic and conventional herds is forage intake (grazing pasture in summer, silage in winter) (AHDB, 2012; Soil Association, 2018). Organic standards (at least in Europe) state cows ought to consume at least 60% of their diet (dry matter) as forage although, in practice, this is usually considerably higher because of the price differential between relatively inexpensive forage, especially grazing (with no fertiliser costs), and expensive concentrate feeds, is very much wider for organic than conventional farms (AHDB, 2012).

In theory, diets high in leafy vegetation ought to produce milk with more n-3 compared with milk from grain feeding, which is higher in n-6 fatty acids (Elgersma et al., 2006; Stergiadis et al., 2015; Stergiadis et al., 2018a). Consequently, higher forage intakes by organic cows raise PUFA and ALA consumption and this (Collomb et al., 2004; Destaillats et al., 2005):

- increases ALA in milk, as more escapes rumen hydrogenation and reaches the udder
- raises VA in milk, an intermediate of ALA hydrogenation
- increases CLA synthesis in the udder, due to the action of Δ^9 desaturase enzyme on VA

High grazing intakes are also associated with long-chain n-3 (EPA and DPA) in milk, possibly due to a higher dietary supply of these FA, or to the increased transformation of ALA by Δ^5 and Δ^6 desaturases enzymes (Barcelo-Goblijn and Murphy, 2009). Since ALA, EPA and DPA are the main n-3 PUFA in milk, the overall n-3 content of organic milk is reported to be higher than conventional milk in many studies (O'Donnell et al., 2010; Butler et al., 2011c; Capuano et al., 2015; Stergiadis et al., 2019; Davis et al., 2020a).

In winter, when cows tend not to graze, concentrate feeding on organic farms still tends to be lower than on conventional farms, although differences might be less marked than in summer (AHDB, 2012, Soil Association, 2018). This is reflected in milk quality. Retail surveys in the UK report total and individual n-3 contents to be higher in organic than conventional milk throughout the year although differences are lower in winter (Butler et al., 2011c; Stergiadis et al., 2019). Ensiling forage reduces its PUFA content (Glasser et al., 2013), although grass or grass/clover silage still contains substantially more n-3 PUFA compared with typical concentrate feeds or maize silage and this may explain (a) the higher n-3 PUFA in milk even when forage-fed cows are not grazing, and (b) the effect of season on the extent of the findings.

Another major difference between organic and conventional diets is the inclusion of clover and other forage plants in pastures and silages. Clover passes through the rumen faster than grass, potentially as a result of its higher degradability and lower fibre content (Beever et al., 1986; McDonald, 2011; Stergiadis et al., 2018b). In the case of red clover, its breakdown is protected by polyphenol oxidase in the plants (Lee et al., 2007). Consequently, hydrogenation is lower, allowing more ALA to escape and pass into milk (Collomb et al., 2004; Destaillats et al., 2005). The proportionate transfer of ALA from cow diet to milk, and its eventual concentration in milk, is greater when clover is available compared to solely feeding grass (O'Callaghan et al., 2016; Stergiadis et al., 2018b). This effect of clover has been confirmed both with fresh (Stergiadis et al., 2018b) and ensiled forages (Dewhurst et al., 2003), enhancing the theory of lower rumen hydrogenation.

Grazing diverse swards further enhances milk FA profiles. Milk from cows grazing Alpine swards had 63% more n-3 compared to milk from simpler valley vegetation (Leiber et al., 2005). A similar response has been reported for cows with access to chicory – Mangwe et al. (2020) reported 19–26% more ALA, 17–21% more n-3 and 10–12% more PUFAs in milk when cows accessed chicory swards for 5.5 hours each day, in addition to grazing grass swards. Milk from Jersey cows grazing a diverse ley in an organic arable rotation (with plantain, chicory, salad burnet, caraway and bird's foot trefoil along with ryegrass, white and red clovers) was higher in PUFA, especially n-3 (7% more PUFA and 13% more n-3) relative to milk from a ryegrass and white clover pasture (Loza et al., 2023). This extra benefit from including herbs in the pasture was surprising since milk from the 'control' cows grazing the grass-clover sward already had 20% more PUFA and almost 90% more n-3, compared with pre-turnout, when cows were on an organic grass/clover silage-based diet.

4.2 Seasonal variation

Changes in seasonal herbage growth and grazing conditions lead to changes in dairy diets throughout the year, as summer grazing is replaced by conserved silage or hay in winter. This inevitably results in seasonal fluctuations in milk quality with winter milk having a poorer FA profile (13% less PUFA, 42% less CLA and 29% less n-3) (Butler et al., 2008, Butler et al., 2011b). To some extent, feeding oil seeds like rape or linseed can replace the unsaturated FA missing from conserved forages and restore milk PUFA and n-3 to 'summer' levels (Glasser et al., 2008; Stergiadis et al., 2014). In this respect, organic cows consuming grass/clover silage, not only produce milk higher in PUFA and n-3 than non-organic cows eating grass-dominated silage but appear more responsive to oilseed supplementation (milk n-3 and PUFA output shown to be 33% and 34% higher respectively, compared with non-organic cows given the same level of supplementation) (Butler et al., 2019).

4.3 Genetic influences

Organic farms are less likely to use Holstein/Friesian genetics than more intensive, non-organic farms. Also, a preventative approach to health management means organic herds rely more on good hygiene and robust cow genotypes (Van Diepen et al., 2007). Pasture-based production systems, as on many organic farms, tend to use crossbred cows, incorporating robust breeds better able to convert a forage diet into milk, such as British and New Zealand Friesian, Jersey, Guernsey, Ayrshire, Brown Swiss and Shorthorn – possibly also involving high-yielding Holstein genetics (to maintain productivity) (Davis et al., 2020b; Ormston et al., 2022). However, the genetic influence over differences in milk FA is unclear and is highly likely to be overridden by the effects of differing diets in organic and non-organic production (Stergiadis et al., 2015, 2018a; Davis et al., 2020b).

4.4 Scope for improvement

Unlike protein composition, steps to improve the nutritional quality of milk fat need to focus on feeding management (rather than breeding decisions), maximising both herbage intake, through judicious grazing management and sward diversity. In situations where grazing is not possible, especially in winter, silage from diverse swards could be supplemented with organic oilseeds although this perhaps needs further research to validate.

5 Mineral composition

Milk and dairy foods have for generations been considered a key source of minerals, particularly calcium (Ca). However, minerals collectively form less than 0.8% of milk by weight (McDonald, 2011). Dairy foods are also an important source of iodine and, to a lesser extent, phosphorous and magnesium, especially for children with high dairy intakes (Newton et al., 2023). The mineral content of milk does vary to some extent, mostly reflecting mineral supply in dairy diets. However, there is some metabolic control in the udder over mineral transfer from blood to milk to ensure neonate nutrition and avoid potential toxicity.

The review by Palupi et al. (2012) comparing the nutritional quality of conventional and organic dairy products makes no mention of the mineral content of milk, whereas the subsequent Średnicka-Tober et al. (2016) meta-analysis identified consistent differences reflected in the title: "...higher iron, but lower iodine and selenium concentrations in organic milk." There has been another meta-analysis considering minerals and heavy metals in milk from cows in different production systems and countries (Zwierzchowski and Ametaj, 2018), covering toxic as well as essential elements from 72 studies. The data collected for organic milk is more limited than for 'commercial' milk or that produced under non-organic management. For instance, results for calcium content state 'n=' for conventional milk as 2791 but only 93 for organic milk, and many of the trace elements reported have not been evaluated in organic products.

The meta-analysis (Zwierzchowski and Ametaj, 2018) is a thorough investigation into the mineral content of milk from various origins, reporting concentrations of 7 macro-, 41 trace- and 7 toxic elements in milk classed as 'commercial' as well as from conventional and organic production. It confirms the lower iodine concentrations reported by Średnicka-Tober et al. (2016) but contradicts the iron finding, reporting it to be lower in organic milk, while Se concentrations did not appear to differ between production systems. Results from Zwierzchowski and Ametaj (2018) for organic milk found lower levels of most major minerals Ca, Mg, P, K and Na, with concentrations ranging from 30% to 40% less than non-organic or commercial milk. Microminerals and trace elements were highly variable in the papers reviewed and only 16 (of the 41) elements were listed for organic milk.

Organic milk was found to be significantly lower in cobalt, barium, boron, silicon, vanadium, copper, manganese, iodine, iron, zinc and molybdenum than milk from conventional production and/or commercial milk. However, except for iodine, dairy foods are not considered an important source of these minerals. Without exception, organic milk had lower concentrations of all heavy metals reported compared with results for conventional production – relative values were only 27% arsenic, 24% lead, 15% aluminium, 14% chromium, 11% nickel and cadmium levels found in non-organic milk, down to undetectable levels of mercury. Unfortunately, the chapter does not clearly discuss the origin of this contamination or steps to reduce their appearance in milk. However, a few years prior to both these meta-analysis (Średnicka-Tober et al., 2016, Zwierzchowski and Ametaj, 2018), the organic dairy sector (in the UK) suspected poor iodine status was leading to health problems with cows and calves and iodine supplementation was introduced on many farms – which would increase the iodine content of organic milk (Flachowsky et al., 2014) although this was subsequent to the research included in both reviews.

More recent research investigating the mineral content of UK organic and non-organic milk has been reported by groups at the University of Reading (Qin et al., 2021; Stergiadis et al., 2021; Newton et al., 2023) and Newcastle

University (Butler et al., 2019), with interesting if mixed findings. Many differences identified between the two production systems may not be physiologically relevant or important for 'quality' – for several reasons. Some are:

- relatively small (organic milk is 5% higher in Ca @ 1.05 vs 0.96 g/kg (Qin et al., 2021))
- inconsistent and conflicting; Butler et al. (2019) found organic milk to be 47% lower in iodine (152 vs 287 μg/litre, $p < 0.001$), in contrast to a study by Qin et al. (2021) finding 10% higher concentrations with organic milk (365 vs 331 μg/kg) and Stergiadis et al. (2021) found no significant difference in the iodine content of organic and non-organic milk from herds with varying reliance on grazing – all relatively high, ranging from 301 μg/kg to 395 μg/kg),
- discuss elements for which milk is a notoriously poor source compared with other foods, such as copper or iron, along with other elements analysed by Zwierzchowski and Ametaj (2018).

If the mineral composition of organic milk is to be optimised, a key step is to identify underlying trace element status and possible deficiencies on production farms. As well as influencing milk composition or quality, these could have a negative impact on herd health (Enjalbert et al., 2006; McDonald, 2011). Corrective inputs can then target known deficiencies and avoid oversupply or potential interactions between elements associated with blanket supplementation.

Care is needed, however, especially with iodine since it is transferred readily from dairy diets to plasma and into milk, as well as from ionophores used for sanitating teats during milking (Flachowsky et al., 2014). Although dairy foods are an important source of iodine for consumers, and organic milk can be relatively low in iodine, the difference for consumers between deficiency and toxicity is relatively narrow, especially for children (with an 'adequate intake' at 100 μg/day and 'tolerable upper limit' of only 300 μg/day (EFSA, 2014)). For instance, feeding 170 g seaweed/cow/day (*Ascophyllum nodosum;* a rich source of iodine, and occasionally used as an organic supplement) to Jersey cows raised milk iodine concentrations to dangerous levels of 1370 μg/litre, (Antaya et al., 2015) – *breaching the child upper tolerable limit with only 220 ml daily milk consumption – ignoring intake from any other dairy products in their diet.*

6 Drug and pesticide residues

With the relatively high use of pesticides and veterinary drugs in many non-organic production systems, consumers may choose organic food as a more acceptable alternative. Recent evidence (Penagos-Tabares et al., 2023; Wei et al., 2023) tends to support this approach to some degree. An Austrian group investigated the presence of pesticides and drug residues in diets and feeds on Austrian organic and non-organic dairy farms (Penagos-Tabares et al., 2023). Total mixed diets were appropriate for housed cows. All were silage based (grass, maize or other cereals) with hay, straw with a range of cereal grains and by-product concentrate feeds – with the notable exclusion of grass or other fresh forages.

Two drug residues and 13 pesticides (mostly fungicides and insecticides) were identified in feeds, including 5 reported to exceed the EU maximum permitted residue concentrations and 4 products not authorised for use in the EU. At least 1 pesticide product was identified in diets on almost all (91 of the 92) non-organic farms, 60% had more than 1 residue while, on 9 of the farms, 5 or 6 different products were detected in feed. Permitted medicated feed additives (monenesin and nicarbazin) were found on 8 farms, all within the EU permitted limit. Results for the nine organic farms showed much lower contamination although worryingly, one feed sample was contaminated by a pesticide and another had evidence of a feed additive – against most, if not all, certification standards.

Another recent study looking at neonicotinoid insecticides in retail milk available in China (Wei et al., 2023) also produced worrying results. Over 200 milk samples were screened for residues, with 134 from domestic production and 71 from imported samples, mostly from Europe and Australasia. Sixteen different residues were identified in the milk samples. Like the Austrian feed study, all samples, including organic milk, contained at least 1 type of residue, although contamination was considerably lower in the organic milk.

The authors modelled typical daily intakes and 'relative potency factors' using the measured neonicotinoid concentrations and potency along with typical dairy consumption data. Looking at the figures for children (at greater risk than adults due to their relatively high dairy consumption and lower body weight), both maximum calculated values for organic milk were an order of magnitude less than the non-organic brands (0.056 vs 0.709 μg/kg body weight/day and 0.366 vs 1.66 ng/ml, respectively). However, on the whole, organic milk appeared only marginally *better* considering the similarity in median values for each milk type (0.004 vs 0.005 μg/kg body weight/day neonicotinoid consumption and 0.011 vs 0.015 ng/ml for the relative potency factor).

Interestingly, almost all the pesticides identified in these studies (Penagos-Tabares et al., 2023; Wei et al., 2023) were fungicides and insecticides, designed to target arable crops. As such these residues persist in cereals, cereal byproducts, oilseeds and other ingredients used in concentrate feeds (as well as direct human consumption (Rempelos et al., 2022)). The risk of feed contamination and hence transfer into milk would appear to be lower for dairy cows fed diets dominated by home-grown forages, avoiding such products. Dairy diets dominated by forages are therefore likely to enhance milk quality, by reducing the risk of pesticide residues, as well as enhancing milk fat quality.

7 The influence of cow–calf contact on milk composition

Concerns about the negative welfare effects of early separation of cows and calves have generated research to identify ethical alternatives. However, very few studies consider the impact of partial suckling on milk composition. Johnsen et al. (2016) and Barth (2020) both report the fat content of machine-collected milk from dairy cows suckling calves to be lower than that from non-nursing cows. Unfortunately, neither paper gives details of the magnitude of these differences or any impact on fat composition.

More detail is given in an earlier paper by Cozma et al. (2013), identifying differences in FA profiles in milk from cows with and without calves present at milking. Although the cows and calves in this study were not always together, suckled calves accessed dams for 1 min before clusters were applied in the parlour and then suckled all residual milk after each milking. As above, allowing calves to suckle the residual (alveolar) milk, depressed the fat content of harvested milk and altered its composition, although differences were less marked (but still significant) for Holstein cows compared with dual-purpose Saler cows (which produced 60% more total milk when calves were present at milking). As stated, the fat content was lower in harvested milk from suckled cows (31.8 vs 33.3 g/l for Holsteins and 26.8 vs 35.3 g/l from Salers) and it had a significantly lower concentration of linoleic acid (LA, 6–7% less), ALA (4–9% less) and PUFA (4–7% less). This confirms the residual milk, held back for calves, is not only higher in fat than that let-down during milking (Ontsouka et al., 2003) and suggests it is disproportionally higher in PUFA. This is only a single study but it suggests cow–calf contact (CCC) systems might produce milk with lower concentrations of beneficial fatty acids compared with more conventional management, although this an area ripe for further research.

8 Conclusion

As mentioned, under current processing methods and payment schemes, the incentive to enhance the quality of organic milk is quite limited. Farmers might be rewarded for raising the total fat and protein content of the milk produced and cheese makers will make more cheese from each litre of milk if it is relatively high in casein rather than whey proteins. The evidence of consumer benefits is mixed. In theory, their health will be enhanced by selecting an organic alternative if the milk and dairy products consumed have:

- a better fat profile (less SFA and more n-3 PUFA)
- an adequate but not excessive iodine content
- has been produced by cows fed predominantly forage diets, without concentrate feeds which could be contaminated with pesticides

Unfortunately, there is presently little incentive to achieve these goals and the organic sector might need to evaluate realistic rewards for 'quality'.

9 References

Aertsens, J., Verbeke, W., Mondelaers, K. and Van Huylenbroeck, G. 2009. Personal determinants of organic food consumption: a review. *British Food Journal* 111(10):1140–1167.

AHDB. 2012. *Feeding+. 8. Managing organic feeding*. p. 14. AHDB, Kenilworth, UK.

AHDB. 2023. *Profitable Lifetime Index £PLI*. Vol. 2023. https://ahdb.org.uk/knowledge-library/profitable-lifetime-index-pli, May 2023.

Antaya, N. T., Soder, K. J., Kraft, J., Whitehouse, N. L., Guindon, N. E., Erickson, P. S., Conroy, A. B. and Brito, A. F. 2015. Incremental amounts of Ascophyllum nodosum meal do not improve animal performance but do increase milk iodine output in early lactation dairy cows fed high-forage diets1. *Journal of Dairy Science* 98(3):1991–2004.

Barcelo-Goblijn, G. and Murphy, E. J. 2009. Alpha-linolenic acid and its conversion to longer chain n-3 fatty acids: benefits for human health and a role in maintaining tissue n-3 fatty acid levels. *Progress in Lipid Research* 48:355–374.

Barth, K. 2020. Effects of suckling on milk yield and milk composition of dairy cows in cow-calf contact systems. *Journal of Dairy Research* 87(S1):133–137.

Beaulieu, A. and Palmquist, D. 1995. Differential effects of high fat diets on fatty acid composition in milk of Jersey and Holstein cows. *Journal of Dairy Science* 78:1336–1344.

Beever, D. E., Dhanoa, M. S., Losada, H. R., Evans, R. T., Cammell, S. B. and France, J. 1986. The effect of forage species and stage of harvest on the processes of digestion occurring in the rumen of cattle. *British Journal of Nutrition* 56(2):439–454.

Benbrook, C. M., Butler, G., Latif, M. A., Leifert, C. and Davis, D. R. 2013. Organic production enhances milk nutritional quality by shifting fatty acid composition: a United States-wide, 18-month study. *PLoS One* 8(12). https://doi.org/10.1371/journal.pone.0082429

Benbrook, C. M., Davis, D. R., Heins, B. J., Latif, M. A., Leifert, C., Peterman, L., Butler, G., Faergeman, O., Abel-Caines, S. and Baranski, M. 2018. Enhancing the fatty acid profile of milk through forage-based rations, with nutrition modeling of diet outcomes. *Food Science and Nutrition* 6(3):681–700.

Brodziak, A., Wajs, J., Zuba-Ciszewska, M., Król, J., Stobiecka, M. and Jańczuk, A. 2021. Organic versus conventional raw cow milk as material for processing. *Animals* 11(10). https://doi.org/10.3390/ani11102760

Butler, G. 2014. Manipulating dietary PUFA in animal feed: implications for human health. *Proceedings of the Nutrition Society* 73(1):87–95.

Butler, G., Nielsen, J. H., Larsen, M. K., Rehberger, B., Stergiadis, S., Canever, A. and Leifert, C. 2011a. The effects of dairy management and processing on quality characteristics of milk and dairy products. *NJAS - Wageningen Journal of Life Sciences* 58(3–4):97–102.

Butler, G., Nielsen, J. H., Slots, T., Seal, C., Eyre, M. D., Sanderson, R. and Leifert, C. 2008. Fatty acid and fat-soluble antioxidant concentrations in milk from high- and low-input conventional and organic systems: Seasonal variation. *Journal of the Science of Food and Agriculture* 88(8):1431–1441.

Butler, G., Stergiadis, S., Chatzidimitriou, E., Franceschin, E., Davis, H. R., Leifert, C. and Steinshamn, H. 2019. Differing responses in milk composition from introducing rapeseed and naked oats to conventional and organic dairy diets. *Scientific Reports* 9(1). https://doi.org/10.1038/s41598-019-44567-8

Butler, G., Stergiadis, S., Seal, C., Eyre, M. and Leifert, C. 2011b. Fat composition of organic and conventional retail milk in northeast England. *Journal of Dairy Science* 94(1):24–36.

Butler, G., Stergiadis, S., Seal, C. J., Eyre, M. D. and Leifert, C. 2011c. Fat composition of organic and conventional retail milk in northeast England. *Journal of Dairy Science* 94:24–36.

Capuano, E., Gravink, R., Boerrigter-Eenling, R. and van Ruth, S. M. 2015. Fatty acid and triglycerides profiling of retail organic, conventional and pasture milk: implications for health and authenticity. *International Dairy Journal* 42:58–63.

Chilliard, Y., Glasser, F., Ferlay, A., Bernard, L., Rouel, J., and Doreau, M. 2007. Diet, rumen biohydrogenation and nutritional quality of cow and goat milk fat. *European Journal of Lipid Science and Technology* 109:828–855.

Collomb, M., Sieber, R. and Butikofer, U. 2004. CLA isomers in milk fat from cows fed diets with high levels of unsaturated fatty acids. *Lipids* 39:355–363.

Cozma, A., Martin, B., Guiadeur, M., Pradel, P., Tixier, E., and Ferlay, A. 2013. Influence of calf presence during milking on yield, composition, fatty acid profile and lipolytic system of milk in Prim'Holstein and Salers cow breeds. *Dairy Science and Technology* 93(1):99–113.

Davis, H., Chatzidimitriou, E., Leifert, C. and Butler, G. 2020a. Evidence that forage-fed cows can enhance milk quality. *Sustainability (Switzerland)* 12(9). https://doi.org/10.3390/su12093688

Davis, H., Stergiadis, S., Chatzidimitriou, E., Sanderson, R., Leifert, C. and Butler, G. 2020b. Meeting breeding potential in organic and low-input dairy farming. *Frontiers in Veterinary Science* 7. https://doi.org/10.3389/fvets.2020.544149

Destaillats, F., Trottier, J., Galvez, J. and Angers, P. 2005. Analysis of α-linolenic acid biohydrogenation intermediates in milk fat with emphasis on conjugated linolenic acids. *Journal of Dairy Science* 88:3231–3239.

Dewhurst, R. J., Evans, R. T., Scollan, N. D., Moorby, J. M., Merry, R. J. and Wilkins, R. J. 2003. Comparison of grass and legume silages for milk production. 2. In vivo and in sacco evaluations of rumen function. *Journal of Dairy Science* 86(8):2612–2621.

EFSA. 2014. Scientific Opinion on Dietary Reference Values for iodine. in *EFSA Panel on Dietetic Products, Nutrition and Allergies (NDA)*. Vol. 2023. European Food Safety Authority (EFSA), Parma, Italy.

Elgersma, A., Tamminga, S. and Ellen, G. 2006. Modifying milk composition through forage. *Animal Feed Science and Technology* 131:207–225.

Enjalbert, F., Lebreton, P. and Salat, O. 2006. Effects of copper, zinc and selenium status on performance and health in commercial dairy and beef herds: retrospective study. *Journal of Animal Physiology and Animal Nutrition* 90(11–12):459–466.

Flachowsky, G., Franke, K., Meyer, U., Leiterer, M. and Schöne, F. 2014. Influencing factors on iodine content of cow milk. *European Journal of Nutrition* 53(2):351–365.

Gai, N., Uniacke-lowe, T., O'regan, J., Faulkner, H. and Kelly, A. L. 2021. Effect of protein genotypes on physicochemical properties and protein functionality of bovine milk: a review. *Foods* 10(10). https://doi.org/10.3390/foods10102409

Glasser, F., Doreau, M., Maxin, G. and Baumont, R. 2013. Fat and fatty acid content and composition of forages: a meta-analysis. *Animal Feed Science and Technology* 185:19–34.

Glasser, F., Ferlay, A. and Chilliard, Y. 2008. Oilseed lipid supplements and fatty acid composition of cow milk: a meta-analysis. *Journal of Dairy Science* 91(12):4687–4703.

Griinari, J., Corl, B., Lacy, S., Chouinard, P., Nurmela, K. and Bauman, D. 2000. Conjugated linoleic acid is synthesized endogenously in lactating dairy cows by Δ^9-Desaturase. *Journal of Nutrition* 130:2285–2291.

Haug, A., Høstmark, A. T. and Harstad, O. M. 2007. Bovine milk in human nutrition: a review. *Lipids in Health and Disease* 6:1–16.

Johnsen, J. F., Zipp, K. A., Kälber, T., Passillé, A. M. D., Knierim, U., Barth, K. and Mejdell, C. M. 2016. Is rearing calves with the dam a feasible option for dairy farms?–current and future research. *Applied Animal Behaviour Science* 181:1–11.

Larsen, M. K., Nielsen, J. H., Butler, G., Leifert, C., Slots, T., Kristiansen, G. H. and Gustafsson, A. H. 2010. Milk quality as affected by feeding regimens in a country with climatic variation. *Journal of Dairy Science* 93(7):2863–2873.

Lee, M. R. F., Parfitt, L. J., Scollan, N. D. and Minchin, F. R. 2007. Lipolysis in red clover with different polyphenol oxidase activities in the presence and absence of rumen fluid. *Journal of the Science of Food and Agriculture* 87(7):1308–1314.

Leiber, F., Kreuzer, M., Nigg, D., Wettstein, H. R. and Scheeder, M. R. L. 2005. A study on the causes for the elevated n-3 fatty acids in cows' milk of alpine origin. *Lipids* 40(2):191–202.

Loza, C., Davis, H., Malisch, C., Taube, F., Loges, R., Magistrali, A. and Butler, G. 2023. Milk fatty acids: the impact of grazing diverse pasture and the potential to predict rumen-derived methane. *Agriculture (Switzerland)* 13(1). https://doi.org/10.3390/agriculture13010181

Mangwe, M., Bryant, R. and Gregorini, P. 2020. Rumen fermentation and fatty acid composition of milk of mid lactating dairy cows grazing chicory and ryegrass. *Animals* 10(1). https://doi.org/10.3390/ani10010169

McDonald, P., Edwards, R. A., Greenhalgh, J. F. D., Morgan, C. A., Sinclair, L. A. and Wilkinson, R. G. 2011. *Animal Nutrition*. 7th ed. Pearson Education Harlow, UK.

Newton, E. E., Pétursdóttir, Á. H., Beauclercq, S., Clarke, J., Desnica, N. and Stergiadis, S. 2023. Variation in macrominerals and trace elements in cows' retail milk and implications for consumers nutrition. *Food Chemistry* 418. https://doi.org/10.1016/j.foodchem.2023.135809

O'Callaghan, T. F., Hennessy, D., McAuliffe, S., Kilcawley, K. N., O'Donovan, M., Dillon, P., Ross, R. P. and Stanton, C. 2016. Effect of pasture versus indoor feeding systems on raw milk composition and quality over an entire lactation. *Journal of Dairy Science* 99(12):9424–9440.

O'Donnell, A. M., Spatny, K. P., Vicini, J. L. and Bauman, D. E. 2010. Survey of the fatty acid composition of retail milk differing in label claims based on production management practices. *Journal of Dairy Science* 93:1918–1925.

Ontsouka, C. E., Bruckmaier, R. M. and Blum, J. W. 2003. Fractionized milk composition during removal of colostrum and mature milk. *Journal of Dairy Science* 86(6):2005–2011.

Ormston, S., Davis, H., Butler, G., Chatzidimitriou, E., Gordon, A. W., Theodoridou, K., Huws, S., Yan, T., Leifert, C. and Stergiadis, S. 2022. Performance and milk quality parameters of Jersey crossbreds in low-input dairy systems. *Scientific Reports* 12(1). https://doi.org/10.1038/s41598-022-10834-4

Palladino, R., Buckley, F., Prendiville, R., Murphy, J., Callan, J. and Kenny, d. 2010. A comparison between Holstein-Friesian and Jersey dairy cowsand their F1 hybrid on milk fatty acid composition under grazing conditions. *Journal of Dairy Science* 93:2176–2184.

Palupi, E., Jayanegara, A., Ploeger, A. and Kahl, J. 2012. Comparison of nutritional quality between conventional and organic dairy products: a meta-analysis. *Journal of the Science of Food and Agriculture* 92(14):2774–2781.

Penagos-Tabares, F., Sulyok, M., Faas, J., Kriska, R., Khiaosa-ard, R. and Zebeli, Q. 2023. Residues of pesticides and veterinary drugs in diets of dairy cattle from conventional and organic farms in Austria. *Environmental Pollution* 316. https://doi.org/10.1016/j.envpol.2022.120626

Qin, N., Faludi, G., Beauclercq, S., Pitt, J., Desnica, N., Pétursdóttir, Á., Newton, E. E., Angelidis, A., Givens, I., Juniper, D., Humphries, D., Gunnlaugsdóttir, H. and Stergiadis, S. 2021. Macromineral and trace element concentrations and their seasonal variation in milk from organic and conventional dairy herds. *Food Chemistry* 359. https://doi.org/10.1016/j.foodchem.2021.129865

Rempelos, L., Wang, J., Barański, M., Watson, A., Volakakis, N., Hoppe, H. W., Kühn-Velten, W. N., Hadall, C., Hasanaliyeva, G., Chatzidimitriou, E., Magistrali, A., Davis, H., Vigar, V., Średnicka-Tober, D., Rushton, S., Iversen, P. O., Seal, C. J. and Leifert, C. 2022. Diet and food type affect urinary pesticide residue excretion profiles in healthy individuals: results of a randomized controlled dietary intervention trial. *American Journal of Clinical Nutrition* 115(2):364–377.

Schwendel, B. H., Wester, T. J., Morel, P. C., Tavendale, M. H. Deadman, C., Shadbolt, N. M. and Otter, D. E. 2015. Invited review: organic and conventionally produced milk-an evaluation of factors influencing milk composition. *Journal of Dairy Science* 98(2):721–746.

Slots, T., Butler, G., Leifert, C., Kristensen, T., Skibsted, L. H. and Nielsen, J. H. 2009. Potentials to differentiate milk composition by different feeding strategies. *Journal of Dairy Science* 92(5):2057–2066.

Soil Association. 2018. *Soil Association organic standards: farming and growing*. Soil Association, Bristol, UK.

Średnicka-Tober, D., Barański, M., Seal, C. J., Sanderson, R., Benbrook, C., Steinshamn, H., Gromadzka-Ostrowska, J., Rembiałkowska, E., Skwarło-Sońta, K., Eyre, M., Cozzi, G., Larsen, M. K., Jordon, T., Niggli, U., Sakowski, T., Calder, P. C., Burdge, G. C., Sotiraki, S., Stefanakis, A., Stergiadis, S., Yolcu, H., Chatzidimitriou, E. Butler, G., Stewart, G. and Leifert, C. 2016. Higher PUFA and n-3 PUFA, conjugated linoleic acid, α-tocopherol and iron, but lower iodine and selenium concentrations in organic milk: A systematic literature review and meta- and redundancy analyses. *British Journal of Nutrition* 115(6):1043–1060.

Stergiadis, S. 2009. Effects of dairy management on milk quality. in *Nafferton Ecologival Farming Unit*. Vol. PhD. Newcastle University, Newcastle upon Tyne.

Stergiadis, S., Berlitz, C. B., Hunt, B., Garg, S., Ian Givens, D. and Kliem, K. E. 2019. An update to the fatty acid profiles of bovine retail milk in the United Kingdom: implications for nutrition in different age and gender groups. *Food Chemistry* 276:218–230.

Stergiadis, S., Bieber, A., Chatzidimitriou, E., Franceschin, E., Isensee, A., Rempelos, L., Baranski, M., Maurer, V., Cozzi, G., Bapst, B., Butler, G. and Leifert, C. 2018a. Impact of US Brown Swiss genetics on milk quality from low-input herds in Switzerland: interactions with season. *Food Chem* 251:93–102.

Stergiadis, S., Hynes, D. N., Thomson, A. L., Kliem, K. E., Berlitz, C. G. B., Gunal, M. and Yan, T. 2018b. Effect of substituting fresh-cut perennial ryegrass with fresh-cut white clover on bovine milk fatty acid profile. *Journal of the Science of Food and Agriculture* (Under review).

Stergiadis, S., Leifert, C., Seal, C. J., Eyre, M. D., Larsen, M. K., Slots, T., Nielsen, J. H. and Butler, G. 2015. A 2-year study on milk quality from three pasture-based dairy systems of contrasting production intensities in Wales. *The Journal of Agricultural Science* 153:708–731.

Stergiadis, S., Leifert, C., Seal, C. J., Eyre, M. D., Nielsen, J. H., Larsen, M. K., Slots, T., Steinshamn, H. and Butler, G. 2012. Effect of feeding intensity and milking system on nutritionally relevant milk components in dairy farming systems in the north east of England. *Journal of Agricultural and Food Chemistry* 60(29):7270–7281.

Stergiadis, S., Leifert, C., Seal, C. J., Eyre, M. D., Steinshamn, H. and Butler, G. 2014. Improving the fatty acid profile of winter milk from housed cows with contrasting feeding regimes by oilseed supplementation. *Food Chemistry* 164:293–300.

Stergiadis, S., Qin, N., Faludi, G., Beauclercq, S., Pitt, J., Desnica, N., Pétursdóttir, Á. H., Newton, E. E., Angelidis, A. E., Givens, I., Humphries, D. J., Gunnlaugsdóttir, H., and Juniper, D. T. 2021. Mineral concentrations in bovine milk from farms with contrasting grazing management. *Foods* 10(11). https://doi.org/10.3390/foods10112733

Stergiadis, S., Seal, C. J., Leifert, C., Eyre, M. D., Larsen, M. K. and Butler, G. 2013. Variation in nutritionally relevant components in retail Jersey and Guernsey whole milk. *Food Chemistry* 139(1–4):540–548.

Stoop, W. M., Bovenhuis, H., Heck, J. M. L. and van Arendonk, J. A. M. 2009. Effect of lactation stage and energy status on milk fat composition of Holstein-Friesian cows. *Journal of Dairy Science* 92(4):1469–1478.

Van Diepen, P., McLean, B. and Frost, D. 2007. *Livestock breeds and organic farming systems*. ADAS, Cardiff, UK.

Walker, G. P., Dunshea, F. R. and Doyle, P. T. 2004. Effects of nutrition and management on the production and composition of milk fat and protein: a review. *Australian Journal of Agricultural Research* 55(10):1009–1028.

Wei, X., Pan, Y., Tang, Z., Lin, Q., Jiang, Y., Chen, J., Xian, W., Yin, R., Li, J. and Qiu, R. 2023. Neonicotinoids residues in cow milk and health risks to the Chinese general population. *Journal of Hazardous Materials* 452. https://doi.org/10.1016/j.jhazmat.2023.131296

Zwierzchowski, G. and Ametaj, B. N. 2018. Minerals and heavy metals in the whole raw milk of dairy cows from different management systems and countries of origin: a meta-analytical study. *Journal of Agricultural and Food Chemistry* 66(26):6877–6888.

Part 2

Improving health, welfare and life opportunities

Chapter 5

The value of One Health and One Welfare approaches in promoting the health and welfare of organic dairy cattle

Lindsay Whistance, The Organic Research Centre, UK

1 Introduction

There is a growing recognition that human, animal and environmental health should not and cannot be managed in isolation from one another. The recent development of 'One Welfare' as a holistic approach to health and well-being came about largely as a response to the perceived limitations of the integrated 'One Health' approach which, is itself, partly developed from the earlier 'One Medicine' approach. All of these approaches continue to be pursued alongside others including 'Eco Health' and 'Planetary Health', indicating the ongoing challenges faced for truly integrative research and inclusive health-directed action.

Within the wider body of published 'integrated planetary health' literature, references to dairy farming are limited and even scarcer for organic farming. Despite this, an investigation of the various integrated approaches to health and welfare, their development and the limitations they face, as well as how they compare to core organic principles, is key to assessing their value to, and potential influence on, organic farming systems in general, and organic dairy

http://dx.doi.org/10.19103/AS.2024.0138.06

farming in particular. The aim of this chapter is firstly to give an overview of how the concepts of One Health and One Welfare have evolved and been debated over time, and how they seem to be practiced. Secondly, the chapter serves as an opinion paper on how the current use of the two concepts can conflict with the aims of organic dairy farming as defined by the four organic principles.

2 The development of the One Health approach

A belief in the integrated nature (of, at least, some elements) of human, animal and environmental health has existed from ancient times in philosophical discussion, physiological research and local practice. Historically, the division between students of animal health and human health was less distinct with much overlap between physiological research (cadavers and vivisections) as well as diseases and their treatment. In time, this overlap led to the development of the 'One Medicine' approach to integrated human and domestic animal health management. However, these early approaches were typically confined to the treatment and control of disease with the major actors limited to physicians and veterinarians.

The One Health approach, or simply 'One Health', was established more formally in the early 2000s as a direct response to the risk of a human pandemic resulting from highly pathogenic avian influenza (HPAI). The World Health Organisation (WHO) e.g. defines One Health as: 'an integrated, unifying approach that aims to sustainably balance and optimise the health of people, animals and ecosystems' (https://www.who.int/health-topics/one-health). The need for wider interdisciplinary collaboration was recognised as important in order to offer the strongest way to address this specific threat alongside those arising from other emerging (and re-emerging) zoonotic diseases (Marbán-Castro et al., 2019). In 2005, a set of four fundamental principles (the Beijing Principles) were developed for HPAI control (Gibbs, 2014):

- To promote a multi-sectoral approach;
- To support individual countries in developing and owning their own control strategies;
- To consider both short- and long-term actions for rapid response and capacity building; and
- To ensure efficiency through monitoring and evaluation.

Collaborations were to include governments, non-governmental organisations (NGOs), researchers and specialists in, e.g. wildlife, the environment, sociology, the economy and anthropology, alongside human and animal health scientists. It was recognised that, whilst each profession and discipline brought different expertise to the table, each could help in developing a more holistic and enduring solution to threats from zoonotic disease. The One Health Initiative developed an umbrella concept to illustrate the scope of One Health (Figure 1).

The 'One Health' concept is now widely recognised and its language has been adopted by governments, international health and research bodies along with human and veterinary medicine organisations. For example, in 2019, the European Food Safety Authority (EFSA) and the European Centre for Disease Prevention and Control (ECDPC) produced a 276-page 'One Health' report on zoonoses monitoring activities in humans, food and farmed animals carried out in Europe in 2018 (EFSA and ECDPC, 2019). The World Organisation for Animal Health (OIE) has also highlighted how health is based on the interdependence of human, animals and plant health (https://www.woah.org/) (Figure 2).

3 One Health and organic dairy farming: synergies and potential conflicts

Organic dairy farms and farmers do not operate in a spatial, social, economic or ecological vacuum. They are necessarily part of the wider health – or absence of disease – narrative highlighted by One Health approaches and therefore have responsibilities beyond the health of their own farm. This responsibility is made explicit in the core organic Principle of Health, where 'the health of individuals

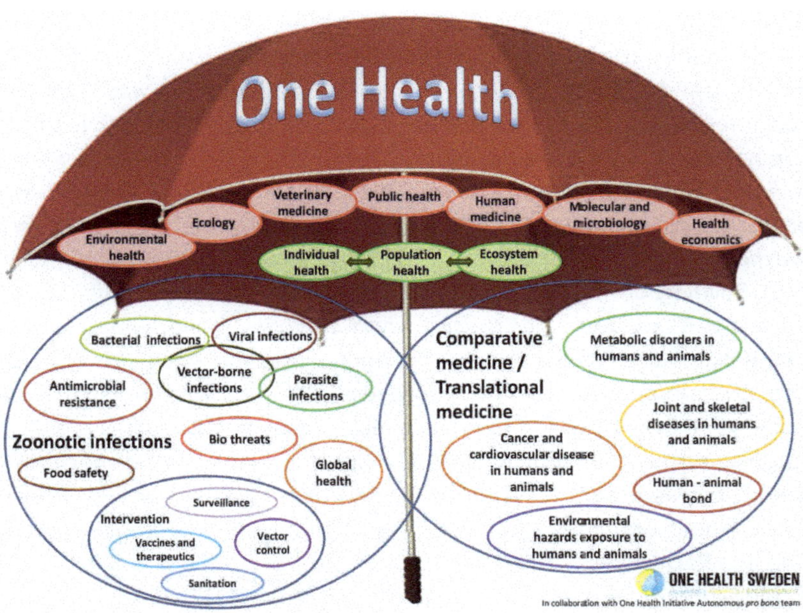

Figure 1 Focus areas included in the One Health approach from www.onehealthintiative .com (Gibbs, 2014).

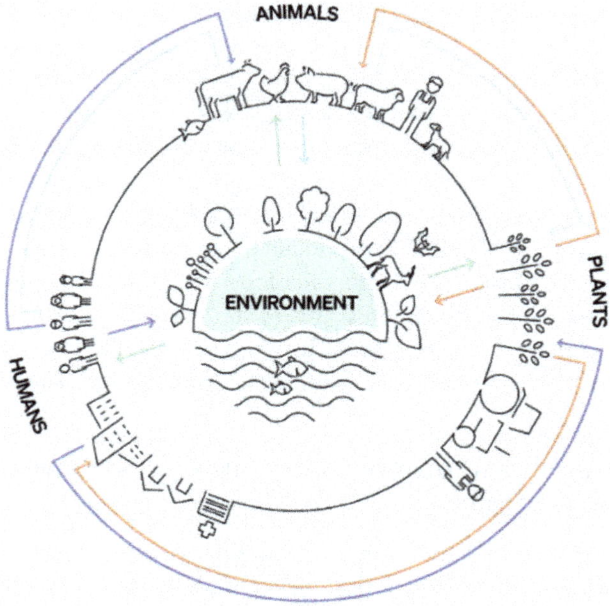

Figure 2 WOAH illustration of the interdependence of health between the different actors (OIE, n.d.).

and communities cannot be separated from the health of ecosystems' (IFOAM, 2005).

The dominant focus in organic food systems is to foster health through 'wholeness and integrity of living systems' (IFOAM, 2005). This was highlighted in early research into health within an organic context which took a different approach from conventional veterinary medicine: *Instead of the contemporary obsession with disease and its causes, they set out to discover the causes of Health. This led inevitably to an awareness of wholeness (the two words after all, have the same origin) and to a gradual understanding that all life is one* (Balfour, 1978). Building on this, Balfour (1978) stated, *the biota is a whole, of which we are a part, and the other species which compose it and helped to create it, are entitled to existence in their own right. This is the wholeness approach.* Part of this approach includes building a relationship between farming and the natural world (e.g. by giving farmed animals access to outdoors and grazing). In organic systems, animals are viewed not just in terms of what they provide humans, but as a part of Nature with their own intrinsic value in a holistic approach that promotes good health alongside other desirable outcomes (Lund and Röcklinsberg, 2001).

More broadly, the four core organic Principles of Health, Ecology, Fairness and Care echo the inclusive and precautionary approach implicit in One Health, describing a way to produce food designed to 'sustain and enhance the health of soil, plant, animal, human and planet as one and indivisible' (Principle of Health) and to act in a 'precautionary and responsible manner to protect the health and well-being of current and future generations and the environment' (Principle of Care). The interdependence of human, animal, plant and broader ecosystem health highlighted in One Health reflects the writings (published in 1943) of Lady Eve Balfour, a pioneer organic farmer in England, who wrote *'the health of man, beast, plant and soil is one indivisible whole'* (Balfour, 2006, p. 21).

However, there are important potential differences between the One Health approaches and organic principles. Given its origins, initial One Health approaches were focussed primarily on the control of specific diseases and primarily from the point of view of human health. Other than to acknowledge the role of human activity (e.g. bush meat markets and deforestation) as a factor in the rise and spread of diseases, One Health does not necessarily offer a more holistic world view for how humans should produce food in a more sustainable way which limits the (re)emergence and spread of these diseases. Moral and practical challenges may then arise where these higher-level disease-control strategies conflict with organic standards or principles.

These challenges may be exacerbated if One Health approaches prioritise certain objectives (e.g. prevention of human disease) over others (e.g. animal or broader ecosystem health). It is noticeable in Fig. 1 e.g. that, while environmental and public health are highlighted, animal health as an objective is only referred to indirectly under veterinary medicine. This focus on certain priorities can, for example, be seen in a review of the quantitative outcomes of One Health programmes by Falzon et al. (2018). This identified environmental 'interventions' as an important measure to limit the impact of existing diseases on humans rather than its wider benefits for animals and ecosystems too. More broadly, the ecosystem services provided by organic and other agroecological farming systems are often categorised and valued as 'public goods' benefitting all members of (human) society (Baldock et al., n.d.). However, despite its focus on fairness and ecology, this is still a largely anthropocentric position compared to the eco-centric position which, it can be argued, lies closer to the principles underlying organic food systems.

In addition, the One Health focus on national and international integrated, multi-disciplinary approaches to disease prevention and management (as important as these are) also has the potential, if not well managed, to implement strategies which do not align with, or are even hostile to, the aims of organic farming. Finally, with the reductionist approach to disease management, it is important to note that, in addition to potential tensions between One Health and organic principles, some of these tensions also affect the organic movement itself in how best to balance responses to the presence of health threats and the core organic principles of Health, Ecology,

Fairness and Care. There are two examples below that illustrate some of these tensions.

The first relates to ongoing practices used to control bovine tuberculosis (bTB) in UK cattle herds. These have included the (initially compulsory) 'proactive' culling of wild badgers in risk areas, since they are potential vectors of the disease, regardless of the disease status of the badgers, or the herd. By addressing the role of wider environmental factors in the spread of disease, it has been argued that this approach to disease control can be seen to conform with and be justified by, One Health principles (Lederman, 2016). In practice, the culling of badgers to control bTB has proved controversial on both ethical and scientific grounds with conflicting evidence on its effectiveness (Langton et al., 2022; Birch et al., 2024; Torgeson et al., 2024).

In assessing the practice against the organic Principles of Health, Ecology and Care, this approach is not consistent with the principles of Ecology or Care. From the point of the Principle of Care, badgers and all other wildlife vectors of bTB should be recognised as victims of this disease. The Principle of Ecology also states that organic farming should 'protect and benefit the common environment including landscapes, ...habitats, ... (and) biodiversity' (IFOAM, 2005). From an ecological perspective, whilst the intention of culls is to reduce bTB incidence, they can also increase incidences of infection through social perturbation, so that disease-free badger setts and cattle herds then become exposed to the disease (Bielby et al., 2014). Such disturbances can persist for up to eight years after a cull (Carter et al., 2007), weakening the integrity of the whole system.

Some of the tensions arising from the mismatch between current national practice and the organic principles can be seen in the policy of the Soil Association, one of the bodies responsible for organic standards and certification in the United Kingdom (UK). On the one hand, it has advocated that alternative measures, such as adequate biosecurity and vaccines, should be preferentially employed for both cattle and wildlife alike. On the other hand, in recognition of the risk to animal health and farm livelihoods under current national farming and disease control practices, the Soil Association has allowed its organic cattle farmer members to make their own decisions on participation in the (now voluntary) culling programme, without affecting their organic certification (Case, 2013). Nevertheless, they made it clear that they consider the development and use of vaccinations as the long-term solution (Case, 2013). A decade later a trialled vaccine is now available and being used for badgers, although it has not (yet) replaced culling as the dominant response on conventional or organic farms.

The potential for better alignment with the organic principles is now on the horizon, not just with the development of a Bacillus Calmette-Guérin (BCG) vaccine for cattle, but also a DIVA (Differentiating Infected from Infected Animals) test capable of identifying infected cattle from vaccinated animals (APHA, 2021), increasing the likelihood that the wider industry, across all countries, will adopt this approach for cattle and wildlife vectors alike (Woodroffe et al., 2024). Indeed, on 30 August 2024, the recently elected UK Government announced a new bTB eradication strategy which will include the use of badger vaccinations at a national scale and bring an end to the badger cull in England (Defra and Zeichner, 2024). The change in strategy at national level should help to support farming practices which better align with all four organic principles.

The second example refers to animal movement practices. Analysis of official animal movement records in Britain shows that cattle movements 'consistently outperform environmental, topographic and other anthropogenic variables as the main predictor of disease occurrence', particularly in areas where bTB has already been reported (Gilbert et al., 2005). This is an issue that affects all dairy farmers, including organic farmers. While operating a closed milking herd is typical on organic dairy farms, this practice is rarely extended to male calves or beef cross calves which are commonly sold (Herrler et al., 2023). One of the ongoing challenges for organic dairy farming is to balance the need for cows to have calves in order to produce milk and what to do with those calves not required as replacement heifers.

Additionally, whilst a low replacement rate is a good reflection of cow longevity, which in turn is an indicator of good health status and therefore desirable in organic dairy systems, these systems largely adhere to a 365-day calving interval, as in conventional dairy systems. This increases the number of animals considered as 'surplus' to the milking system and thus perceived as a burden on the farm's available resources. In practice, replacement rates in organic dairy herds are not consistently lower than in conventional herds, even if the reasons for culling between conventional and organic herds may be different (Ahlman et al., 2011). The issue of unwanted calves presents a moral dilemma for organic producers and is one which has come under increasingly critical scrutiny (Nielsen and Thamsborg, 2002). In all systems, the outcomes and challenges for these, so-called 'surplus' calves is further compounded by the continued selection of animals for a single purpose, i.e. either beef or dairy (Bolton and von Keyserlingk, 2021).

Both of these examples highlight the challenges of balancing potentially competing priorities (e.g. relating to economic performance versus welfare and broader environmental needs) both within the One Health movement and within the organic dairy farming sector.

4 One World, One Health™

Through a recent development itself, the One Health approach rather paradoxically now encompasses two distinct approaches:

- 'One Medicine, One Health'; and
- 'One World, One Health™'.

As Bresalier et al. (2021) noted in their review of the development of One Health, narrowing the concept of health primarily to disease control was supported by the successful development of vaccines and antimicrobials which enabled targeted control of infectious diseases and largely ignored the significant environmental factors influencing both the presence of infectious agents and their impacts on populations. These environmental factors were more evident in poorer countries with limited access to medication during infectious disease outbreaks.

Whilst One Medicine, One Health continues with human and animal medical research as the principal focus, One World, One Health™ (a trademark of the Wildlife Conservation Society) has expanded the focus to more fully cover the environmental and life sciences. The One World, One Health™ approach recognises the limitation of health research that does not consider and engage with the environment as a key actor. It provides a more holistic approach and a potential route to balance the core organic principles of Health, Ecology, Fairness and Care.

In 2005, the same year in which the Four Organic Principles were published, One World, One Health™ produced 12 detailed recommendations (The Manhattan Principles) *for establishing a more holistic approach to preventing epidemic/epizootic disease and for maintaining ecosystem integrity for the benefit of humans, their domesticated animals, and the foundational biodiversity that supports us all*. Though the wording of the principles is more detailed, they can be summarised as follows:

1. Recognise the essential link between human, domestic animal and wildlife health.
2. Recognise that decisions regarding land and water use have real implications for health.
3. Include wildlife health science as an essential component of global disease prevention.
4. Recognise that human health programmes can greatly contribute to conservation efforts.
5. Devise adaptive, holistic and forward-thinking approaches to the prevention of emerging and resurging diseases.
6. Seek opportunities to fully integrate biodiversity perspectives when developing solutions to infectious disease threats.
7. Reduce the demand for and better regulate the international live wildlife and bushmeat trade.
8. Restrict the mass culling of free-ranging wildlife species for disease control.

9 Increase investment in the global human and animal health infrastructure.
10 Form collaborative relationships among governments, local people and the private and public (i.e. non-profit) sectors.
11 Provide adequate resources and support for global wildlife health surveillance networks.
12 Invest in educating and raising awareness.

The 12 principles 'urge' world leaders, society, health providers and researchers to recognise essential links between health and functioning ecosystems including

- land and water use; and
- wildlife health and conservation.

The principles advocate for the implementation of interconnected approaches to address the threat of disease which fully integrate biodiversity, domestic animal and human perspectives. These include regulating the bushmeat trade and restricting mass culling of wild species for disease control. Furthermore, these principles urge an increase in investment in the global health infrastructure including surveillance and sharing of information and medicines. More broadly, they call for collaborative relationships between all sectors and an investment in educating and raising awareness *to increase recognition that we must better understand the relationships between health and ecosystem integrity to succeed in improving prospects for a healthier planet* (Principle 12) (WCS, 2005). The 12 Manhattan Principles also expand upon the Beijing Principles by adding a list of focus areas to support a multi-sectoral approach as well as calling for education of, and collaboration with, local peoples.

Nevertheless, the Manhattan Principles still remain largely focussed on disease control. The detailed wording of Manhattan Principle Six e.g. urges the seeking of 'opportunities to fully integrate biodiversity conservation perspectives and human needs (including those related to domestic animal health) when developing solutions to infectious disease threats'. The inclusion of domestic animals in parenthesis risks them being perceived merely as a sub-category of disease control to meet human needs, rather than as a key part of ecosystems which should be protected in their own right.

Despite the call for participatory approaches articulated in Principle 10, One World, One Health™-inspired programmes have also often been driven in practice by a relatively narrow group of higher-level professionals and experts (e.g. scientists in government or in national research bodies) with responses developed and actioned by governments and, notably, by the pharmaceutical industry, but with minimal use of participatory approaches. This has then meant a failure to develop resilient and sustainable practices at ground level which considers the health of each 'stakeholder' (including the landscape, wild and farmed animals) at the farmer/citizen level. Top-level understanding is, of course, necessary to understand the causes and consequences of disease and to develop 'best practice' in appropriate treatment and control of any given disease. However, top-down approaches risk failing if there is insufficient consultation with and participation by local populations on the ground. They also risk failure if they do not address broader issues of developing sustainable practices to optimise agroecosystem health.

Some of these problems were identified in a review of One Health initiatives in the UK by Kock (2015). The author identified that most activity was from veterinarians with a dearth of published material from other specialisms such as the medical and public health communities. This and other studies (e.g. Wallace et al., 2015) also highlighted a very narrow definition of health, namely, the absence of disease. This narrow definition risks restricting the focus of research and action to a focus on the presence versus absence of disease, rather than moving towards systems of holistic health, building system resilience to prevent disease from occurring in the first place.

As this suggests, weaknesses such as limiting the focus of health to an absence of disease persist in the wider One Health literature. This can be seen e.g. Nguyen et al. (2022) who use a One Health lens to evaluate the sustainability of dairy farming in the United States (US). From a primarily veterinarian perspective, they argue for the benefits to planetary health of modern breeding and intensive management systems which, over 30 years, they claim has resulted in a 24% reduction in methane emissions per unit of milk produced through increases in feed conversion efficiency and milk yields per cow. These increases in efficiency have also resulted in a reduced land footprint and water use in milk production. They use measures such as a reduction in somatic cell counts to show overall improvements in cow health. Whilst this perspective could be seen as consistent with One Health approaches, and not necessarily excluded by the Beijing or Manhattan Principles, it is clearly completely at odds with the principles and practices of organic farming (Bonney, 2008; Potterton et al., 2011; Bürow et al., 2011).

These problems in defining and applying One Health in a more holistic way that might be consistent with organic principles are also shown e.g. in Garcia et al. (2019) who apply a One Health perspective to measures to improve dairy food safety in the US. They focus on identifying critical points throughout the current, largely conventional dairy supply chain in the US in the prevention, detection and management of both zoonotic pathogens and diseases such as mastitis, including a continuing, more selective therapeutic use for antibiotics (after their use as a growth promotor was banned in regions such as the European Union in 2006 (EC Reg. 1831/2003/EC)). This study assumes conventional, more intensive production systems to be the norm and makes no reference to the potential role of the dairy production system itself (including breeding, housing and feeding practices) in unintentionally promoting sub-clinical disease. The study goes on to recommend current best practice in the US dairy supply chain as a management template for 'low- and middle-income countries to improve public health, nutrition and economic benefits to farmers'. Here, animal and planetary health are not included.

5 From One Health to EcoHealth and Planetary Health

In 2019, at the 'One Planet, One Health, One Future' Conference in Berlin, the Manhattan Principles were reviewed, updated and re-named the Berlin Principles. Though, as with the Manhattan Principles, the wording is more detailed, the 10 Berlin Principles can be summarised as follows:

1. Recognise and take action to retain the essential health links between human, wildlife, domestic animal and plants, and all nature.
2. Take action to develop strong institutions that integrate understanding of human and animal health with the health of the environment.
3. Take action to combat the current climate crisis, which is creating new severe threats to human, animal and environmental health.
4. Recognise that decisions regarding land, air, sea and freshwater use directly impact health and wellbeing of humans, animals and ecosystems.
5. Devise adaptive, holistic and forward-looking approaches to the detection, prevention, monitoring and control of emerging and resurging diseases.
6. Take action to meaningfully integrate biodiversity conservation perspectives and human health and wellbeing when developing solutions for disease threats.
7. Increase cross-sectoral investment in the global human, livestock, wildlife, plant and ecosystem health infrastructure.
8. Enhance capacity for cross-sectoral and trans-disciplinary health surveillance.
9. Form participatory, collaborative relationships among governments, NGOs, and indigenous peoples and local communities.
10. Invest in educating and raising awareness for global citizenship and holistic planetary health approaches.

The need for an update was based on *the massive and existential changes – notably increasing climate change and biodiversity loss* (Gruetzmacher et al., 2021). The updates were intended to strengthen the focus on overall ecosystem health and integrity as well as take account of pressing issues such as increasing antimicrobial resistance and climate change impacts.

Whilst the Berlin Principles strengthen the focus on a more integrated and participatory approach, they do not fully resolve earlier problems with One Health agendas. There could still be said to be an underlying anthropocentric focus on identifying and managing disease risks that particularly affect the human population. There is also a recognition of the enormous challenges in bringing together many different disciplines and interest groups, each with its own perspectives, goals and suggested solutions (Gibbs, 2014). This has led to the call for 'One Communication' programmes to address gaps in communication, coordination and collaboration across different sectors (Mateus et al., 2022). As just one aspect of this broader challenge, it has been pointed out that traditional models of funding research will need to adapt to the challenges and requirements of more collaborative approaches (Queenan et al., 2017; Falzon et al., 2018).

Some of these issues are reflected in the development of further approaches, each reflecting a similar awareness of the need for interconnectedness but with a different focus. These approaches include (Beasley, 2009)

- 'One Toxicology';
- 'Planetary Health'; and
- 'Eco-Health'.

Lerner and Berg (2017) have compared 'One Health', 'EcoHealth' and 'Planetary Health' approaches, focussing on definitions, core values and contributing sciences (Table 1). They found not only similarities but also important differences in definitions, core values and focus areas of the principle science disciplines driving different approaches. These again highlight the difficulty in attaining a true 'Oneness', at least under the current way in which political, social, medical, agricultural and research groups are organised and interact.

In the comparison shown in Table 1, the approach that appears to most closely align with an organic perspective is 'Eco-Health' (or 'Ecohealth') with its focus on ecological health. This also most closely aligns with the definition of health developed by the WHO in 1948: *Health is a state of complete physical, mental and social well-being and not merely the absence of disease or infirmity* (WHO, 2020). The inclusion of well-being in the WHO definition makes it clear that health is more than the absence of disease. By expanding this definition to include animals as well as humans, Eco-Health envisages attaining *sustainable human and animal health and well-being, through healthier ecosystems* (Waltner-Toews, 2009, p. 519). Ecohealth also distinguishes itself by stating that it can be defined as a *systemic, participatory approach to understanding and promoting health and wellbeing in the context of social and ecological interactions* (Waltner-Toews, 2009, p. 520) and that *whilst other schools of thought define fields of inquiry, ecohealth defines an area of practice* (p. 519).

Table 1 A comparison of definitions, focus areas and values in different approaches to health.

		One Health		EcoHealth	Planetary Health	
		Narrow	Wide		Narrow	Wide
Core contributing sciences	Human	Public health	Public health Human medicine Molecular and microbiology Health economics Social status	Public health Human medicine Rural and urban development and planning Social sciences Anthropology	Public health Human medicine	Human medicine Economy Energy Natural resources
	Animal	Veterinary medicine	Veterinary medicine	Veterinary medicine	–	Agricultural sciences (including plant and animal sciences)
	Ecosystem	–	Environmental health Ecology	Conservation and ecosystem management	–	Ecology Other environmental sciences (including climate and biodiversity research) Marine sciences
Knowledge base		Western scientific	Western scientific	Western scientific Indigenous Knowledge	Western scientific	Western scientific
Core values	Health	Individual health	Individual and population health	Population health	Individual and population health	Individual and population health
	Groups	HumansAnimals	Humans Animals Ecosystems	Humans Animals Ecosystems	Humans	Humans
	Other			Biodiversity Sustainability (for humans, animals and ecosystems)	Sustainability (for humans)	Sustainability (for humans)

(Lerner and Berg, 2017; modified. Note: see the original article for associated references).

This most closely aligns with organic farming and complements an on-farm focus on health within a wider context. However, neither Eco-Health or any of the other global One Health approaches makes any reference to the organic principles of Health, Ecology, Fairness and Care, despite some commonalities in intention.

6 One Welfare approaches

In response to some of these problems, the concept of One Welfare was introduced as a recognition that the existing One Health approaches can be limited in their focus. One Welfare is not intended to replace but to complement One Health by recognising that human and animal health are closely linked to human wellbeing and animal welfare (Garcia Pinillos et al., 2016). A second umbrella diagram was created to illustrate the scope of One Welfare (Figure 3). As Lindenmayer and Kaufman (2022) stated, 'One Health and One Welfare together can establish a basic ethic of respect for other living and non-living elements of our planet, provide a foundation for the science of One Health and replace exploitation and annihilation with stewardship and responsibility

Figure 3 The scope of One Welfare (Garcia Pinillos et al., 2016).

for our planet and all its life forms'. A specific intention of developing the One Welfare approach was to *empower the animal welfare and human wellbeing fields to address the connections between science and policy more effectively in various areas of human society, including environmental science and sustainability*. When arguing the necessity for One Welfare, Garcia Pinillos et al. (2022) also refer to an important point made by Colonius and Earley (2013), who stated that the *separation between human, social and animal welfare is an artificial compartmentalisation. These disciplines... ...heavily depend on each other in an ecological context*. Nevertheless, these authors also make clear that any proposed improvements in animal welfare *must not be addressed without ensuring that changes meant to improve animal welfare do not result in a net decrease in welfare in humans*, without any specific guiding pillars for action and making it clear that the 'One' is not indicative of any equality between the three key partners (human, animals and nature).

Whilst livestock welfare is explicitly mentioned, alongside farmer wellbeing, it is notable that the majority of named focus areas in the umbrella are directly related to areas of concern resulting in poor human and animal welfare, rather than a focus on positive practices which lead to good welfare. Looking at published One Welfare literature indicates a dominant focus on poor welfare and its causes for both humans and animals, limiting research to what to do when things go wrong in existing systems rather than how to develop high welfare systems where resilience and robustness are built in (Squance et al., 2021; Garcia Pinillos, 2022). Just as health is more than the absence of disease, good welfare is also much more than the absence of active maltreatment or the result of negligence. A potential area of concern is the inclusion of 'Increased Productivity' as an independent measure of improved welfare. This may be acceptable if it is interpreted as meaning that high welfare results in higher productivity. According to Garcia Pinillos et al. (2016), the inclusion of 'Increased Productivity' in the One Welfare focus themes refers to the loss of production related to poor welfare. The authors state that *practical evidence demonstrates that improved farm animal welfare results in superior yields (e.g. better hoof care improves the health of cows and healthy cows have a better milk yield)*. However, this position is itself misleading since milk yield is not 'superior' with good welfare but rather inferior with poor welfare where energy is spent on coping. This focus could become further problematic if it encourages too much of a focus on increasing productivity as a goal in its own right. After all, one of the reasons for the development of organic animal farming was because of the concern about the negative impact on both animal welfare and farmer well-being of intensive systems focussing on maximising yields (Waiblinger et al., 2002; Hansen et al., 2020). A key weakness in One Health and One Welfare approaches, from an organic farming viewpoint, is the lack of a distinction between systems which build in good welfare and systems which focus on the mitigation of poor welfare outcomes (Lerner and Berg, 2017).

7 New approaches to complex problems: post-normal science

Post-normal science (PNS) was developed in 1980s as a problem-solving framework to address sustainable governance of natural systems, recognising the need for 'new tools for new problems'. It identified that traditional research methods (pure and applied) were struggling to manage three key challenges (Funtowicz and Ravetz, 2003):

- Uncertainty;
- Value loading; and
- Diverse legitimate perspectives.

Post-normal science is based on distinguishing between three different levels of inquiry (Figure 4):

- Academic applied science;
- Professional consultancy; and
- Post-normal science.

Where both uncertainty and stakes are low, 'normal' applied scientific approaches remain appropriate. Research settings tend to involve suggesting a specific hypothesis and then testing its validity in carefully controlled

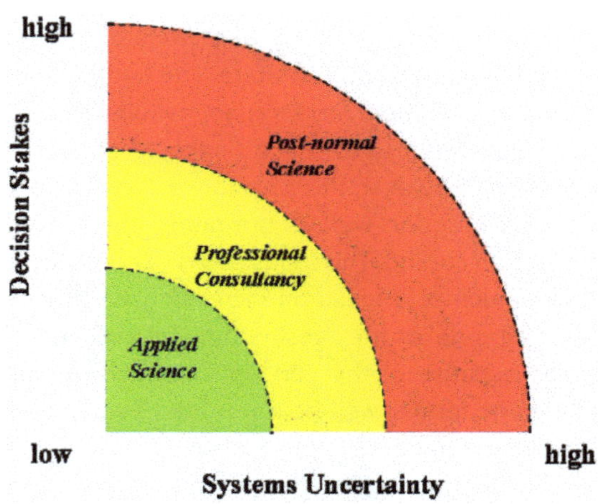

Figure 4 Post-normal science (PNS) diagram (Funtowicz and Ravetz, 2003).

experimental conditions (e.g. field trials). To ensure that findings are objective, repeatable and robust, these settings deliberately avoid the complexity and the myriad economic and other pressures that characterise real-world farm conditions. However, where stakes and uncertainty increase to medium, further expertise is needed including 'skill, judgement and even courage' of professionals working with individual farming systems (Funtowitz and Ravetz, 2003). These two levels can be appropriate for investigating levels of welfare, including risks to health, at the farm level.

However, as stakes and uncertainty increase further, these competencies are not sufficient in themselves and require the inclusion of an 'extended peer community' which involves a broader notion of 'facts' (Kønig et al., 2017). The extended peer community helps to prevent errors such as the narrow framing of a problem to make it fit a given disciplinary framework as well as avoiding Type III errors (where errors in problem framing lead to the wrong problem being assessed).

Post-normal science helps to reposition the goal of scientific enquiry away from simple (and at times naïve assumption of) objective 'truth' to 'quality' and the development of information considered more fit for sustainable decision-making in real-world conditions. The term quality includes not only the product (knowledge) but also processes, people and purposes (Funtowicz and Ravetz, 1993). An acceptance of uncertainty and unpredictability as facts of life is also critical and, in so doing, forces us to understand the limitations of a research approach focussed on (top down) control and predictability when dealing with highly complex systems. At its root, this acceptance demands a democratisation of responsibility.

As with organic farming, PNS is based on norms and values. In a review of published PNS research, Kønig et al. (2017) identified 33 different norms and values (Figure 5). The five norms in bold have been highlighted as components in the acronym 'TRUST'. These are as follows:

- Transparency;
- Robustness;
- Uncertainty management;
- Sustainability; and
- Transdisciplinarity.

Highlighting these norms, the authors intend them to be used to help regain public trust in scientific advice indicating the value of a general change in approach beyond the focus here on organic dairy farming. The inclusion of prototypical norms including 'holism', 'precaution', 'integrity' and 'empowerment' indicate that health and welfare of farmed animals (including organic dairy cattle) would be better served using this approach compared to the more reductionist

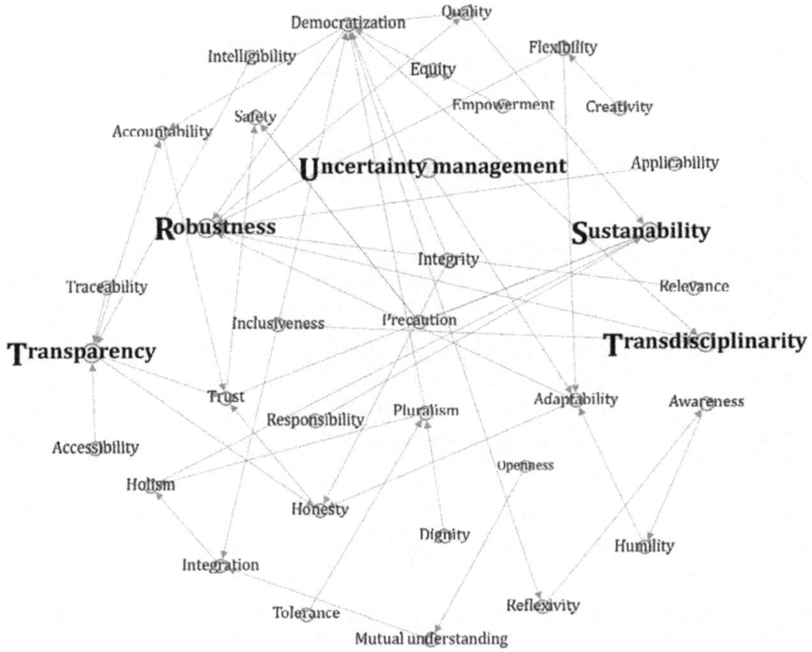

Figure 5 The ethos of PNS (Kønig et al., 2017).

One Health/One Welfare approach that seems to have occurred in practice. Indeed, whilst the value of norms is made explicit, they are not fixed. As Kønig et al. (2017) state, *we do not claim that the ethos of PNS will be the only norms and values that are brought into play in these practices, on the contrary, the ethos of PNS is better understood as a potential or a tool from normative meaning making.*

The recognition of the need for more inclusive, transdisciplinary research e.g. is not new. For example, at a 1970 meeting of the International Society for Applied Ethology (ISAE), it was agreed to *promote exchange of information between those concerned with behaviour and well-being of animals* and that there should be a *stimulation of enquiry and research into the behaviour of animals among agricultural scientists, veterinarians and zoologists* (Petherick and Duncan, 1991). However, at a wider level, with the increasing recognition of uncertainty of facts, differing (and sometimes competing) values, high stakes and the need for urgent solutions to real-world problems, PNS argues that the quality of the process and product are not well served by limiting any inquiry to a community of disciplinary peers (Funtowicz and Ravetz, 2008).

8 Conclusion

This chapter has explored the development of One Health and its successive variants (including One Word, One Health™ and One Welfare) in both principles articulated and in reported practice. These approaches have been compared against the four core organic Principles of Health, Ecology, Fairness and Care. The chapter has identified a number of tensions and potential weaknesses both within and between these approaches, as well as how they relate to organic principles. These tensions include

- a narrow focus on dealing with disease rather than promoting health;
- a short-term focus on management rather than a longer-term focus on prevention;
- a failure to take into account how different production practices and systems may impact health and welfare; and
- a prioritisation of benefits for human health over animal and environmental health and welfare considerations.

In addition, top-down framing and implementation of these approaches by a relatively narrow group of high-level experts (e.g. from government agencies, academic researchers and veterinarians) has sometimes reinforced these potential areas of weakness. As the examples of badger culling and movement of surplus calves in the UK demonstrate, balancing the principles of Health, Ecology, Fairness and Care has also been a challenge with organic dairy farming itself. Addressing these challenges, achieving the ambitions of the One Health and Welfare movement, and achieving alignment with the organic principles of Health, Ecology, Fairness and Care will require new approaches such as post-normal science which take into account not just product (knowledge) but also processes, people and purposes in tackling complex problems.

9 Where to look for further information

- A quick guide to post-normal science: https://i2insights.org/2021/10/19/guide-to-post-normal-science/
- Science for the Post Normal Age: https://commonplace.knowledgefutures.org/pub/6qqfgms5/release/1

10 References

Ahlman, T. et al. (2011). Culling reasons in organic and conventional dairy herds and genotype by environment interaction for longevity. *Journal of Dairy Science*. 94: 1568–1575.

APHA. (2021). The potential role for BCG vaccination in global efforts to control and eradicate bovine tuberculosis. A technical discussion paper compiled by the OIE Reference Laboratory for Bovine Tuberculosis based at the United Kingdom's Animal and Plant Health Agency (Weybridge). Submitted to WOAH 29 June, 2021. https://rj8a5f.n3cdn1.secureserver.net/wp-content/uploads/2022/12/APHA_technical

_discussion_paper_WOAH_-BCG_vaccination_cattle.pdf Online: [Accessed 06/04/2023].

Baldock, D. et al. (n.d.). Public Goods and public intervention in agriculture. European Network for Rural Development. Online: https://ec.europa.eu/enrd/enrd-static/fms/pdf/260BDE6D-0066-3464-FD34-E3BB6AD3BB51.pdf [Accessed 24/05/2022].

Balfour, E. B. (1943). *The living soil* (2nd ed. 2006). Soil Association, Bristol.

Balfour, E. B. (1978). Towards a sustainable agriculture: the living soil. In: Besson, J.-M. and Vogtmann, H. (eds.) *Towards a sustainable agriculture. Proceedings of the IFOAM International Conference, Sissach, Switzerland, 1977.* Verlag Wirz: Aarau.

Beasley, V. (2009). 'One Toxicology', 'Ecosystem Health' and 'One Health'. *Veterinaria Italiana*. 45: 97–110.

Bielby, J. et al. (2014). Badger responses to small-scale culling may compromise targeted control of bovine tuberculosis. *PNAS*. 111: 9193–9198.

Birch, C. P. D. et al. (2024). Difference in differences analysis evaluates the effects of the badger control policy on bovine tuberculosis in England. *Nature Scientific Reports*. 14: Article 4849.

Bolton, S. E. and von Keyserlingk, M. A. G. (2021). The dispensible surplus dairy calf: is this issue a "Wicked Problem" and where do we go from here? *Frontiers in Veterinary Science*. 8: 660934. https://doi.org/10.3389/fvets.2021.660934

Bonney, R. (2008). The business of farm animal welfare. In: Dawkins, M. S. and Bonney, R. (eds.) *The future of animal farming. Renewing the ancient contract*. Blackwell, Oxford, UK, pp. 63–72.

Bresalier, M. et al. (2021). One Health in history. In: Zinstag, J., et al. (eds.) 2nd ed. *One Health: The Theory and Practice of Integrated Health Approaches*. CABI, Wallingford, pp. 1–14.

Bürow, E. et al. (2011). The effect of grazing on cow mortality in Danish dairy herds. *Preventive Veterinary Medicine*, 100: 237–241.

Carter, S P. et al. (2007). Culling-induced social perturbation in Eurasian badgers Meles meles and the management of TB in cattle: an analysis of a critical problem in applied ecology. *Proceedings of the Royal Society B*. 274: 2769–2777.

Case, P. (2013). Soil association rejects call to oppose badger cull. *Farmer's Weekly* 8th April, 2013. https://www.fwi.co.uk/livestock/health-welfare/livestock-diseases/bovine-tb/soil-association-rejects-call-to-oppose-badger-cull

Colonius, T. J. and Earley, R. W. (2013). One welfare: a call to develop a broader framework of thought and action. *Journal of the American Veterinary Medical Association*. 242: 309–310.

Defra and Zeichner, D. (2024). Press release: Government to end badger cull with new TB eradication strategy. https://www.gov.uk/government/news/government-to-end-badger-cull-with-new-tb-eradication-strategy. Online [Accessed 02/09/2024].

EC Regulation 1831/2003/EC on additives for use in animal nutrition, replacing Directive 70/524/EEC on additives in feeding-stuffs.

EFSA and ECDC. (2019). The European Union One Health 2018 Zoonoses Report. https://doi.org/10.2903/j.efsa.2019.5926. Online: [Accessed 05/06/2021].

Falzon, L. C. et al. (2018). Quantitative outcomes of a One Health approach to study global health challenges. *EcoHealth*. 15: 209–227.

Funtowicz, S. O. and Ravetz, J. R. (1993). Science for the post-normal age. *Futures*. 25: 739–755.

Funtowicz, S. O. and Ravetz, J. R. (2003). Post-normal science. Internet Encyclopedia of Ecological Economics. Chapter 3. International Society for Ecological Economics. https://www.isecoeco.org/pdf/pstnormsc.pdf. Online: [Accessed 23/12/21].

Funtowicz, S. O. and Ravetz, J. R. (2008). Uncertainty and quality in science for policy. In: Leinfellner, W. and Eberlein, G. (eds.) *Philosophy and methodology of the social sciences*. Vol. 15. Kluwer Academic Publishers, Dordrecht, pp. 17–31.

Garcia Pinillos, R. (ed.). (2022). *One Welfare: a framework to improve animal welfare and human welfare*. CABI, Wallingford (Oxford), UK.

Garcia Pinhillos, R., et al. (2016). One Welfare – a platform for improving human and animal welfare. *Veterinary Record*. 179: 412–413. supplementary material. vetrec-2016-october-179-16-412-inline-supplementary-material-1.pdf (onewelfareworld.org) [Accessed 04/05/2021].

Garcia, S. N. et al. (2019). A one health perspective on dairy production and dairy food safety. *One Health*. 7: Article 100086.

Gibbs, E. P. J. (2014). The evolution of One Health: a decade of progress and challenges for the future. *Veterinary Record*, 174: 85–91. https://doi.org/10.1136/vr

Gilbert, M. et al. (2005). Cattle movements and bovine tuberculosis in Great Britain. *Nature*. 435: 491–496. https://doi.org/10.1038/nature03548

Gruetzmacher, K. et al. (2021). The Berlin principles on one health – Bridging global health and conservation. *Science of the Total Environment*. 764: 142919. https://doi.org/10.1016/j.scitotenv.2020.142919

Hansen, B. G. et al. (2020). Automatic milking systems and farmer wellbeing – exploring the effects of automation and digitalisation in dairy farming. *Journal of Rural Studies*. 80: 469–480.

Herrler, M. et al. (2023). Public awareness, attitude and empathy regarding the management of surplus dairy calves. *Journal of Agricultural and Environmental Ethics*. 36: 11. https://doi.org/10.1007/s10806-023-09905-x

IFOAM Organics International. (2005). https://www.ifoam.bio/sites/default/files/2020-03/poa_english_web.pdf Online: [Accessed 04/04/2021].

Kock, R. (2015). Structural One Health – are we there yet? *Veterinary Record*. 176: 140–142. https://doi.org/10.1136/vr.h193

Kønig, N. et al. (2017). The ethos of post-normal science. *Futures*. 91: 12–24.

Langton, T. E. S. et al. (2022). Analysis of the impact of badger culling on bovine tuberculosis in cattle in the high-risk area of England, 2009–2020. *Veterinary Record*. 190(6): e1384.

Lederman, Z. (2016). One health and culling as a public health measure. *Public Health Ethics*, 9(1): 5–23.

Lerner, H. and Berg, C. (2017). A comparison of three holistic approaches to health: One Health, EcoHealth, and Planetary Health. *Perspective*, 4: 163. https://doi.org/10.3389/fvets.2017.00163

Lindenmayer, J. M. and Kaufman, G. E. (2022). One health and one welfare. In: Stephens, T. (ed.) *One Welfare in practice, the role of the veterinarian*. CRC Press, Oxford, pp. 1–30.

Lund, V. and Röcklinsberg, H. (2001). Outlining a conception of animal welfare for organic farming systems. *Journal of Agricultural and Environmental Ethics*, 14, 391–424.

Marbán-Castro, E. et al. (2019). Re-emerging zoonoses with "one health" approach. *Revista MVZ Cordoba*. 24: 7280–7284. https://doi.org/10.21897/rmvz.1777

Mateus, T. L., et al. (2022). We know One Health, but we also need One Communication. In: Filho, W. L., et al. (eds.) *Sustainable policies and practices in energy, environment and health research*. Springer, Germany, pp. 245–259. https://doi.org/10.1007/978-3-030-86304-3_15

Nguyen, B. T. et al. (2022). Dairy production sustainability through a one-health lens. *Journal of the American Veterinary Medical Association*. 261: 12–16. https://doi.org/10.2460/javma.22.09.0429

Nielsen, B. and Thamsborg, S. M. (2002). Dairy bull calves as a resource for organic beef production: a farm survey in Denmark. *Livestock Production Science*, 75(3), 245–255. https://doi.org/10.1016/ S0301-6226(01)00322-0

Petherick, J. C. and Duncan, I. J. H. (1991). Society for veterinary ethology 1966–1991, 25th anniversary review. In: Appleby, M. C., et al. (eds.) *Applied animal behaviour, past, present and future*. Proceedings of the ISAE International Congress. 1991, Edinburgh. Hertfordshire, UK. Universities Federation for Animal Welfare, pp. 11–16.

Potterton, S. L., et al. (2011). Risk factors associates with hair loss, ulceration and swelling at the hock in free stall-housed UK dairy herds. *Journal of Dairy Science*. 106: 173–177.

Queenan, K., et al. (2017). Roadmap to a One Health agenda 2030. *CAB Reviews: Perspectives in Agriculture, Veterinary Science, Nutrition and Natural Resources*. 12: 14. https://doi.org/10.1079/PAVSNNR201712014

Squance, H. et al. (2021). Strategies for implementing a One Welfare framework into emergency management. *Animals*. 11(11): Article 3141.

Torgeson, P. R. et al. (2024). Absence of effects of widespread badger culling on tuberculosis in cattle. *Nature Scientific Reports*. 14: Article 16326.

Waiblinger, S., et al. (2002). The relationship between attitudes, personal characteristics and behavior and production of dairy cows. *Applied Animal Behaviour Science*. 79: 195–219.

Wallace, R. G., et al. (2015). The dawn of structural One Health: a new science tracking disease emergence along circuits of capital. *Social Science and Medicine*. 129: 68–77.

Waltner-Toews, D. (2009). Eco-health: a primer for veterinarians. *Canadian Veterinary Journal*. 50: 519–521.

WCS, Wildlife Conservation Society. (2005). The Manhattan Principles. https://oneworldonehealth.wcs.org/About-Us/Mission/The-Manhattan-Principles.aspx Online: [Accessed 21/12/2022].

WHO. (2020). *Basic documents* (49th ed.). World Health Organisation, Geneva. Licence: CC BY-NC-SA 3.0 IGO. https://apps.who.int/gb/bd/pdf_files/BD_49th-en.pdf Online: [Accessed 21/01/23].

Woodroffe, R. et al. (2024). Farmer-led badger vaccination in Cornwall: epidemiological patterns and social perspectives. *Nature and People*. https://doi.org/10.1002/pan3.10691

World Organisation for Animal Health (OIE). (n.d.-b). https://www.woah.org/en/what-we-do/global-initiatives/one-health/ Online: [Accessed 15/12/2022].

Chapter 6

The lives of calves from organic dairy farms

Kerstin Barth, J. H. von Thünen Institute, Federal Research Institute for Rural Areas, Forestry and Fisheries, Germany; Julie Føske Johnsen, Norwegian Veterinary Institute, Norway; Marie Haskell, SRUC, UK; Silvia Ivemeyer, J. H. von Thünen Institute, Federal Research Institute for Rural Areas, Forestry and Fisheries, Germany; and Mette Vaarst, Aarhus University, Denmark

1 Introduction

For many years, the rearing of healthy, vital calves for the future dairy herd has been the major focus for farmers, advisors, and researchers. Evidence of the effects of calf feeding, particularly the concept of metabolic programming (Soberon *et al.*, 2011; reviewed by Hammon *et al.*, 2020), on animal productivity later in life provided the impetus for a more detailed examination of the methods of calf rearing applied in current practice on farms and the potential for improvement. Furthermore, efficient rearing (referring to fast growth without becoming diseased), was seen as an option to reduce the age at first calving and thus to increase economic sustainability, although longevity is not necessarily correlated to a low age at first calving (Krpálková *et al.*, 2014). However, ad libitum liquid feeding, thereby providing milk-fed calves with adequate nutrition, has been demonstrated to improve their development and resilience (e.g. Curtis *et al.*, 2018).

http://dx.doi.org/10.19103/AS.2024.0138.07

In organic herds, this milk should preferably come from the dam (*Regulation (EU) 2018/848)*, but organic milk powder is also accepted, as it is interpreted as 'natural milk' (e.g. Anon. 2024). Many practices based on research outcomes focused on growth, future production and efficiency of calves that were intended to become dairy cows. Although organic farming and regulation gave some consideration on early contact between dam and calf, group housing and outdoor stay for young calves, the focus on growth and productivity has also been strong in organic dairy herds, where the milking herd and milk production were the main focus.

The organic dairy sector has been strong and successful in many countries, and organic dairy products have generally been one of the main drivers of organic markets in many European countries (Willer *et al.*, 2023). At the farm level, ruminants are good agroecological system partners because they can use all types of grassland and utilize nitrogen fixers such as clover, as well as deliver manure. The positive environmental profile complements the interest in the nutritional benefits and the absence of chemicals in organic dairy products, especially from many families with children (Riefer and Hamm, 2008). However, this emphasis on milk production keeps a strong focus on the dairy cows and those female calves which the farmers intend to include in the dairy herd, as much in organic dairy farming, as in conventional.

According to the organic principle of fairness (IFOAM, 2020), '... animals should be provided with the conditions and opportunities of life that accord with their physiology, natural behaviour and well-being'. This means that all animals should be treated equally well and have the opportunity for a good life, irrespective of species or sex, so already here, a dilemma emerges. The intention in organic farming to combine animal-friendly husbandry with sustainable food production poses some challenges, but on the other hand favors the calves' behavioural repertoire to a comparatively high degree, for example in terms of group housing, space requirements and outdoor access.

However, over the last decade, citizens have also become increasingly aware of practices in animal husbandry, and general concerns about the ethical treatment of farm animals have been raised (Cardoso *et al.*, 2016) that contribute to the erosion of trust in the dairy sector (Cook and von Keyserlingk, 2023). Killing newborn male calves just after birth was one of the practices that highlighted the imbalanced value of milk and calves, as well as the imbalanced view of different "classes" of animals on the farm. One procedure that is also widely applied in organic dairy production is the early separation of cow and calf, which has been increasingly criticized by participants in opinion polls (reviewed by Placzek *et al.*, 2021). Dairy farmers in general have given many good reasons for early separation, mainly with the starting point of the high value of milk versus the low value of calves (especially male calves), in combination with the reduced volume of saleable milk in such systems, because calves are allowed to consume more milk (Johnsen *et al.*, 2016; Meagher *et al.*, 2019). Another reason given for early cow-calf separation was that it reduced the risk of transmission of infectious diseases between cow and calf, but depending on which infection is in question, it is not always supported by literature as strongly and uniformly, as some assume (Beaver *et al.*, 2019). Organic dairy farmers also referred to practical difficulties with dam-rearing, such as the requirements of housing systems, which for decades have been built only for cows (Bertelsen and Vaarst, 2023). However, as relevant as dairy farmers' arguments for early cow-calf separation are from an economic and practical farming systems point of view, no matter whether they are organic or conventional, studies have shown that citizens confronted with these arguments do not necessarily change their attitudes (Busch *et al.*, 2017). This also applies to other aspects of animal farming, e.g. the use of cages for hens (Weary *et al.*, 2016).

In light of an increasing focus from citizens and farmers on letting calves staying longer with their dams, it is only logical that research on this topic has increased significantly in recent years, addressing not only the fundamentals of animal behaviour and physiology but also increasingly the practical implementation of the procedures, e.g. how to organize cow-calf contact (CCC) systems during the grazing season. However, CCC systems often seem like significant contrasts to an industrialized dairy industry, which also includes some organic dairy farming, and usually focuses on the maximum performance of the dairy cows or the pasture/land used. However, the organic principles as formulated by IFOAM (2020) insist on organic farming meeting more natural needs, e.g. the principles of fairness mentioned above, in combination with the principle of care, which insists that organic farming '... *can enhance efficiency and increase productivity, but this should not be at the risk of jeopardizing health and well-being*'. This could be an explanation for why a reintroduction and reinforcement of contact between cow and calf is given more attention in organic farming.

The pioneers in this field have often been organic farmers, who have developed quite individual approaches for themselves, or farmers in low-input farming systems, where calves are traditionally raised by the milking cows, e.g. with indigenous breeds like Modicana in Sicily (Erikson *et al.*, 2022). An increasing body of knowledge is being developed both in practice and in science, where experiments are conducted to know more about animal behaviour and motivation. It must be expected to help the development of CCC systems a lot, if there is an increasing body of knowledge and experience to build on, also in many different types of circumstances and local conditions.

Another challenge for organic dairy production is to develop strategies for handling the calves that are not needed for herd replacement. This requires that milk and meat production should both be given value. It is not in accordance with the principles to 'export' a number of the calves from organic farms to conventional veal or beef farms. All animals born on an organic farm should be given value and the opportunity to live a life of good quality, and it is the responsibility of the sector and surrounding human society to ensure this.

2 Management of calves under organic farming conditions

2.1 Regulations

The regulations for keeping calves in organic farming vary widely across the world, but the supply of natural milk is fixed, although some countries allow organic milk powder as feed for calves. Roughage must be offered even to young calves (e.g. *Regulation (EU) 2018/848, NOP eCFR:: 7 CFR 205.239 – Livestock living conditions* 2023). In the EU, maternal milk is encouraged even more than in other regions. However, there are great differences between countries in the regulations on the age at which the animals are required to be kept in groups, as well as when they must have access to pasture. Group housing is mandatory in the EU from the 8th day of life and rearing systems should be based on maximizing grazing time (*Regulation (EU) 2018*). In the USA, however, calves can be confined until they are 6 months old (*NOP eCFR:: 7 CFR 205.239 – Livestock living conditions* 2023).

Furthermore, organic calves are influenced by other legislation, such as welfare regulations which also set rules, and environmental regulations, which influence some farm structures and practices, including the way in which organic regulations are interpreted at the country level. Many countries or regions have specific husbandry regulations or standards for the handling of calves irrespective of whether they are organic or conventional, so that often only additional criteria are included in the regulations for organic farming (Grundin *et al.*, 2019).

2.2 Colostrum and milk feeding

Health promotion and disease prevention are the keystones for the success of organic systems. Ensuring that the cow is strong and healthy and can produce high-quality colostrum, which can be fed to the calf in adequate quantities within the first hours after birth is essential for the immediate protection as well as the further physical development of the newborn calf (reviewed by Hammon *et al.*, 2020).

In the meantime, further studies also show that the transition milk, i.e. the milk produced by the cow in the days after calving, also has an impact on calf growth beyond the period of transition milk feeding (Kargar *et al.*, 2021). The requirement to provide calves with mother's milk, as formulated, e.g. in the EU or Swiss guidelines for organic farming (*Bio Suisse Richtlinien 2023*; Regulation

(EU) 2018/848), could potentially therefore lead to better calf performance if implemented literally, provided that the mother cow produces high-quality colostrum. However, studies concerning the effects of prolonged feeding of milk from their own dams on calves' physiology and health are still lacking, although the increase in dam-calf contact systems offers the option for more comprehensive studies of the subject.

In Europe, organic calves have to be provided with natural milk for at least 90 days (*Regulation (EU) 2018/848*). This period goes far beyond the milk-feeding phase in conventional dairy production today but without reaching the natural weaning age for cattle that starts around 8 months (Reinhardt, 1980). Furthermore, the amount of milk that has to be offered to the calves is not clearly defined, and even on some organic dairy farms, e.g. Norway: Johnsen *et al.* (2021a), Germany: Dachrodt *et al.* (2022), calves are provided with less milk per day than current recommendations prescribe based on scientific studies (reviewed by Khan *et al.*, 2011).

Although the use of larger quantities of high-quality organic milk in the diet of any calf may appear to be economically disadvantageous in the short term, in the long term, it can be expected to have a positive effect on animal health, which will lead to an improvement in environmental sustainability. Milk from healthy cows that does not contain mastitis pathogens (Abb-Schwedler *et al.*, 2014) or drug residues promotes calf intestinal health (Malmuthuge and Le Guan, 2017) while reducing the risk of the development of pathogenic microorganism resistance (Aust *et al.*, 2013). Thus, improving the general health situation in the dairy herd and not using antibiotics and anthelmintics preventively, ensures healthy milk and has a direct impact on calf health and well-being on the farm.

2.3 Roughage feeding and grazing

The calf develops from a monogastric to a ruminant over the first weeks of life. This process should be encouraged and initiated as early as possible, not least because ruminants can, generally live on feed that cannot be consumed by humans, thus contributing to the sustainability of the entire farming and food system (Leiber *et al.*, 2019).

The quality of the roughage must of course meet the calves' needs to support their gut health. Webb *et al.* (2013) showed that straw can cause damage to the abomasum, and recommended feeding other types of roughage to young animals. Calves that follow their dams on grass, are observed to graze at an age of 1 week and show marked grazing behaviour immediately after weaning (Nicolao *et al.*, 2020). Another advantage of calves grazing with their dams is the reduction of parasite pressure on the calves, which have yet to develop immunity to pasture parasites (Constancis *et al.*, 2021, 2022). Older cattle that

Published by Burleigh Dodds Science Publishing Limited, 2025.

have a higher feed intake, simply eat more of the infectious parasite larvae, and since they are immune, they excrete fewer parasite eggs and thereby 'dilute' the infection pressure on the paddock, as discussed by e.g. Nansen *et al.* (1990).

2.4 Birthplace and housing

Before giving birth, cows search for a place where they can calve undisturbed and then establish the bond with their own calf (reviewed by Rørvang *et al.*, 2018). Under semi-natural conditions, cows with their newborn calves do not fully reintegrate into the herd until 1–3 weeks after birth (Reinhardt, 1980). The *Regulation (EU) 2018/848* takes this behaviour into account by allowing individual housing of calves only during the first week of life. However, this specification also allows for complete isolation from conspecifics, as the calves usually have no contact with their dams either.

Single housing for a few days after birth is justified for the purpose of improving animal health, although Chua *et al.* (2002) have shown that at least pair housing at this early stage does not affect it. Nevertheless, early group housing might pose risks for transmission of infection, especially in large or dynamic groups (Pedersen *et al.*, 2009) where younger calves are constantly being introduced, such as in smaller herds when calving is not seasonal.

2.5 Particular constraints in rearing offspring

The principles and guidelines for organic calf rearing do not distinguish between calves intended as replacement for dairy cows, and calves for meat production (male calves, and some female calves). However, organic farming has generally become increasingly large-scale and specialized in many European countries (European Commission, 2023) as well as in the USA (Natzke, 2023), where organic products go into a supermarket-based food system, and where the competition on price is strong. While organic milk has been successful and competitive in the market, the demand for organically produced beef has been lower. Many organic dairy farms are built in similar ways as many non-organic dairy farms, e.g. cubicle housing and milking by robots (Wallenbeck *et al.*, 2019), and although group housed and have more space than many conventional calves, most young organic calves are housed in similar housing systems. Many organic farms are not built to house all the calves for more than a few weeks, and specialized dairy farmers are not experienced in rearing calves to slaughter, or marketing beef meat.

3 Cow-calf contact during the milk-feeding period

In suckler beef systems, the cows rear their calves during the pasture period until weaning and separation, often 4–8 months after birth. In some tropical and sub-tropical traditional dairy farming systems, dairy cows spend the first days with their calves, and are often still stimulated by the presence of calves as a normal part of the milking routine (Froberg *et al.*, 2007, Eriksson *et al.*, 2022).

To obtain as much milk as possible for human consumption, it has been standard practice that calves in intensive dairy farming are separated from their mothers shortly after birth for more than a century (Schlipf, 1908). This is also the case in organic herds. For example, in Danish organic dairy farming, it was a legal requirement for two decades that a calf should stay with its mother for a minimum of 24 h (which is already short), but from January 2023, this stipulation was removed because it was seen as over-implementation of the EU-regulation. Since then, a calf only needed to stay with the mother for 12 h like in conventional herds, unless the dairy company has agreed with the farmers that they keep a minimum of 24 h (Anonymous, 2023).

As already mentioned, calf-rearing systems allowing calves to have contact with a cow during the milk feeding period, are currently emerging, and research addressing the early separation of the dairy cow and calf is now a spiralling research area globally (Johnsen *et al.*, 2016; Beaver *et al.*, 2019; Meagher *et al.*, 2019).

CCC systems entail that calves have physical contact and behavioural interaction with either their own dam (dam-calf contact) or a foster cow, where the calf is reared by a cow that also rears other (alien) calves (Sirovnik *et al.*, 2020). Reflecting different farmers' resources, aims, values and motivations, multiple different CCC systems exist (Fig. 1), including combinations of the ones mentioned above (e.g. Eriksson *et al.*, 2022; Neave *et al.*, 2022; Johanssen *et al.*, 2023; Bertelsen and Vaarst, 2023).

The time span of CCC usually reflects the pre-weaning period and includes suckling. Although a minimal duration has not yet been clearly determined as part of a definition of what constitutes a CCC system in general, in a first step a contact duration of at least 14 days has been suggested (Sirovnik *et al.*, 2020). In Norway, around 3% of all dairy farmers already practice CCC for a minimum of 14 days (Hansen *et al.*, 2023). In the organic certification for CCC systems in Germany, a minimum contact time of 3 months is defined, based on the minimum requirement for a milk feeding period of 90 days *(EU 2018/848).*

3.1 Organizing CCC systems

3.1.1 Housing and contact times

The great variety of CCC systems illustrated in Fig. 1, reflects the fact that they were developed directly on farms and available resources have a major influence on the choice of how the CCC systems were organized (Eriksson *et al.*, 2022; Johanssen *et al.*, 2023; Neave *et al.*, 2022) So far, only very few housing systems have been built and developed to house dairy cows with

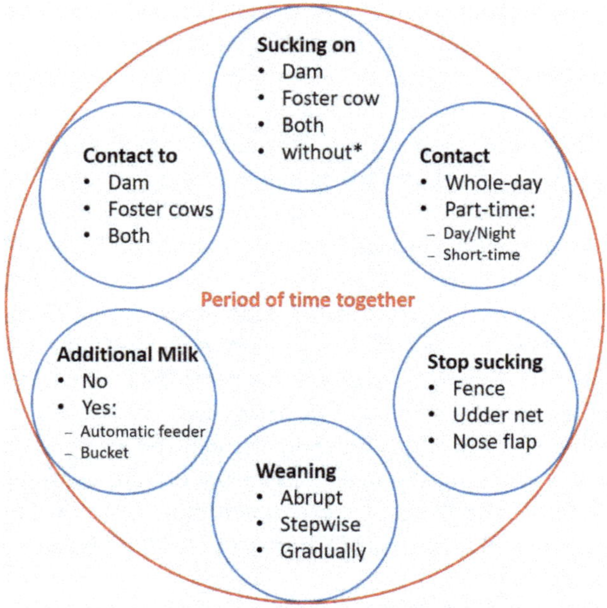

Figure 1 CCC systems are characterized by many different aspects, which are also combined in different ways, resulting in a wide variety of systems. (©Thünen Institute)

calves. Therefore, farmers develop their own CCC system using the buildings and resources available, which will determine the design of the systems.
In general, three main types of dam-calf contact exist: whole-day systems, where calves freely roam around among dairy cows indoors and/or on pasture, and part-time systems that include half-day systems, either over day, or night (Eriksson *et al.*, 2022), or short-time CCC for two or more short periods per day (Sirovnik *et al.*, 2020). The majority of 104 farms with CCC monitored in six European countries, allowed whole-day CCC (48 farms), followed by restricted contact around milking (37 farms), and less often half-day contact (5 farms, Eriksson *et al.*, 2022).

Systems with permanently keeping the calves together with the dams within the group of lactating cows, require barns without slatted alleys (due to differences in minimum slat width by legislation), with enough space and without risk of injuries for the calves. Furthermore, dam-rearing systems require proximity to the milking area, especially when automatic milking systems are used. Separate barns or sections with deep bedding are often used for dams or foster cows if the milking herd barn is not suitable for keeping calves and cows together. Alternatively, there may be a contact area where cows and calves meet only a few times a day.

Space allowance guidelines for CCC systems are not yet well-developed. However, independent of the duration of daily contact allowed between cow and calf, when keeping cows and calves together more space and resources are required, compared to systems designed for cows only. For example, calves need safe lying areas (e.g. a hide), and place for their feed, such as concentrate and hay, and water troughs in their height. If cows and calves can stay in each their own housing system during the majority of the day and meet for some time in a suitable contact area (for suckling and social contact), e.g. in the outdoor yard or a part of the cow stable or in the calf stable, less space is needed. Furthermore, calf creeps providing an area of special feeding and resting, encourage calves to spend time away from the cows, and can make the weaning and separation process easier.

Part-time contact may entail practical benefits related to separate housing of cow and calf. If a system has different 'departments'. contact and suckling can be facilitated by allowing cows to visit their calves in a designated area (cow-driven CCC) or calves to visit the cows (calf-driven CCC, Sirovnik *et al.*, 2020). The access to the calves may be facilitated by a mechanical or push-gate or a computer-controlled smart-gate responding to the cows' Radio Frequency Identification ear-tag (Johnsen *et al.*, 2021b, 2021c, Fig. 2). Calf-driven contact may also be facilitated through manually opening gates between the calf and cow pen (Veissier *et al.*, 2013; Johnsen *et al.*, 2015b) or a computer-controlled push gate (Johnsen *et al.*, 2016).

Figure 2 A computer-controlled smart gate controls the access of cows to an area where they can interact with their calves (© Camilla Mellemstrand)

Published by Burleigh Dodds Science Publishing Limited, 2025.

3.1.2 Grazing in CCC systems

Pasture access during the summer is mandatory for organic dairy cows. Therefore, research on CCC systems also includes pasture management with cows and calves together. Mac *et al.* (2023) studied the impact of CCC at pasture in a whole-day contact system. The authors reported that the calves seemed to habituate well to the transient separation from the dams during milking. However, farmers also have expressed legitimate concerns (such as weather conditions and colostrum supply) regarding rearing calves outdoors with dams (Neave *et al.* 2022; Johannsen *et al.*, 2023). Danish farmers, e.g. have found that it requires extra fencing to avoid calves escaping under the typical, single-strand wire fence and out onto nearby roads (Vaarst *et al.*, 2023).

Infobox 1. Examples of cow-calf contact systems, all of them are organic dairy farms, and links to further information is given in section 7.

From Germany (see Barth *et al.*, 2022):

Combined rearing by the dam and foster cows with short-time contact:

The first three weeks after calving, all cows of the 40-strong herd (German Holstein x German Black Pied) stay together with their own calf in a separate area before the calves move into the calf group. In this group, the calves are kept until the 13th to 15th week of life when they are going to be weaned. All cows were milked twice per day, but the udder of cows that are suckling the calves will not be emptied completely. These cows meet the calf group in the cows' lying area while the other cows are milked. Cows and calves (1.3 to 1.8 calves per cow on average) stay together for about 40 minutes to ensure that all calves suckle and receive affection, like social licking. After the suckling udder and teats of the cows as well as the vitality and health of calves are checked before the calves are moved to their own area adjacent to the cows' area. At the end of the suckling period, the calves are allowed to suckle only once per day and then stay in the calves group without being suckled for a few more weeks till they are moved to the youngstock barn. Most of the milk is turned into various dairy products, e.g. raw milk cheese and yogurt, and marketed directly in the own or other farm shops, and on farmer's markets. A large part of the male and female calves that will not be needed for replacement are fattened on the farm also marketed directly.

Dam rearing with whole-day contact

Cow and calf stay together in the calving area for the first five to seven days post calving to establish a strong bond before they join the dairy herd consisting of 60 Holstein cows. The cows are milked two times per day and the calves are allowed to suckle as much as they like till they will be weaned using a nose flap starting at an age of 91 days.

During summer, the calves stay with the herd on pasture for more than four weeks after weaning, in winter they are separated but can get in touch with their dams without the opportunity to suckle. All calves stay at least three months on the farm. Together with four other farms the farm formed a limited liability company, that is a member of a dairy cooperative, and they market their milk under the label 'Four-Seasons-Milk'" with the addition 'Parenty time for our cows'. They are developing their own strategies, such as selling veal, to increase the sales opportunities for calves and meat. That is why they are now used also a dual-purpose breed for cross-breeding.

Rearing by foster cows in a large farm

This is one of the few examples of a CCC system for which a new barn was specially designed. The dairy herd consists of 1500 Holstein cows resulting in 120 calvings per month. The group calving pen has space for five cows and so the calves can also get to know their future foster cows right from birth. After five days the selected foster cows and all calves move into a larger pen where they meet other cows with calves. From the third week onwards, 16 cows with 48 calves were kept in a large pen which has a separate calf area. Till the end of the 3rd month, calves have access to their foster cow group interrupted only for a few hours, during which bedding, veterinary and udder checks of the cows take place. After that the contact times are reduced and the number of calves per cow is increased - one foster cow group suckles two groups of calves till they are 4 months old. Then the cows are moved to another barn where they are milked regularly two times per day in a rotary parlor. The calves stay behind. All calves are reared on the farm and used for replacement or fattening, respectively. The milk and meat are sold through a large organic supermarket.

From Denmark (see video given in section 7)

Two Danish dairy herds with 320 and 150 dairy cows, respectively, initiated their CCC systems in 2020, letting the calves stay with their mothers during the first 2–3 weeks, after which they separated them and let the calves stay with a foster cow, with 2–4 calves per cow. Both of these herds had the initial intention to let calves stay with their mother throughout the 3 months of milk feeding, but left this system, because they had problems with alveolar milk ejection. They experienced that the calves quite quickly calmed down and seemed content in the foster cow environment, so it was 'only' the cow who called for her calf. Some major learning points were the importance of giving enough space indoor to ensure that the calves stayed free of diarrhea, and the advantage of separate the foster cows from the calves gradually by taking them out one by one, which kept the calves more quiet. On the positive side, seeing how calves learned to move around in systems with cows and follow the group on grass was a great pleasure as a farmer and calf carer. The system was calm and easy and both cows and calves stayed indoor during nights and outdoor during days throughout the summers. This was an argument in favor of the foster cow system, because the foster cows did not need milking twice or three times daily (as was the practice in the herd of 320 cows), and therefore was calmer.

3.1.3 Viewing the CCC systems from different perspectives

Vaarst *et al.* (2020) identified four different angles, from which a CCC system could be viewed, based on interviews and a review of research results from France, Norway, The Netherlands and Denmark. The four perspectives were (1) the calf, (2) the cow, (3) the human (the caregiver) and (4) the system (the given and/or development of the structures and practices on the farm). From the interviews, it became clear that different interviewees had different priorities, which guided their choices on how to organize the CCC system. Furthermore, it became obvious that they gave different weight to three different purposes of CCC systems, seen from an animal point of view: (1) nutrition, (2) care and (3) learning. These purposes addressed whether and how a given system could provide nutrition to the calf in a way that satisfied the natural needs, whether it would allow mutual care between the cow and the calf, and finally, whether it could convey learning between cow and calf in ways that prepared the calves for future group and outdoor life. These criteria could guide future efforts to develop and optimize an array of different CCC systems. As an example, if we view the part-time systems described above, some systems will only allow them to meet for the calf to drink milk, which means that cow and calf will not have much opportunity to lay down together (see the example in the infobox 1), and the calf cannot follow the mother to learn from and get to know the system, and the mother cannot defend and care for the calf during the day, when they are in each their system. On the other hand, some of these systems allow for a gradual weaning and separation, which could stimulate the calf to increased solid feed intake and thereby easier weaning.

3.2 Foster cow systems: practices and arguments for and against

Rearing 2–4 calves with a foster cow, who is alien to the calves, is perceived as a practical solution for CCC by some farmers (Bertelsen and Vaarst, 2023). The foster cows are often not part of the milking herd, and may thus be housed together with the calves in a separate barn compartment or on grazing areas during summer, where they stay with the calves 24 h per day. This makes it a more viable and feasible option for many, especially larger dairy farms (see infobox).

A foster cow system needs a process of careful and supervised bonding between the foster cow and the calves that need to be used to suckling. Furthermore, even when this bond is established, it needs constant monitoring to ensure that all calves drink, the cows continue to accept their calves, and that disease, e.g. mastitis or severe diarrhoea, does not occur. Studies show that the foster cows' acceptance, care and nutrition towards the calf is not always consistent (Loberg and Lidfors, 2001), and may lead to large between-calf variation in growth and affiliative behaviours in such groups (Loberg, 2007; Wieczorreck and Hillmann, 2022).

A recent study indicated a low societal acceptance of foster cow systems (vs. dam-rearing) by consumers (Sirovica *et al.*, 2022). Seen from the animals' point of view, it is an important consideration that foster cow systems still entail dam-calf separation, and hence only partly meet the needs of the calf, and not of the mother-cow. In mother-bonded systems, both cow and calf can perform natural behaviours, and the dam's strong motivation to take care of her own calf is met (Wenker *et al.*, 2020; Constancis *et al.*, 2022), which also is more in accordance with the organic principles of health and fairness (IFOAM, 2020). Thus, it is crucial that future research actually emphasizes these key motivations and behaviours, whilst at the same time accounting for practical and economic feasibility.

3.3 Animal welfare in CCC systems

Cows and calves are highly motivated to be together (Wenker *et al.*, 2020), and the complex bond between cow and calf extends beyond the nutritional relationship (Johnsen *et al.*, 2015a). Natural, maternal behaviours resulting from forming a bond with the calf include licking, suckling and reciprocal social behaviours. Allowing for these behaviours follows the IFOAM principles and the organic concept of naturalness (IFOAM, 2020; Lund, 2006). The link between affiliative behaviours and long-term positive affective states in farm animals is established (Boissy *et al.*, 2007). Because natural behaviour and positive affective states are inherent concepts of animal welfare, this rearing method has the potential to advance the welfare of cows and calves in dairy production. Using the four different viewpoints outlined above, the calves may get many of their needs fulfilled in the dam as well as foster cow systems, but seen from the point of view of the cow, in foster cow systems she is still not able to meet her natural needs, being separated from her calf at an early age.

In addition to performing natural behaviour, it is key to animal welfare that the animals stay healthy and don't become diseased or die. Compared to calves kept in individual pens or smaller groups, where feed intake and faecal consistency can be easily observed, monitoring of calves in CCC systems requires a different approach to ensure that all animals are inspected. A recent Norwegian observational study of 39 CCC herds matched with 70 herds not practising CCC found no differences in calf mortality and overall calf morbidity (Johnsen *et al.*, 2022). The study revealed that the relative treatment risk of omphalitis was higher among CCC calves while that of enteric disease tended to be lower for CCC calves. This study indicated that the prevention of diseases in calves is not a valid argument for the early separation of cow and calf. However, some country-specific considerations regarding endemic cattle

Published by Burleigh Dodds Science Publishing Limited, 2025.

diseases (BVD, Johne's disease, etc.) may require different approaches to the implementation of CCC, e.g. whether all cows must be tested before calving and, if positive, contact with the calf may not be possible.

Even though farmers often report that calf disease has been reduced after the introduction of the CCC, especially regarding diarrhoea (e.g. Eriksson *et al.*, 2022), the contact between calf and cow does not necessarily guarantee an adequate calf health status, as found by e.g. Wenker *et al.* (2022) in a controlled study. The majority of CCC studies addressing the relationship between CCC and cow-calf health and diseases have so far been inconclusive (reviewed by Beaver *et al.*, 2019), calling for more research.

3.4 Impacts on milking and milk yield

Performance of dairy cows is often measured through milk yield and milk composition. Time and duration of contact between calf and dam per day affect the amount of saleable machine milk yield during the suckling period (Nicolao *et al.*, 2022) and beyond (Barth, 2020; Churakov *et al.*, 2023). However, in CCC systems the cow yields milk both to the calf (usually not quantified) and to the milking machine. The amount of milk that can be milked by machine is determined by two factors: the amount that the calf has already drunk and the amount that is made available due to alveolar milk ejection (Tančin *et al.*, 1995). The latter is subject to large individual variations, the cause of which is not yet clear and needs further investigation.

Quantifying the performance of cows in CCC systems needs a more nuanced approach than just talking about saleable milk. Brombin *et al.* (2019) argued that one-welfare and sustainability interests could support a transition towards investing more milk in healthy calves through CCC systems and balancing the focus between milk for human consumption and for the calves, maybe in combination with more involvement in dual-purpose breeds. In particular, the extensive use of milk in veal production and the often still very low quantities of milk used in the rearing of female calves need to be addressed. Both are detrimental to the health of the animals concerned and can result in consequential costs that may even outweigh the presumed milk losses in cow contact rearing. Unfortunately, however, there are still no practical surveys that put the reduction in saleable milk in CCC systems into a wider context.

3.5 Gentle weaning and separation

In European farms, the age at weaning in CCC systems reflects the organic standards, which is a minimum 90 days with whole milk. However, it varies considerably from 12 to 30 weeks (Eriksson *et al.*, 2022). In these cases, weaning and separation often happen at the same time, whereas separation from cows earlier will mean that calves need to learn how to drink milk from buckets or automatic milk feeders.

A major criticism of CCC is the stress caused by the separation of cow and calf after their time together. As most CCC farmers will need to break the cow-calf bond prematurely in order to still obtain enough saleable milk, strong behavioural responses of cow and calf are to be expected at separation if precautions are not taken. These reactions are one reason why some farmers quit CCC after trying it out (Hansen *et al.*, 2023).

Separation from the cow and weaning off milk should be done step by step. Gradual separation and weaning will help to keep the stress for animals and humans as low as possible, to prevent weight losses of calves and milk ejection problems in cows. This might be achieved by gradually reducing the amount of milk for the calf, gradually reducing the contact time between cow and calf, and by implementing a time lag between weaning and separation or vice versa (Schneider *et al.*, 2021).

Encouraging calf nutritional independence seems key to reducing behavioural stress, since calves vocalize in response to the hunger for milk. However, even gradual weaning and separation might cause significant behavioural responses in calves (Vogt *et al.*, 2024). The option of offering calves a different milk source can also be challenging, as relearning drinking from the udder to an artificial teat (automatic feeder or milk bar) takes some time.

Co-creation of new knowledge through investigation of on-farm developed gradual weaning and separation systems needs to be developed, to mitigate adverse effects of separation distress for mother and young. However, given that separation is still premature, we must expect some level of response from both cow and calf, with respect to the cows' motivation to care for the offspring and the calf-filial aspects of the cow-calf bond. Nonetheless, an extended suckling and contact period up to the time of natural weaning (Reinhardt, 1980) could also be an option if cows with good lactation persistence are used, enough resources, like space in barns, are available, and when the extra effort is reflected in the price of milk and meat.

3.6 Case studies: collaborations between research and practice regarding CCC systems

During the early 2000s, research was centred on why there may be benefits of allowing contact between the dairy cow and her calf (Flower and Weary, 2003; Lund, 2006; Newberry and Swanson, 2008). In contrast, several projects around the world now focus on <u>how</u> to implement CCC, on-farm, or in pasture systems. Interestingly, several CCC projects aim to learn about how CCC systems comply with the natural needs of cow and calf. In addition, there is a broad effort to collate farmer experiences with CCC systems, as some farms practice CCC systems already for several decades successfully. As stated by the European

Food Safety Authority (EFSA) Panel on Animal Health and Animal Welfare on request from the European Commission, prolonged CCC in dairy production should increasingly be implemented (Nielsen *et al.*, 2023). Organic farmers are usually pioneers in this respect.

In some countries like Austria, Switzerland and Germany, the majority of farm raising their dairy calves with cows are organic farms, while in France, Sweden and Italy the percentage of organic farms out of the farms practicing CCC is lower: 58%, 42% and 21%, respectively (Eriksson *et al.*, 2022). In Norway, up to 15% of all dairy farmers responded that they planned to implement CCC on their farms in near future (Hansen *et al.*, 2022). As this interest increases, so does the need for recommendations on how to implement CCC in practice on farms, based (a) on experiences and knowledge of farmers, who have already implemented CCC systems, and (b) on results from research projects, several leaflets and extension material have been developed.

3.6.1 Extension material from a survey in different European countries (project 'ProYoungStock') and from a farmer group in Switzerland and South Germany

In the framework of the European CORE Organic Cofund project 'ProYoungStock', more than one hundred farmers practising cow-bonded calf rearing in different European countries were interviewed with standardized closed and open questions regarding their CCC system, their motivation and their experiences with challenges and solutions. One of the main challenges named by the farmers was weaning and separation of cows and calves. Based on the results of the survey, a technical note about the possibilities of gentle and gradual separation and weaning of calves reared in CCC systems has been published (Schneider *et al.*, 2021). Additionally, an existing leaflet based on a farmer-driven process together with FiBL, with general recommendations and several farm examples (with different kinds of CCC systems) in Switzerland and Germany has been translated into several languages and is now available in English, German, French, Italian and Polish (e.g. Spengler Neff *et al.*, 2023). See the list of additional material to read, below.

3.6.2 Cow-bonded calf rearing in dairy farming – a practical guide

In a German EIP project, farmers, scientists and advisors developed a comprehensive guide to provide beginners, experienced farmers, and all those interested in the topic with background knowledge, recommendations and farm examples of possible ways to implement CCC systems on dairy farms (Barth *et al.*, 2022). The guide is based on current scientific as well as experience-based knowledge and an attempt was made to cover all aspects relevant to the topic from the management of systems to economics, and included several inputs from stakeholder experts. See links and references under additional material to read, below.

3.6.3 Standard development and certification of CCC systems in Germany

In Germany, an initiative of organic dairy farmers together with representatives from research, organic label organizations, and animal protection organizations was started to develop criteria for CCC systems in organic dairy farming, which were published in 2021 (IG kuhgebundene Kälberaufzucht 2022; https://ig-kalbundkuh.de/). The criteria refer to calves from dairy cows and are based on the requirements of the organic label organization. Therefore, criteria such as space allowance, feeding and general husbandry were already regulated and their compliance is checked in the yearly organic certifications according to Regulation (EU) 2018/848.

The main criteria of CCC systems have been defined as follows: (a) calves must be suckled by their own mothers or by foster cows, (b) a minimum period of cow-bound rearing of 90 days on the birth farm (or on a nearby partner farm with foster cows), (c) a maximum of 15% of the born calves (females and males) may leave the farm as breeding or fattening animals after only 4 weeks, (d) weaning and separation of cow and calf must not be done abruptly, but gradually and stepwise for both calf and cow. Besides the criteria, in 2022 a checklist for a certification of CCC systems according to the criteria has been developed, that can be carried out voluntarily within the framework of the organic certification since 2023.

As a next step, a mentoring project has been started in 2023: a group of farmers, experienced in cow-bonded calf rearing, offer a mentoring process to farmers that are interested to convert from artificial to cow-bonded calf rearing. In addition to their own experience, mentoring farmers have access to selected extension material regarding CCC systems that have been compiled by researchers.

3.6.4 Stable school for common social learning on CCC systems in Denmark

In Denmark, a project focusing on mother-bonded calf rearing facilitated a process among farmers to learn together about CCC systems (Vaarst *et al.*, 2019), while establishing and developing them on the participating farms using the so-called Stable School approach (Vaarst *et al.*, 2007). Farms were very different, but still found it interesting to learn from each other: Herd sizes ranged from 7 to 600 cows, the farmers came from 5 different dairy companies, and the CCC systems were a mix of mother-bonded and foster cow systems.

They discussed themes chosen by the host farmer of each meeting, and challenges associated with a CCC system, such as space requirements or pasture access, were often on the agenda. They collectively identified potential solutions for implementing various types of CCC on the participating farms, and collectively gained experience from their own farms as well as from each other. Over time, some of the concerns raised in the beginning, became much smaller, such as vocalization at separation or 'wild calves' because of less direct human contact, others remained challenges, such as the amount of saleable milk, which dramatically decreased in mother-bonded CCC systems. However, this leads to the interesting aspect, how they raised ethical questions to their previous and their emerging systems. Many acknowledged their pleasure and joy in seeing the CCC system work and the calves play or learn from the cows, and the articulation of 'losing milk' because the calf drank a lot, was changed into a discussion on 'how to share the milk'. However, many participants viewed foster cow systems as a more feasible – and sometimes the only possible – option for providing CCC on their farm, both because of the available housing facilities, and because of the issues of calves drinking relatively much milk. One major learning concerning foster cow systems was that it required careful bonding processes, even though the calves often cross suckled, and that the persons taking care of the system, really needed to develop skills in observing calves and cows in such a system. This Stable School experience is described in Vaarst *et al.* (2023) and made it quite clear that socially situated learning through this type of emerging Communities of Practice (CoP) is helpful for farmers, who are interested in developing, implementing and improving CCC systems in different dairy farming contexts.

4 Non-replacement calves born in organic dairy herds

As mentioned above, female calves destined for future replacement animals in the dairy herd are usually kept within the herd. Calves of some specialized dairy breeds often have low economic value and, in some farms and countries, have been killed as newborns. For instance, it is estimated that 15% of male dairy calves were killed on their farm of birth in the UK in the period 2016 to 2018 (CHAWG, 2020), while a survey in Canada suggested that 5% of dairy farmer respondents killed between 1% and 100% of their male calves (Renaud *et al.*, 2017). However, organic regulations, or private agreements, e.g. between shareholders of dairy companies in many countries, actively discourage or prohibit the routine killing of healthy newborn calves (e.g. Denmark: Sector recommendations, 2019; UK: Soil Association, 2023). In other countries such as New Zealand, calves may be transported to slaughterhouses at around one week of age (Boulton *et al.*, 2020), where they are processed for veal, or the hides used for leather goods (Dairy Australia, 2021). These animals are referred to as 'bobby' or 'bob' calves. Industry representatives and animal welfare groups are also increasingly concerned about the practice of early slaughter of bull calves and likely consumer backlash (Dairy Australia, 2019; GB Dairy Calf Strategy, 2020; Bolton and von Keyserlingk, 2021).

4.1 Rearing calves to leave the birth farm

Studies showed that only about 10–20% of bull calves born on organic dairy farms were raised on these farms (Germany: Ivemeyer *et al.*, 2018; Denmark: Nielsen and Thamsborg, 2002), and the rest is sold to other (mainly conventional) farms or animal dealers. Regarding female calves on conventional farms, it has been estimated that 60–70% of them were needed for replacement (de Vries *et al.*, 2008; Edwards-Callaway *et al.*, 2019). Since improved longevity is an aim for organic farmers, and is achieved on some farms (Ahlman *et al.*, 2011), the proportion of the female calves not required as replacement becomes even larger.

The most common practice in most European countries is that calves born on organic dairy farms are sold at 2–4 weeks of age into the veal or dairy-beef supply chains, when markets allow this. In countries such as France, the Netherlands and Italy, the majority of all calves are reared for veal (Sans and de Fontguyon, 2009). This is also the case in the United States (Renaud *et al.*, 2017; Creutzinger *et al.*, 2021). Most bull calves from organic farms in Germany (Ivemeyer *et al.*, 2018) and Switzerland (Rell *et al.*, 2020, 2022) go into conventional veal or beef production systems.

White veal is produced from calves that are fed a predominantly milk-based diet with small amounts of roughage, to keep the meat pale. They will normally be individually housed, to restrict movement and thereby muscle development, and they will be slaughtered at 20–26 weeks of age. Nielsen *et al.* (2023) point to insufficient space, individual housing and restriction of iron and fibre in the diet as severe welfare issues associated with white veal production. Rosé veal can be defined as meat from dairy calves below the age of 12 months, which have received a varied diet including roughage (Skelhorn *et al.*, 2020; RSPCA, 2023). Rosé veal is the most commonly reared in the UK where white veal is not consumed.

There is a risk that calves which are intended to leave the birth farms will be given lower priority, compared to the calves which are intended to stay on the farm and are perceived as high value future milking cows. These non-replacement calves may receive poorer quality care on the dairy farm of origin, such as not being fed the same amount and quality of colostrum, receiving inferior quality feed and less attention and care compared to the animals that are meant to stay on the farm (Renaud *et al.*, 2017). It is also possible that less or no bonds are established between caregivers and calves, since the caregivers know that the calves will leave the farm within weeks.

Published by Burleigh Dodds Science Publishing Limited, 2025.

4.2 Transport and mixing of young calves

A number of studies have shown that there are significant welfare challenges associated with early transportation including higher mortality, dehydration and fatigue (Boulton *et al.*, 2020; Roadknight *et al.*, 2021; England *et al.*, 2023; Marcato *et al.*, 2022). These practices are very likely to result in poor welfare, health and growth outcomes, when these calves are moved on to other farms for rearing, and they need more antibiotic treatments, than if they had stayed in the birth farm (Sandelin et al., 2021). It is a well-established fact that mixing of young calves from many herds, gives them a high risk of making them diseased, leading to high antibiotic use (Pardon *et al.*, 2012), because they come from widely different herds and backgrounds (Santman-Berends *et al.*, 2018), increasing social stress and disease risk. To ensure that all these calves receive good standards of care and housing, the calves must provide an economic return to the farmer (Reiber *et al.*, 2020).

4.3 The life of calves seen from an organic principles' viewpoint

There is a strong conflict with organic principles in the practices of early separation from the dam as well as the sale to conventional farms of calves that are not intended to remain on the dairy farm. As stated in the introduction, letting the animals meet their natural needs is explicitly mentioned in the fairness principle, but violated through the early separation from the dam (also leaving the natural needs of the dam unfulfilled), as well as the potentially early transport and mixing with others, being fed with unnatural diets and living short lives under potentially stressful conditions. This way of handling the young calves are also in conflict with the Health principle, which states that *'Health is the wholeness and integrity of living systems. It is not simply the absence of illness, but the maintenance of physical, mental, social and ecological well-being. Immunity, resilience and regeneration are key characteristics of health'* (IFOAM, 2020). Furthermore, it is argued that all animals born under organic conditions should be allowed to live under these conditions during their entire lifetime.

4.4 What are potential solutions?

4.4.1 Prolonged lactation

Reducing the number of calves born each year will reduce the number that need to be reared. Increasing the length of the cow's lactation increases the calving interval and thus reduces the number of calves born overall. Increasing the length of lactation is somewhat controversial as the period of late lactation, when milk yield is relatively low, is extended, which may have negative economic consequences for farmers. A review of the topic by van Knegsel *et al.* (2022)

concluded that while farmers were concerned about a reduction in overall milk yield, a reduction in yield was not always observed, and depended on the age of the cow and the length of the dry period. Additionally, there are many positive outcomes of extended lactations for cow health and fertility (van Knegsel *et al.*, 2022), suggesting that it could also be a good option for organic dairy systems. However, this approach may not be appropriate for cows producing lower yields due to the risk of over-conditioning at the end of the lactation, which increases the risk of dystocia and related diseases in the subsequent lactation (Roche *et al.*, 2009). Furthermore, it could be feasible on farms aiming at a high grazing performance to have spring calvings. This may conflict with the aim of prolonged lactations as annual calvings are then required.

4.4.2 Focus on pasture-based milk and meat

Pasture-based dairy farming, calf rearing and fattening of young animals, using less or no concentrate feed, should be further developed as it has the potential to combine better animal welfare (e.g. Simon *et al.*, 2023) with less greenhouse gas emissions (Murphy *et al.*, 2017) and thus to meet societal expectations for environmentally sound and more animal-friendly meat production. Using grassland to produce high-quality food is highly consistent with the principle of ecology (IFOAM, 2020) aiming for low input and efficient use of resources. Furthermore, consumers show preferences for pasture-raised products (Stampa *et al.*, 2020), which presents opportunities for value-added marketing.

4.4.3 Support rearing the calves from dairy farms for high-value organic beef

At a systemic level, the balance between milk and meat focus on dairy farms, and between beef cattle and calves born on dairy farms and reared as dairy beef, should be reconsidered. The rearing of calves from dairy farms should be prioritized, seen from a life-cycle and environmental perspective (Tichenor *et al.*, 2017). Comparisons of meat quality suggest that dairy beef is as high quality as beef meat (Foraker *et al.*, 2022), and Bonny *et al.* (2016) showed that the eating quality of some of the muscle groups from pure dairy breeds (e.g. Holstein Friesians) was superior to that of pure beef breeds or crosses. It requires a collective effort to find solutions, as e.g. suggested by Proudfoot *et al.* (2022), and since many studies on dairy beef have been conducted under non-organic and non-European conditions, more efforts should be made specifically to solve these challenges at the systemic level for European organic dairy farming, as it is not acceptable that organic calves are not given more value. It would be an attractive development strategy for the organic dairy sector to exploit the options for producing high-value organic beef, preferably on the birth farm. Currently, this happens in some cases, where non-replacement calves are reared on the birth farm until they are of saleable age, instead of suffering from transport and regrouping stress, which can result in high rates of mortality.

Sandelin *et al.* (2021) observed a mortality rate of 2.5% among calves on Finnish fattening farms, when the calves were bought as weaned calves, compared to a mortality rate between 4.6 and 5.3% for calves which were bought in as milk-fed calves. This study was based on conventional farming, but is relevant for calves born on organic dairy farms, since far the most organic calves go to conventional veal or fattening farms.

Prices of calves being raised on milk in organic dairy herds are completely out of proportion, compared to current market prices for beef meet. In addition, organic dairy farms do not have facilities or staff to raise calves until slaughter. Furthermore, environmental legislation restricts the possible number of animal units per farm in many European countries, hence rearing of calves will be economically unviable, as shown, for example, in a survey of all Irish dairy farms (Maher *et al.*, 2021).

The British Calf Strategy aims to encourage responsible breeding practices that produce dairy-origin calves that are suitable for the beef supply chain (GB Calf Strategy, 2020). Larger organic dairy farms in particular, often base their breed selection on milk yield. For example, more than 90% of German organic farms with more than 70 cows (milk yield >7000 kg per year) keep Holstein cows (Ivemeyer *et al.*, 2018). Calves from such specialized dairy breeds tend to have lower meat-to-bone ratios at slaughter than calves from beef breeds because of the selection for higher milk yield (Berry, 2021), which is a main challenge when rearing calves from dairy farms, even if crossed with heavier breeds. Some dairy breeds, particularly Jerseys and their crosses, are also slower growing with a smaller mature size (Coleman *et al.*, 2016), which makes them less economically viable to rear as beef animals, and maybe even problematic to deliver to an abattoir, because the equipment is of a standardized size. In addition, the fat of Jersey animals is yellow, which makes the meat unattractive for some consumers (Dunne *et al.*, 2006).

Some have also pointed to the use of sexed semen as a solution, and the use has increased over the last decade. This is allowed in organic farming in Europe according to Regulation (EU) 2018/848, although some organic farmers' organizations, e.g. BioSuisse (2023) in Switzerland or Demeter in many countries, do not allow this and question this in relation to the organic principles. This use of sexed semen has allowed farmers to be selective regarding which of their cows should give birth to the new generations of herd-replacement female calves (de Vries et al., 2008). Although this might facilitate the use of beef-breed semen, and in this way get better prices for calves exported from the farm, but it cannot be a stand-alone strategy, not least because it is still considered conflicting to the organic principles by many.

Published by Burleigh Dodds Science Publishing Limited, 2025.

5 Conclusion

Ensuring that calves are treated in an ethical manner and in accordance with the principles of organic farming has been one of the biggest challenges faced by organic dairy farmers for many years. While on the one hand, it was mostly organic farmers who initially re-introduced the contact between cow and calf on their farms, on the other hand, many organic calves are exported into the conventional veal or beef industry. To change the latter, the expansion of organic fattening and marketing structures and changes on the farms themselves are necessary. Research can support this process by developing holistic approaches that value all calves equally, regardless of their intended use and for reasons of sustainability. The coupling of milk and meat production should become a stronger focus of research. However, the exchange between practice and research is particularly important. Examples in terms of case studies are given to demonstrate how pressing issues have been tackled jointly and fostered development in different settings.

6 Future trends in research

For CCC systems, the effects of CCC on the health of cows and calves will remain in focus. In addition to issues related to individual disease risk, such as the incidence of umbilical infections, those related to the management of infectious diseases, e.g. Johne's disease, will become more important. From the perspective of animal welfare, gentle weaning and separation methods have to be (further) developed. Technical solutions could support this, and the application of methods of so-called precision livestock farming in CCC systems should be explored. However, much of the available knowledge on raising healthy, vigorous calves is based on studies of calves raised on milk replacers and often with no or only social contact with peers. There is a clear knowledge gap regarding the effect of contact with the dam or an alien cow on the development of calves. Although consumers prefer mother-bonded rearing and cows should have the opportunity to live out their natural maternal behaviour, the use of foster cows can be a first step in this direction. However, foster cow, as well as dam-rearing systems, must achieve at least the quality of well-managed artificial calf-rearing methods using bucket or automatic milk feeding.

The high variability of the CCC systems inevitably leads to a large variation and sometimes contradictory results. In order to ensure the comparability of studies and to enable later evaluation in meta-analyses, a group of scientists has already developed initial recommendations on the methodology of studies in this field (de Oliviera *et al.*, 2020, Ferneborg *et al.*, 2020) that should be applied on a broad scale and, of course, further developed.

To improve the situation for calves that will not replace cows of the dairy herds, research should focus on the conditions that will promote the expansion of organic fattening, processing and marketing structures. Structural changes in terms of shared responsibility for environmentally and animal ethically friendly solutions which takes equally care of dairy cows and their calves, should be supported by research e.g. scenario development and economic as well as environmental and agricultural systems research.

In order to strengthen the social acceptance of organic milk and meat production, the effects of the new farming methods on the environment should be studied more closely, in particular, the consequences of increased calf milk intakes in CCC and other rearing systems.

7 Where to look for further information

A good introduction to the topic of CCC is provided by the following reviews:

- Meagher, R.K., Beaver, A., Weary, D.M. and von Keyserlingk, M.A.G. (2019), 'Invited review: a systematic review of the effects of prolonged cow-calf contact on behavior, welfare, and productivity', *J Dairy Sci*, 102(7), 5765–5783.
- Johnsen, J.F., Zipp, K.A., Kälber, T., Passillé, A.M. de, Knierim, U., Barth, K. and Mejdell, C.M. (2016), 'Is rearing calves with the dam a feasible option for dairy farms? –Current and future research', *Appl Anim Behav Sci*, 181, 1–11.

Definitions and recommendations for the design and reporting of investigations can be found here:

- Sirovnik, J., Barth, K., de Oliveira, D., Ferneborg, S., Haskell, M.J., Hillmann, E., Jensen, M.B., Mejdell, C.M., Napolitano, F., Vaarst, M., Verwer, C., Waiblinger, S., Zipp. K. and Johnsen, J.F. (2020), 'Methodological terminology and definitions for research and discussion of cow-calf contact systems', *J Dairy Res*, 87S1, 108–114.
- de Oliveira, D., Barth, K., Haskell, M.J., Hillmann, E., Jensen, M.B., Johnsen, J.F., Mejdell, C.M., Waiblinger, S. and Ferneborg, S. (2020), 'Methodology for experimental and observational animal studies in cow-calf contact systems', *J Dairy Res*, 87S1, 115-121.
- Ferneborg, S., Napolitano, F., Vaarst, M., Mejdell, C.M., Waiblinger, S. and de Oliveira, D. (2020), 'Methodology for studying human attitudes and behaviour to cow–calf contact systems', *J Dairy Res*, 87S1, 122-127.

Instructions for the practical implementation of CCC on the dairy farm:

Published by Burleigh Dodds Science Publishing Limited, 2025.

- Abstract about separation and weaning of calves reared in CCC systems: https://orgprints.org/id/eprint/42549/
- Leaflet with general recommendations and several farms examples (with different kind of cow-calf contact systems): https://www.fibl.org/de/shop/1660-mother-bonded-calf-rearing.
- Cow-bonded calf rearing in dairy farming – a practical guide (with decision trees, calculation sheets and farm descriptions): https://www.kuhgebundene-kaelberaufzucht.de/wp-content/uploads/Leitfaden-kuhgebundene-Kaelberaufzucht_englische-Uebersetzung.pdf.
- Some videos with English subtitles describing the development of a few CCC-systems in Denmark and Germany and with focus on the development process and the different perspectives of these systems, can be found here:
  - https://www.youtube.com/watch?v=y3A3F8sHd3o.
  - https://www.youtube.com/watch?v=f4Kubh4MX70.
  - https://www.youtube.com/watch?v=VjBWjtc-UIA.
- (subtitles possible to click under 'settings'/the flower where 'undertekster' / 'Untertitel' (=subtitles) can be set to English).

8 References

Abb-Schwedler, K., Maeschli, A., Boss, R. et al. (2014), 'Feeding mastitis milk to organic dairy calves: effect on health and performance during suckling and on udder health at first calving', *BMC Vet Res*, 10, 267.

Ahlman, T., Berglund, B., Rydhmer, L. and Strandberg, E. (2011), 'Culling reasons in organic and conventional dairy herds and genotype by environment interaction for longevity', *J Dairy Sci*, 94(3), 1568–1575.

Anon. (2023), Brancheanbefalinger om økologisk mælk og kød (The sector's recommendations to organic milk and meat. In Danish) https://lf.dk/viden-om/oekologi/lovgivning-og-kontrol/brancheanbefalinger-oekologisk-maelk-og-koed (accessed 18 November 2023)

Anon. (2024) Regler for økologisk kvæghold. Innovationscenter for Økologisk Landbrug (Rules for organic cattle farming. In Danish). https://icoel.dk/media/aijhz2yj/regler-for-oekologisk-kvaeghold-2024.pdf (accessed 24 February 2024)

Aust, V., Knappstein, K., Kunz, H.-J. et al. (2013), 'Feeding untreated and pasteurized waste milk and bulk milk to calves: effects on calf performance, health status and antibiotic resistance of faecal bacteria', *J Anim Physiol Anim Nutr*, 97(6), 1091–1103.

Barth, K. (2020), 'Effects of suckling on milk yield and milk composition of dairy cows in cow–calf contact systems', *J Dairy Res*, 147, 133–137.

Barth, K., Bock, A., Breden et al. (2022) et al., 'Cow-bonded calf rearing in dairy farming - A practical guide.' https://www.kuhgebundene-kaelberaufzucht.de/wp-content/uploads/Leitfaden-kuhgebundene-Kaelberaufzucht_englische-Uebersetzung.pdf. (accessed 18 November 2023)

Beaver, A., Meagher, R.K., von Keyserlingk, M.A.G. and Weary, D.M. (2019), 'Invited review: A systematic review of the effects of early separation on dairy cow and calf health', *J Dairy Sci*, 107, 5784–5810.

Berry, D.P. (2021), 'Invited review: beef-on-dairy–The generation of crossbred beef × dairy cattle', *J Dairy Sci*, 104(4), 3789–3819.

Bertelsen, M. and Vaarst, M. (2023), 'Shaping cow-calf contact systems: farmers' motivations and considerations behind a range of different cow-calf contact systems', *J Dairy Sci*, 106(11), 7769–7785.

BioSwiss - Standards for the Production, Processing and Trade of "Bud" Products. https://www.bioaktuell.ch/fileadmin/documents/ba/Bioregelwerk-2023/deutsch/bs_all_d/rili_e.pdf (accessed 3 July 2023).

Boissy, A., Manteuffel, G., Jensen, M.B. et al. (2007), 'Assessment of positive emotions in animals to improve their welfare', *Physiol Behav*, 92(3), 375–397.

Bolton, S.E. and von Keyserlingk, M.A.G. (2021), 'The dispensable surplus dairy calf: is this issue a "wicked problem" and where do we go from here?', *Front Vet Sci*, 8. https://doi.org/10.3389/fvets.2021.660934

Bonny, S.P.F., Hocquette, J.-F., Pethick, D.W. et al. (2016), 'The variation in the eating quality of beef from different sexes and breed classes cannot be completely explained by carcass measurements', *Animal*, 10(6), 987–995.

Boulton, A.C., Kells, N.J., Cogger, N. et al. (2020), 'Risk factors for bobby calf mortality across the New Zealand dairy supply chain', *Prev Vet Med*, 174. https://doi.org/10.1016/j.prevetmed.2019.104836

Brombin, A., Pezzuolo, A. and Brščić, M. (2019), 'Are we ready for the big change in the dairy production system?', *Res Vet Sci*, 126, 17–19.

Busch, G., Weary, D.M., Spiller, A. and von Keyserlingk, M.A.G. (2017), 'American and German attitudes towards cow-calf separation on dairy farms', *PLoS One*, 12(3), e0174013.

Cardoso, C.S., Hötzel, M.J., Weary, D.M. et al. (2016), 'Imagining the ideal dairy farm', *J Dairy Sci*, 99(2), 1663–1671.

CHAWG (2020) *Fifth Report: GB Cattle Health & Welfare Group*. ©Cattle Health and Welfare Group of Great Britain 2020.

Chua, B., Coenen, E., van Delen, J. and Weary, D.M. (2002), 'Effects of pair versus individual housing on the behavior and performance of dairy calves', *J Dairy Sci*, 85(2), 360–364.

Churakov, M., Eriksson, H. K., Agenäs, S. and Ferneborg, S. (2023), 'Proposed methods for estimating loss of saleable milk in a cow-calf contact system with automatic milking', *J Dairy Sci*. https://doi.org/10.3168/jds.2022-23099.

Coleman, L.W., Hickson, R.E., Schreurs, N.M. et al. (2016), 'Carcass characteristics and meat quality of Hereford sired steers born to beef-cross-dairy and Angus breeding cows', *Meat Sci*, 121, 403–408.

Constancis, C., Ravinet, N., Bernard, M. et al. (2021), 'Rearing system with nurse cows and risk factors for Cryptosporidium infection in organic dairy calves', *Prev Vet Med*, 190. https://doi.org/10.1016/j.prevetmed.2021.105321

Constancis, C., Chartier, C., Leligois, M. et al. (2022), 'Gastrointestinal nematode and lungworm infections in organic dairy calves reared with nurse cows during their first grazing season in western France', *Vet Parasitol*, 302. https://doi.org/10.1016/j.vetpar.2022.109659

Cook, N. B. and von Keyserlingk, M.A.G. (2023), 'Perspective: Prolonged cows calf contact – A dilemma or simply another step in the evolution of the dairy industry?', *J Dairy Sci*. https://doi.org/10.3168/jds.2023-23840

Creutzinger, K., Pempek, J., Habing, G. et al. (2021), 'Perspectives on the Management of Surplus Dairy Calves in the United States and Canada', *Front Vet Sci*, 8. https://doi.org/10.3389/fvets.2021.661453

Curtis, G., McGregor Argo, C., Jones, D. and Grove-White, D. (2018), 'The impact of early life nutrition and housing on growth and reproduction in dairy cattle', *PLoS One*, 13(2), e0191687.

Dachrodt, L., Bartel, A., Arndt, H. et al. (2022), 'Benchmarking calf health: Assessment tools for dairy herd health consultancy based on reference values from 730 German dairies with respect to seasonal, farm type, and herd size effects', *Front Vet Sci*, 9, 990798. https://doi.org/10.3389/fvets.2022.990798

Dairy Australia (2021) https://www.dairy.com.au/dairy-matters/you-ask-we-answer/what-are-bobby-calves (accessed 17 November 2023).

de Oliveira, D., Barth, K., Haskell, M. et al. (2020), 'Methodology for experimental and observational animal studies in cow-calf contact systems', *J Dairy Res*, 87S1, 115-121.

de Vries, A., Overton, M., Fetrow, J. et al. (2008), 'Exploring the impact of sexed semen on the structure of the dairy industry', *J Dairy Sci*, 91(2), 847–856.

Dunne, P.G., Monahan, F.J., O'Mara, F.P. and Moloney, A.P. (2009), 'Colour of bovine subcutaneous adipose tissue: A review of contributory factors, associations with carcass and meat quality and its potential utility in authentication of dietary history', *Meat Sci*, 81(1), 28–45.

Edwards-Callaway, L.N., Walker, J. and Tucker, C.B. (2018), 'Culling Decisions and Dairy Cattle Welfare During Transport to Slaughter in the United States', *Front Vet Sci*, 5, 343. https://doi.org/10.3389/fvets.2018.00343

England, Z.A., Maggard, H.L., Fisher, A.D., Roadknight, N.W. and Pempek, J.A. (2023), 'Condition of bob veal calves on arrival at an abattoir in Ohio', *Animal Welfare*, 32, e7,1–8. https://doi.org/10.1017/awf.2022.8

Eriksson, H., Fall, N., Ivemeyer, S. et al. (2022), 'Strategies for keeping dairy cows and calves together – a cross-sectional survey study', *Animal*, 16(9), 100624. https://doi.org/10.1016/j.animal.2022.100624

European Commision 2023. Organic farming in the EU. A decade of organic growth. Agricultural Market Brief No. 20, January 2023, pp.32 https://agriculture.ec.europa.eu/system/files/2023-04/agri-market-brief-20-organic-farming-eu_en.pdf (accessed 24 February 2024)

Ferneborg, S., Napolitano, F., Vaarst, M. et al. (2020), 'Methodology for studying human attitudes and behaviour to cow−calf contact systems', *J Dairy Res*, 87S1, 122–127.

Flower, F.C. and Weary, D.M. (2003), 'The Effects of Early Separation on the Dairy Cow and Calf.' *Anim Welf*, 12(3), 339–348.

Foraker, B.A., Frink, J.L. and Woerner, D.R. (2022), 'Invited review: a carcass and meat perspective of crossbred beef × dairy cattle', *Transl Anim Sci*, 6(2). https://doi.org/10.1093/TAS/TXAC027

Froberg, S., Aspegren-Guldorff, A., Olsson I. et al. (2007), 'Effect of restricted suckling on milk yield, milk composition and udder health in cows and behaviour and weight gain in calves, in dual-purpose cattle in the tropics', *Tropic Anim Health Prod*, 39, 71–81.

GB Calf Strategy (2020). British Calf Strategy Group. https://projectblue.blob.core.windows.net/media/Default/Dairy/Publications/DairyCalfStrategy_200826_WEB.pdf (accessed 17 November 2023)

Grundin, J., Eriksson, H., Fall, N. et al. (2019), 'Review of the regulations concerning organic dairy calf rearing in seven European countries', *Report ProYoungStock Consortium*. https://orgprints.org/id/eprint/39785/ (accessed 17 November 2023)

Hammon, H.M., Liermann, W., Frieten, D. and Koch, C. (2020), 'Review: Importance of colostrum supply and milk feeding intensity on gastrointestinal and systemic development in calves', *animal*, 14, S133–S143. https://doi.org/10.1017/S1751731119003148

Hansen, B.G., Langseth, E. and Berge, C. (2023), 'Animal welfare and cow-calf contact-farmers' attitudes, experiences and adoption barriers', *J Rural Stud*, 97, 34–46.

IFOAM (2020), 'Principles of organic agriculture'. https://www.ifoam.bio/principles-organic-agriculture-brochure, (accessed 17 November 2023)

IG kuhgebundene Kälberaufzucht (2022), Kriterien und Kontrollunterlagen der Interessensgemeinschaft kuhgebundene Kälberaufzucht. https://ig-kalbundkuh.de/kriterien-und-kontrollunterlagen-der-interessengemeinschaft/ (accessed 17 November 2023)

Ivemeyer, S., Brinkmann, J., March, S., Simantke, C., Winckler, C. and Knierim, U. (2018), 'Major organic dairy farm types in Germany and their farm, herd, and management characteristics', *Organic Agric*, 8(3), 231–247.

Johanssen, J.R.E., Kvam, G.-T., Logstein, B. and Vaarst, M. (2023), 'Interrelationships between cows, calves, and humans in cow-calf contact systems – An interview study among Norwegian dairy farmers', *J Dairy Sci*, 106. https://doi.org/10.3168/jds.2022-22999

Johnsen, J.F., Ellingsen, K., Grøndahl, A.M. et al. (2015a), 'The effect of physical contact between dairy cows and calves during separation on their post-separation behavioural response', *Appl Anim Behav Sci*, 166, 11–19.

Johnsen, J. F., Beaver, A., Mejdell, C. M. et al. (2015b), 'Providing supplementary milk to suckling dairy calves improves performance at separation and weaning', *J Dairy Sci*, 98(7),4800–4810

Johnsen, J.F., Zipp, K.A., Kälber, T. et al. (2016), 'Is rearing calves with the dam a feasible option for dairy farms? – Current and future research', *Appl Anim Behav Sci*, 181, 1–11.

Johnsen, J.F., Holmøy, I.H., Mejdell, C.M. et al. (2021a), 'A cross-sectional study of associations between herd-level calf mortality rates, compliance with legislation on calf welfare, and milk feeding management in Norwegian dairy herds', *J Dairy Sci*, 104(1), 839–848.

Johnsen, J.F., Kischel, S.G., Sætervik Rognskog, M. et al. (2021b), 'Investigating cow-calf contact in a cow-driven system: performance of cow and calf'. *J Dairy Res*, 1–4. https://doi.org/10.1017/S0022029921000200.

Johnsen, J.F., Johanssen, J.R.E., Aaby, A.V. et al. (2021c), 'Investigating cow-calf contact in cow-driven systems: behaviour of the dairy cow and calf', *J Dairy Res*, 88 (1), 52–55.

Johnsen, J.F., Garmo, R. and Holmøy, I.H. (2022), 'Calf mortality, calf treatments and mastitis treatments in Norwegian dairy herds practicing cow-calf contact versus herds practicing artificial calf rearing', *International Symposium of Veterinary Epidemiology and Economic*, Canada, 403.

Just, A., Wellmann, R. and Bennewitz, J. (2018), 'Estimation of relative economic weights and the marginal willingness to pay for breeding traits of Brown Swiss cattle using discrete choice experiments', *J Dairy Sci*, 101(6), 5207–5213.

Kargar, S., Bahadori-Moghaddam, M., Ghoreishi, S.M. et al. (2021), 'Extended transition milk feeding for 3 weeks improves growth performance and reduces the susceptibility to diarrhea in newborn female Holstein calves', *Animal*, 15(3), 100151.

Khan, M.A., Weary, D.M. and von Keyserlingk, M.A.G. (2011), 'Invited review: effects of milk ration on solid feed intake, weaning, and performance in dairy heifers', *J Dairy Sci*, 94(3), 1071–1081.

Krpálková, L., Cabrera, V.E., Kvapilík, J., Burdych, J. and Crump, P. (2014), 'Associations between age at first calving, rearing average daily weight gain, herd milk yield and dairy herd production, reproduction, and profitability', *J Dairy Sci*, 97(10), 6573–6582.

Leiber, F., Müller, A., Maurer, V., Schader, C. and Bieber, A. (2019), 'Organic dairy farming: towards sustainability'. In M. Vaarst and S. Roderick (Eds) *Improving organic animal farming*. Burleigh Dodds Science Publishing Limited, Cambridge, UK, Chapter 11, 225–244.

Loberg, J. and Lidfors, L. (2001), 'Effect of stage of lactation and breed on dairy cows' acceptance of foster calves', *Appl Anim Behav Sci*, 74(2), 97–108.

Loberg, J.M. (2007), *'Behaviour of Foster Cows and Calves in Dairy Production: Acceptance of Calves, Cow-Calf Interactions and Weaning'*, Doctoral Thesis, Swedish University of Agricultural Sciences.

Lund, V. (2006), 'Natural living – a precondition for animal welfare in organic farming', *Livest Sci*, 100(2–3), 71–83.

Mac, S.E., Lomax, S. and Clark, C.E.F. (2023), 'Dairy cow and calf behavior and productivity when maintained together on a pasture-based system', *Anim Bio*, 36(2), 322–332.

Maher, J.W., Clarke, A., Byrne, A.W. et al. (2021), 'Exploring the opinions of Irish dairy farmers regarding male dairy calves', *Front Vet Sci*, 8, 635565.

Malmuthuge, N. and Le Guan, L. (2017), 'Understanding the gut microbiome of dairy calves: opportunities to improve early-life gut health', *J Dairy Sci*, 100(7), 5996–6005.

Marcato, F., van den Brand, H., Kemp, B. et al. (2020), 'Effects of pretransport diet, transport duration, and type of vehicle on physiological status of young veal calves', *J Dairy Sci*, 103(4), 3505–3520.

McCullock, K., Hoag, D.L.K., Parsons, J. et al. (2013), 'Factors affecting economics of using sexed semen in dairy cattle', *J Dairy Sci*, 96(10), 6366–6377.

Meagher, R.K., Beaver, A., Weary, D.M. and von Keyserlingk, M.A.G. (2019), 'Invited review: a systematic review of the effects of prolonged cow-calf contact on behavior, welfare, and productivity', *J Dairy Sci*, 102(7), 5765–5783.

Murphy, B., Crosson, P., Kelly, A.K. and Prendiville, R. (2017), 'An economic and greenhouse gas emissions evaluation of pasture-based dairy calf-to-beef production systems', *Agric Syst*, 154, 124–132.

Nansen, P., Steffan, P., Monrad J., Grønvold, J. and Henriksen, S.A (1990), 'Effects of separate and mixed grazing on trichostrongylosis in first- and second-season grazing calves', *Veterinary Parasitology*, 36(3–4), 265-276.

Natzke, D. (2023), Organic farms get fewer and larger. https://www.agproud.com/ext/resources/2023/01/16/0223pd-infocus.pdf

Neave, H.W., Sumner, C.L., Henwood, R.J.T. et al. (2022), 'Dairy farmers' perspectives on providing cow-calf contact in the pasture-based systems of New Zealand', *J Dairy Sci*, 105(1), 453–467.

Newberry, R.C. and Swanson, J.C. (2008), 'Implications of breaking mother–young social bonds', *Appl Anim Behav Sci*, 110(1–2), 3–23.

Nicolao, A., Coppa, M., Bouchon, M. et al. (2020), 'Early-life dam-calf contact and grazing experience influence post-weaning behavior and herbage selection of dairy calves in the short term', *Front Vet Sci*, 7, 600949.

Nicolao, A., Veissier, I., Bouchon, M. et al. (2022), 'Animal performance and stress at weaning when dairy cows suckle their calves for short versus long daily durations', *Animal*, 16(6), 100536.

Nielsen, B. and Thamsborg, S.M. (2002), 'Dairy bull calves as a resource for organic beef production: a farm survey in Denmark', *Livest Prod Sci*, 75(3), 245–255.

Nielsen, S.S., Alvarez, J., Bicout, D.J. et al. (2023), 'Welfare of calves', *EFSA J*, 21(3), e07896.

NOP eCFR: 7 CFR 205.239 – Livestock living conditions 2023. https://www.ecfr.gov/current/title-7/subtitle-B/chapter-I/subchapter-M/part-205/subpart-C/section-205.239 (accessed 26 February 2023).

Pardon, B., K. de Bleecker, M. Hostens, J. et al. (2012), 'Longitudinal study on morbidity and mortality in white veal calves in Belgium', *BMC Vet Res*, 8(1), 26. https://doi.org/10.1186/1746-6148-8-26

Pedersen, R.E., Sørensen, J.T., Skjøth, F. et al. (2009), 'How milk-fed dairy calves perform in stable versus dynamic groups', *Livest Sci*, 121(2-3), 215–218.

Placzek, M., Christoph-Schulz, I. and Barth, K. (2021), 'Public attitude towards cow-calf separation and other common practices of calf rearing in dairy farming – a review', *Organic Agric*, 11(1), 41–50.

Proudfoot, K. L., Hendricks, J., Higgins, A. et al. (2022), ,The Entrepreneurs: Dairy farmer perspectives on finding an industry solution for the surplus calf issue–A participatory case study', *Front Sustain Food Syst*, 415. https://doi.org/10.3389/FSUFS.2022.961068

Regulation (EU) 2018/848 on organic production and labelling of organic products and repealing Council Regulation (EC) No 834/2007 2018. (accessed 26 February 2023).

Reiber, C., Wollmeister, M., Sommer, T. and Chagunda, M.G.G. (2020), 'Status quo und Determinanten der Kälbervermarktung von ökologischen und konventionellen Milchviehbetrieben in Baden-Württemberg', (in English: Status quo and determinants of calf marketing of organic and conventional farms in Baden-Württemberg). *Züchtungskunde*, 92(5), 320–338.

Reinhardt, V. (1980), '*Investigation into social behaviour of cattle. Two-year observation on a semi-wild herd of zebu (Bos indicus)*', Springer Basel.

Rell, J. Home, R.; Bähler, C. et al. (2022), 'Motivations for Swiss veal farmers to adopt calf health management strategies that enable reduction in antibiotic use', *Anim Prod Sci*, 62, 5(3), 490–500.

Rell, J., Wunsch, N., Home, R. et al. (2020), 'Stakeholders' perceptions of the challenges to improving calf health and reducing antimicrobial use in Swiss veal production.' *Prev Vet Med,* 179, 104970.

Renaud, D.L., Duffield, T.F., LeBlanc, S.J. et al. (2017), 'Management practices for male calves on Canadian dairy farms', *J Dairy Sci*, 100(8), 6862–6871.

Riefer, A. and Hamm, U. (2008), 'Changes in families' organic food consumption', *AgEconSearch*. https://ageconsearch.umn.edu/record/43653/

Roadknight, N., Mansell, P., Jongman, E. et al. (2021), 'Blood parameters of young calves at abattoirs are related to distance transported and farm of origin', *J Dairy Sci*, 104(8), 9164–9172.

Roche, J.R., Friggens, N.C., Kay, J.K. et al. (2009), 'Invited review: Body condition score and its association with dairy cow productivity, health, and welfare', *J Dairy Sci*, 92(12), 5769–5801.

Rørvang, M.V., Nielsen, B.L., Herskin, M.S. and Jensen, M.B. (2018), 'Prepartum maternal behavior of domesticated cattle: a comparison with managed, feral, and wild ungulates', *Front Vet Sci*, 5, 45.

RSPCA. (2023), RSPCA Welfare Standards for Beef Cattle. https://science.rspca.org.uk/documents/d/science/web-beef-welfare-standards-2023-1 (accessed 17 November 2023)

Sandelin, A., Hälli, O., Härtel, H. et al. (2021), 'Effect of farm and animal-level factors on youngstock mortality and growth on calf rearing farms', *Prev Vet Med*, 193, 105416.

Sans, P. and de Fontguyon, G. (2009), 'Veal calf industry economics', *Revue de Médecine Vétérinaire*, 160, 420–424.

Santman-Berends, I. M. G. A., de Bont-Smolenaars, A.J.G., Roos, L. et al. (2018), 'Using routinely collected data to evaluate risk factors for mortality of veal calves', *Prev Vet Med*, 157, 86–93.

Schlipf. (1908), *Handbuch der Landwirtschaft (in English: Handbook of agriculture)*, 16. edition, Verlagsbuchhandlung Paul Parey Berlin

Schneider, C., Bieber, A., Spengler Neff, A. and Ivemeyer, S. (2021), 'Separation and weaning of calves reared in cow-calf contact systems.' *ProYoungStock - Practice Abstract.* https://orgprints.org/id/eprint/43342/1/schneider-etal-2021-42549-CORE_Organic_practice%20abstract_ProYoungStock_EN.pdf (accessed 6 July 2023)

Simon, G., E. Hillmann and Barth, K. (2023), 'Pasture-based fattening does not cause severe non-perforating lesions in veal calves of dairy breeds', *JDS Com*, 4(6),496–501

Sirovica, L.V., Ritter, C., Hendricks, J. et al. (2022), 'Public attitude toward and perceptions of dairy cattle welfare in cow-calf management systems differing in type of social and maternal contact', *J Dairy Sci*, 105, 3248–3268.

Sirovnik, J., Barth, K., de Oliveira, D. et al. (2020), 'Methodological terminology and definitions for research and discussion of cow-calf contact systems', *J Dairy Res*, 87S1, 108–114.

Skelhorn, E.P.J., Garcia-Ara, A., Nova, R.J., Kinston, H., Waapenaar, W. (2020), 'Public perception and opinon of rose veal in the UK', *Meat Sci*, 167, 108032.

Soberon, F., Ryan, C.M., Nydam, D.V. et al. (2011), 'The effects of increased milking frequency during early lactation on milk yield and milk composition on commercial dairy farms', *J Dairy Sci*, 94(9), 4398–4405.

Soil Association (2023). Soil Association Organic Standards for Great Britain. Farming and Growing. Version 1.2: published on 12th January, 2023. https://www.soilassociation.org/media/23372/sa-gb-farming-_growing-standards.pdf (accessed 17 Novmeber 2023)

Spengler Neff, A., Schneider, C. and Ivemeyer, S. (2023), *'Cow-reared Calf Husbandry in Dairy Farming.'* FiBL, Frick, Schweiz. https://www.fibl.org/de/shop/1660-mother-bonded-calf-rearing

Stampa, E., Schipmann-Schwarze, C. and Hamm, U. (2020), 'Consumer perceptions, preferences, and behavior regarding pasture-raised livestock products: a review', *Food Qual Pref*, 82, 103872.

Tančin, V., Harcek, L., Broucek, J. et al. (1995), 'Effect of suckling during early lactation and changover to machine milking on plasma oxytocin and cortisol levels and milking characteristics in Holstein cows', *J Dairy Res*, 62(2), 249–256.

Tichenor, N.E., Peters, C.J., Norris, G.A., Thoma, G. and Griffin, T. S. (2017), 'Life cycle environmental consequences of grass-fed and dairy beef production systems in the Northeastern United States', *J Clean Prod*, 142, 1619–1628.

Vaarst, M., Nissen, T. B., Østergaard, S. et al. (2007), 'Danish stable schools for experiential common learning in groups of organic dairy farmers', *J Dairy Sci,* 90(5), 2543–2554

Vaarst, M., Hellec, F., Sørheim, K. et al. (2019), 'Calves with their dams in dairy cow systems', *CORE organic report*, p. 59. https://orgprints.org/id/eprint/36915/2/fullreport-pdf-version-december-2019.pdf

Vaarst, M., Hellec, F., Verwer, C. et al. (2020), 'Cow calf contact in dairy herds viewed from the perspectives of calves, cows, humans and the farming system. Farmers' perceptions and experiences related to dam-rearing systems', *Landbauforschung - J Sust and Organic Agric. Syst*, 70(1), 49–57.

Vaarst, M. and Christiansen, I.A. (2023), 'Three years of situated social learning and development of diverse cow-calf contact systems in Danish organic dairy farms', *J Dairy Sci*, 106. https://doi.org/10.3168/jds.2022-22755

van Knegsel, A.T.M., Burgers, E.E.A., Ma, J. et al. (2022), 'Extending lactation length: consequences for cow, calf, and farmer', *J Anim Sci*, 100(10), 1–10.

Veissier, I., Caré, S. and Pomiès, D. (2013), 'Suckling, weaning, and the development of oral behaviours in dairy calves', *Appl Anim Behav Sci*, 147(1–2), 11–18.

Vogt, A., Barth, K., Waiblinger, S. and König von Borstel, U. (2024)'Can a gradual weaning and separation process reduce weaning stress in dam-reared dairy calves? A comparison with the two-step method', *J Dairy Sci*. https://doi.org/10.3168/jds.2024-23809.

Wallenbeck, A., Rousing, T., Sørensen, J.T. et al. (2019), 'Characteristics of organic dairy major farm types in seven European countries', *Org Agr*, 9(3), 275–291.

Weary, D.M., Ventura, B.A. and von Keyserlingk, M.A.G. (2016), 'Societal views and animal welfare science: understanding why the modified cage may fail and other stories', *Animal*, 10(2), 309–317.

Webb, L.E., Bokkers, E.A.M., Heutinck, L.F.M. et al. (2013), 'Effects of roughage source, amount, and particle size on behavior and gastrointestinal health of veal calves', *J Dairy Sci*, 96(12), 7765–7776.

Wenker, M.L., Bokkers, E.A.M., Lecorps, B. et al. (2020), 'Effect of cow-calf contact on cow motivation to reunite with their calf', *Sci Rep*, 10(1), 14233.

Wenker, M.L., Verwer, C.M., Bokkers, E.A.M. et al. (2022), 'Effect of type of cow-calf contact on health, blood parameters, and performance of dairy cows and calves', *Front Vet Sci*, 9, 855086.

Wieczorreck, L. and Hillmann, E. (2022), Ist die ammengebundene Aufzucht eine tiergerechte Alternative zur künstlichen Aufzucht von Milchviehkälbern? (In English: Is a foster cow rearing system an animal-friendly alternative to artificial rearing of dairy calves?) In *Aktuelle Arbeiten zur artgemäßen Tierhaltung 2022*, KTBL-Schrift 11530, 90–100.

Willer, H., Schlatter, B. and Trávníček, J. (Eds) (2023), *'The World of Organic Agriculture. Statistics and Emerging Trends 2023'*, Research Institute of Organic Agriculture FiBL, Frick, and IFOAM – Organics International, Bonn (accessed 2 February 2023).

Chapter 7

Improving herd health management and disease prevention in organic dairy cattle farming

Lisa Morgans, Royal Agricultural University/(formerly Innovation for Agriculture), UK; Catarina Svensson, Swedish University of Agricultural Sciences, Sweden; Julie Duval, National Research Institute for Agriculture, Food and the Environment (INRAE), France; and Silvia Ivemeyer, Thünen Insititute - Institute of Organic Farming, Germany

http://dx.doi.org/10.19103/AS.2024.0138.08

1 Introduction

Animal health and welfare are the foundations of thriving and ethical animal farming systems and are essential to ensuring the health and safety of our food. The scale and nature of animal farming globally necessitates a critical approach to how animals are reared, kept and managed within our food system (Acari, 2017). Ensuring animal welfare is at the forefront of our food production systems – is not only an ethical issue but is now considered an inherent part of what defines sustainable farming and key to fulfilment of the United Nations Sustainable Development Goals (Health for Animals, 2021). This has been driven by a growing concern amongst consumers about how we care for farm animals and manage their health and welfare within our food production systems (see Livestock's Long Shadow, FAO; Alonso, 2020).

Animals are recognised as sentient beings in the eyes of European Union Law under Article 13 (https://eur-lex.europa.eu/EN/legal-content/glossary/animal-welfare.html). The way we look after them, and we how plan and implement herd health management strategies, is therefore key. Furthermore, the healthier animals are, the less likely they will need treatment with medicines such as antibiotics or anthelmintics, which is critical in helping us combat the growing threat of Antimicrobial Resistance (AMR) across the globe.

In organic dairy farming, welfare including health is central to a functioning and thriving farm enterprise (see 'Where to find further information' at the end of this chapter for a list of the guiding principles for organic production in Europe). However, welfare, especially health, states are not always as optimal as one might hope (Sanders et al., 2019; Krieger et al., 2017; Sundrum et al., 2001). There are myriad factors that influence animal health and welfare that need to be considered in order to achieve a minimal disease burden and high standards of welfare. Organic dairy farming emphasizes the importance of preventive health as a cornerstone in herd health planning and avoiding reliance on medicines, such as antibiotics, which underpins many poor farming practices in some farming systems worldwide (Manyi-Loh, 2018). Additionally, freedom from disease and good health is only one component of 'good' welfare (Yeates and Main, 2008; Stokes et al., 2020). Although we are aware that health is, of course, just one part of welfare, we will refer to the often-used term 'animal health and welfare planning' (AHWP) throughout this chapter.

Animal Health and Welfare Planning (AHWP) covers a number of approaches and has a number of names. Veterinary Herd Health Management (VHHM) is a similar approach defined by Green et al. (2012) as a method to optimize health, welfare and production in a population of dairy cows. A particular focus is regular objective observations of cows and their environment and the systematic analysis of relevant data, enabling informed, timely decisions to adjust and improve herd management over time. It is generally agreed that systematic analysis of data and assessments of animal-based measures and welfare indicators are necessary ingredients in VHHM/AHWP as a foundation for informed decisions about actions to be taken. However, one can question the emphasis the authors place on objective observation, considering our inherent biases and degree of subjectivity in our perceptions and judgements, the influence of social structures and the importance of diverse perspectives in innovation processes (Darnhofer, 2020; Enticott, 2023).

Many of the challenges to achieving good animal welfare are a result of our management practices, such as breeding techniques (Hartcher and Lum, 2020) and/or our food production systems (Brand et al., 2001). Various approaches exist to help improve animal welfare, as well as prevent disease and reduce reliance on medicine use on dairy farms, from closer collaboration between farmers and their veterinarians to the use of benchmarking and data-driven decision-making (Sumner et al., 2018). In the first part of this chapter, we aim to give an overview of the challenges facing AHWP on organic dairy farms with a brief review of the recent literature. This introduction and 'scene setting' are split into five sections:

- The complexity of animal health
- Reducing reliance on conventional medicine
- Advisory support for AHWP
- Behavioural factors in improving health and welfare
- Successfully implementing herd health management strategies

We believe these are the main types of challenges to overcome in order to improve AHWP on farms. Some practical suggestions to overcome these challenges are then discussed in the second part of this chapter.

2 Challenges: the complexity of animal health as part of animal welfare

> *Health is the wholeness and integrity of living systems. It is not simply the absence of disease, but the maintenance of physical, mental, social and ecological well-being. Immunity, resilience and regeneration are key characteristics of health.*
>
> *(IFOAM, 2005)*

2.1 *Animal health as part of animal welfare: moving beyond the five freedoms*

While animal health is important, it is still only one of several components of animal welfare. According to Fraser (2008), the concept of animal welfare comprises three key points:

- basic health and functioning
- affective states
- the ability to live in a natural way

Animal welfare encompasses the animal's state of being – how it feels. It also includes the context in which it lives and a sense of 'naturalness'. Welfare is more than the absence of pain, suffering and physical harm but also, e.g. the experience of positive emotions (Knierim, 2002) or the quality of an animal's life as it is experienced and valued by the animal (Bracke et al., 1999). A healthy life is described as one of the five pillars of Positive Welfare that, together with a Comfortable, Pleasurable, Confident and Interesting Life, constitute a Good Life or a 'life worth living' for animals under our care. This is something consumers, or food citizens (Doherty et al., 2020), are increasingly wanting to see demonstrated in the products they buy, despite citizens often experiencing cognitive dissonance in their buying choices, with conflicting values and beliefs complicating buying behaviour (Schroder and McEachern, 2004). These welfare concerns, therefore, have implications for how we rear farm animals and therefore the purpose and function of AHWP.

Unhealthy cattle with high levels of infectious or metabolic disease are an economic loss, in terms of production losses over the animal's lifetime and ongoing treatment costs, which can be seen quite clearly with diseases such as mastitis and lameness (van Soest et al., 2019). There have also been studies exploring the link between animal health, welfare and farmer wellbeing and how poor animal welfare affects farmers' wellbeing (Hansen et al., 2019). It would appear that good cattle welfare is a 'win-win' situation for economic, social and environmental reasons. However, we still see a wide range of poor welfare practices across farming systems, from calves to adult cows, and from indoor and housed to extensive outdoor systems. Furthermore, there are still unacceptably high levels of disease present in conventional and organic dairy farming systems, in spite of many initiatives to combat the most prevalent or significant diseases (Krieger et al., 2017).

2.2 Dairy farming, the environment and the role of animal health

Despite growing criticism surrounding the environmental impact of cattle production and the greenhouse gas (GHG) emissions it produces (Steinfeld et al., 2006; Poore and Nemcek, 2018), cattle are well-evolved to produce nutritionally dense meat and milk from grass that we cannot otherwise eat (Cheng et al., 2022). They also have multiple other benefits for our shared environment ranging from:

- Returning nutrients back to the soil from their manure, providing food for insects such as dung beetles (Benyon et al., 2015)
- Trampling carbon-rich plant matter back into the soil whilst grazing
- Through their grazing behaviour, stimulating the natural seedbed, which supports a whole ecosystem and adds to a biodiverse farming system (Dumont et al., 2016).

The challenge here is to consider the role and impact of cattle production, and more specifically dairy farming, from a wider systems-based perspective where GHG emissions are not the only measure of environmental impact. We must also recognise the impacts on biodiversity, soil health and the importance of dairy farming to many communities around the world.

Nevertheless, the impact of cattle farming globally on increasing GHG emissions is substantial. How we farm cattle has significant implications for the climate crisis, regardless of whether certified organic or not. In fact, in some studies, the extensive farming of cattle which is characteristic of organic systems has been shown to have a worse carbon footprint than conventional systems (Poore and Nemecek, 2018). This means that advances by AHWP, so that herds are healthier for longer and use resources more efficiently, may mean reduced emissions overall and have positive benefits for our shared environment.

2.3 Animal health in organic and agroecological farming approaches

Agroecology is increasingly acknowledged as being able to contribute to transforming food systems by applying ecological principles to agriculture and ensuring a regenerative use of natural resources and ecosystem services (HLPE, 2019). Scaling up agroecological practices is considered one potential way to increase the productivity of farms, ensure food security, improve incomes and rural livelihoods and reverse trends in biodiversity loss (De Schutter, 2010). Moreover, agroecology can contribute to the design of farming systems which are more resilient to climate change and reduce the need for purchased inputs, e.g. fossil fuels and agrochemicals (De Schutter, 2010). Animals are important in agroecosystems as they can serve the system in different ways, e.g. by contributing to closed nutrient cycles (with manure helping to fertilise crops) or, more broadly, by shaping the vegetation of ecosystems through grazing (Gliessman 2007; Dumont et al., 2013).

Several farming practices can be regarded as agroecological or regenerative with no current consensus on whether a system is considered agroecological or not (HPLE, 2019). When comparing definitions of agroecology and commonly-accepted agroecological practices to EU regulation on organic agriculture, and the norms of organic farming as produced by IFOAM (International Federation of Organic Agriculture Movements), there are no exclusive requirements related to the use of animals in the former (Migliorini and Wezel, 2017). Currently, no widely used certification schemes or labels exist for agroecological farming. In contrast, EU organic regulation sets down specific requirements for certain husbandry and animal welfare practices e.g. the use of chemically synthesised allopathic medicinal products in animal production (Migliorini and Wezel, 2017). However, organic agriculture and agroecology do share certain principles and practices and both are seen as promising solutions to develop sustainable agri-food systems (FAO, 2017; Migliorini and Wezel, 2017).

As one of five principles for designing sustainable animal production systems based on ecological processes, Dumont et al. (2013) highlighted the need *"to adopt management practices aiming to improve animal health"*. Health plays a central role in organic farming. It is one of the four principles of organic farming defined by IFOAM alongside the principles of ecology, care and fairness. The principle of health (also quoted above) states that "Health is the wholeness and integrity of living systems. It is not simply the absence of illness, but the maintenance of physical, mental, social and ecological well-being. Immunity, resilience, and regeneration are key characteristics of health" (IFOAM). Concepts, such as resilience link to the promotion of health by supporting animals' abilities to fulfil their needs and their capacity to react to disturbances, e.g. more extreme weather due to climate change (Döring et al., 2015).

3 Challenges: reducing reliance on conventional medical treatments

It is widely acknowledged in the veterinary profession and amongst many involved in food production that 'prevention is better than cure' when it comes to herd health and, specifically, antimicrobial stewardship. Preventive practices help minimise disease and maintain health, such as maintenance of clean pasture, bedding and low stocking densities to reduce the build-up of bacteria and other infectious agents. Preventive farm practices help farmers avoid the need to use 'last resort' therapies, such as antibiotics or anthelmintics, where use has consequences such as increasing antibiotic resistance (Rose et al., 2020). However, preventive treatments are sometimes perceived to be expensive by farmers (e.g. vaccines (Richens at el., 2015)) or ineffective by veterinarians and advisors (e.g. herbal remedies (Hellec et al., 2021)). Preventive practices are sometimes perceived to be labour intensive (e.g. regular weighing of animals, handling cattle with enhanced levels of care or thorough bedding management) and are poorly adopted in many areas of cattle farming, including organic dairy farming (Phillips and Heins, 2022).

The challenge of antimicrobial stewardship facing dairy farmers, and veterinarians working with dairy farming clients, is at least in part a combination of the following factors:

- Antimicrobials (which include anthelmintics and antibiotics) have until recently been very effective at treating microbial/parasitic/bacterial diseases in individuals and groups of animals and have become cheap to use for farmers, making them a first-line intervention in many cases in conventional agriculture. Using antibiotics and anthelmintics have become ingrained herd health management practices that are hard to shift (Bellet, 2018).
- These medicines are ubiquitous and widely available to farmers in some countries, depending on the region, sometimes with little oversight or guidance from the prescribing veterinarian (Clifford et al., 2018).
- Levels of knowledge on medicine stewardship vary and is often limited in many places around the world, leading to overuse/misuse.
- Antimicrobials are critical in 'propping up' some farming practices and systems, which cannot function economically without them (Bellet, 2018).
- This overreliance on antimicrobials poses a significant risk for selecting for antimicrobial resistance (Graham et al., 2019).

As a result of issues such as antimicrobial resistance (AMR), medicines such as antibiotics and anthelmintics are strictly controlled in many parts of the world (e.g. only available with a veterinary prescription or from a Suitably Qualified Person in the case of anthelmintics) and avoided in most organic dairy systems. This has been in part driven by consumer demand for organic dairy products free from medicine residues, rather than AMR, but more recently concern about the development of AMR has reinforced this demand (Miftari et al., 2022).

Organic farming certification (Regulation (EU) 2018/848) requires farmers to practice good animal husbandry so that antibiotic treatments are less necessary. There is an emphasis on enhanced stockpersonship, good housing, early detection of disease and prompt treatment with alternatives, e.g. fluid therapy, advanced nutrition, herbal remedies and homeopathy. Organic farming regulation also impacts medicine usage and consequently AHWP. Regulations vary across countries with some organic certifiers requiring farms to have zero antibiotic use in order to supply some markets (Bulut et al., 2021). According to the European Regulation on organic production, chemically synthesised allopathic veterinary medicinal products (including antibiotics) are not allowed for preventive treatments but are permitted for therapeutic use.

EU regulations state:

Chemically synthesised allopathic veterinary medicinal products, including antibiotics, may be used where necessary, under strict conditions and under the responsibility of a veterinarian, when the use of phytotherapeutic, homeopathic and other products is inappropriate (Regulation (EU) 2018/848).

These medicinal products can be used where necessary when the use of phytotherapeutic, homeopathic and other 'natural' products is inappropriate, and with a limitation of a maximum of three courses of treatments per year and a doubled withdrawal period (European Council, 2018). Private organic farming organisations, e.g. 'Bioland' in Germany or 'Demeter' (international) have additional prohibitions of or limitations on use of certain remedies (Bioland, 2023; Demeter International, 2022). These limitations are based on the negative environmental effects of these substances.

The average antimicrobial use on dairy farms varies between and within countries, and between different major organic farm types, but is generally lower than in conventional systems (Wallenbeck et al., 2018; Ivemeyer et al., 2012). The use of homeopathy also differs, especially between countries, with typical organic dairy farms in Nordic countries like Denmark and Sweden using fewer homeopathic mastitis treatments than, e.g. in Switzerland, Austria or Lithuania (Wallenbeck et al., 2018).

Alternatives to conventional medicine are discussed as part of the AHWP process and on organic dairy farms. There are numerous alternatives to antimicrobial and anthelmintic treatments, ranging from herbal remedies to vaccinations, both of which are widely adopted in organic farming (see Wallenbeck et al., 2018). Each treatment has pros and cons, depending on the reason for use and the causative pathogen. Within the veterinary profession in the UK, there is a desire to see more vaccine use to prevent the severity and spread of infectious diseases like pneumonia and bovine viral diarrhoea (BVD) (NOAH, 2022). Conversely, in Sweden, there is a scepticism about the efficacy of available vaccines and concerns that their use could divert attention away from other approaches such as the need to improve housing facilities and management routines (Hoelzer et al., 2018). Furthermore, there is scepticism amongst the veterinary profession about certain other alternatives, i.e. homeopathy and some herbal remedies, which has hindered veterinary involvement on some organic farms and within organic farming more broadly (Hellec et al., 2021, Duval et al., 2016b, 2017b)). Alternative therapies used in organic dairy farming are discussed elsewhere in this book.

The variability on organic dairy farms has led to the establishment of many diverse approaches to AHWP and variation in the provision of advisory services, with many organic farmers less than satisfied with the advice provided, forcing them to source their own information when managing their herds and deciding on alternative therapies (Duval et al., 2016b). Due to this lack of trust in existing support, peer-to-peer learning approaches have become very popular in organic farming to help farmers build their knowledge rather than relying on traditional 'top-down' advisory services (Vaarst et al., 2007).

Nevertheless, AHWP can still play an important role in guiding farmers on when, where and how to reduce reliance on antimicrobials and how to use them responsibly when they are necessary for safeguarding animal welfare. The AHWP process is an opportunity for the collection and analysis of herd-level data to appraise current performance and inform decisions, so that future management of herd health continues to improve (Speksnijder et al., 2017).

4 Challenges: advisory support for Animal Health and Welfare Planning approaches

Veterinarians are key actors in improving animal health and welfare and are often characterised as the trusted advisor who can help farmers optimise animal health and improve their herd's productivity (Enticott et al., 2012). In many countries (e.g. the Netherlands), it is a regulatory requirement that a named veterinarian develops and signs off a herd health plan agreed with the farmer and reviews it regularly (e.g. yearly). However, in some places, the veterinary profession has struggled to optimise the herd health planning process, e.g. in the UK (Mahon et al., 2021). Veterinarians have also struggled to engage and support organic dairy clients in their efforts to improve herd health and avoid/reduce AMU, e.g. through alternative treatments (Brock et al., 2022; Duval et al., 2016b, 2017b; Vaarst et al., 2006, 2007).

The role of veterinarians on organic, as well as conventional, dairy farms varies. Farmers invite veterinarians to have a certain role on their farm, ranging from a 'fire-fighter' managing emergencies, to collaborations where farmers treat veterinarians as partners in herd health management (Duval et al., 2016b; Skjølstrup et al., 2021). It has been found that Swedish organic dairy farmers contacted their veterinarians later than conventional counterparts to help them deal with calves with diarrhoea or when mastitis was suspected. Moreover, they were less likely to have a dedicated farm veterinarian that they used frequently compared to conventional farmers (Emanuelson et al., 2018). In Denmark and France, veterinarians were not always considered by organic dairy farmers as relevant advisors in supporting them in defining animal health strategies (Duval et al., 2016b; Vaarst et al., 2006, 2007). According to Danish organic dairy farmers aiming to phase out antimicrobial use, veterinarians were not considered as supportive in defining appropriate alternative animal health strategies (Vaarst et al., 2006, 2007). This can lead to situations in which interactions between organic farmers and veterinarians are limited to dealing with acute animal health problems and/or situations in which veterinarians are considered merely as technicians (Brock et al., 2022; Duval et al., 2016b, 2017b; Hellec et al., 2021; Vaarst et al., 2006). This has resulted in organic farmers finding different sources of information to support themselves and learn how to prevent disease, reduce antibiotic use and care for their cows (Duval et al., 2017; Hellec et al., 2021).

The organic farming context presents several challenges for farmers and veterinarians to collaborate on the subject of animal health. From a veterinarian's point of view, organic farming principles, regulations and individual farmers' animal health goals and approaches, sometimes interfere with what they consider as 'good animal health management practices'. Veterinarians can question some organic principles, such as prioritizing naturalness at the potential expense of animal health. They can also question organic regulations, e.g. the priority given to the use of alternative medicines over synthetic drugs. Veterinarians are also challenged by certain animal health approaches used by organic farmers (Duval et al., 2016b; Skjølstrup et al., 2021).

Organic dairy farmers have experienced differences between their animal health approaches and those proposed by veterinarians. For example, organic farmers aim to limit the use of synthetic products, especially where harmful side effects have been noted such as anthelmintic resistance or ecological toxicity, but veterinarians often rely on synthetic chemical solutions, and until recently, have a limited knowledge and/or lack of interest in searching for alternative methods. Where this is now changing is around anthelmintic use, targeted treatments based on diagnostic tests (e.g. faecal egg/worm counts) and integrated parasite management. Nevertheless, this is not widespread and many other qualified people who are not veterinarians are also permitted to sell and provide anthelmintics to farmers (i.e. Suitably Qualified Persons), albeit with a prescription from 2025 across the EU (Regulation (EU) 2019/6).

Some farmers have questioned the independence of some veterinarians working within veterinary practices with business models that have been (and many still are) structured around income from medicine sales. This is variable across countries with some places like Denmark decoupling veterinary advice from medicine sales.

The services offered by veterinarians as part of AHWP typically focus on the level of the individual animal, rather than on the level of the herd, with many routine herd visits dealing primarily with care for individual cows or fertility and pregnancy checks (Derks et al., 2013; Ritter et al., 2019; Ries et al., 2022). This is despite efforts by the veterinary profession in some countries to move to more population-level medicine (Sjöström et al., 2019). There are several barriers to veterinarians adopting a preventative role (see Hall and Wapenaar 2012; Ruston et al. 2016; Svensson et al., 2022a, 2022b). Critically, this will require veterinary business models moving away from a reactive model (Aplin et al., 2022) which relies on medicine sales as the main source of income, as seen in the US and the UK, to one where medicine sales are decoupled from prescribing veterinary drugs as seen in Denmark, and veterinarians take on a more population-level consultancy role.

In the case of farmers' long-term health promotion strategies, farmers found some veterinarians were overly focussed on curing disease and lacked a proactive attitude towards finding sustainable, preventative solutions to animal health problems (Duval et al., 2017b). Organic dairy farmers with a focus on reducing the use of antibiotics emphasised that their first priority is to improve herd health more generally. This requires day-to-day routines that limit the risk of developing infectious diseases (e.g. providing sufficient fresh air, clean bedding, as well as monitoring individual animals to detect issues early) (Vaarst et al., 2006).

In certain cases, farmers regretted a lack of dialogue and sharing of experience between themselves and veterinarians on practices that could improve animal health. There are organic farmers who have positive experiences involving veterinarians but often in the context of farmer groups and/or continued education on animal health. Contributing to the positive experiences were the independence of the advisors involved in these groups (i.e. not selling veterinary drugs) and the fact that these exchanges were seen as a source of information on different practices to manage their herd and based on experience sharing. Characteristics of these activities might explain, at least partly, their success and contrast with experiences farmers had experienced during exchanges with veterinarians (Duval et al., 2017b).

5 Challenges: behavioural factors in improving health and welfare

There are copious amounts of information available on how to improve animal health and welfare in organic farming (see https://www.organicseurope.bio/ as an example). However, using and applying this knowledge through the framework of AHWP is still perceived to be a major challenge, making implementing recommended changes on organic dairy farms an area of ongoing study (Grodkowski et al., 2023). Simply having a health plan for the farm, and informing the relevant actors of the changes they need to make on farm is not enough to result in change, as seen in the chronic levels of lameness and calf disease across European dairy production (Thomsen et al., 2023).

Awareness and knowledge are only some of the factors influencing behaviour and farm practices as described in the widely used ADKAR change management tool (https://www.prosci.com/methodology/adkar):

- **Awareness** of the need for change
- **Desire** to support the change
- **Knowledge** of how to change
- **Ability** to demonstrate skills and behaviour
- **Reinforcement** to make the change stick

A further useful tool to understand behaviour and how to influence certain behaviours is the COM-B theoretical framework. The COM-B behaviour change framework emphasizes (Michie et al., 2011):

- **C**apability
- **O**pportunity
- **M**otivation

Human behaviour, and the variation in people's beliefs, attitudes, motivations, personal capacity and experiences, all play a part. On top of this, the specific farming context and wider socio-economic factors influence why improving herd health on farms is an ongoing challenge and needs tailored approaches (Blanco-Penedo et al., 2019; Krieger et al., 2017). Farmers – and the immediate team around them – are the main actors responsible for implementing the herd health plan and improving herd health and welfare. Therefore, their needs, desires, concerns, knowledge and experiences are crucial drivers and shapers of change on their farms. Starting a dialogue and listening to farmers discuss their own experiences and perspectives of herd health on their farm is integral to a process of change. This is a critical first step that needs to be recognised by an advisor or veterinarian aiming to improve AHWP on organic dairy farms.

Once an advisor/veterinarian steps back from giving information (i.e. telling the farmer what to do) and listens to the needs, concerns and experiences of those managing the dairy cattle, then the next step in AHWP (reflecting and evaluating past and current progress) can commence. AHWP can proceed with the following stages:

- (1) being aware and reflecting on any negative trends from on-farm data and experience
- (2) appraising the risk factors affecting animal health and welfare
- (3) identifying farm-specific problems, ideally before they have arisen
- (4) celebrating the farm's successes (so you can do more of that!).

In a collaborative manner, ideally as a multi-actor farm health team (see https://disarmproject.eu/search-resources/farm-health-toolbox/), it is then appropriate to suggest or recommend solutions/actions *with* and *for* the farmer(s) with detailed descriptions of how to implement them in a tailored fashion for the farm. If farmers are not first and foremost in this process, then they arguably have little ownership of the changes and will be left feeling demotivated and disempowered, in a similar way to being told what to do from afar and not being asked for your thoughts and ideas (Šūmane et al., 2018). It would seem that AHWP processes and approaches have been predominantly top-down and advisor-led in the past, with advisor/veterinary knowledge being the main source of recommendations (Ruston et al., 2016). This has contributed to low levels of farmer participation in and utilisation of AHWP (Svensson et al., 2018; De Groot et al., 2021) and a poor perception of it by organic dairy farmers, with consequently limited impact on levels of dairy cow health and welfare (Vaarst et al., 2011). As described by Chambers back in (1994), putting farmers first and last helps increase buy-in but, more importantly, brings valuable and insightful expertise to the AHWP discussions that may otherwise be missing from more traditional, top-down approaches (Bard et al., 2017; Morgans et al., 2021).

6 Challenges: successfully implementing herd health management strategies

Changing on-farm practices around animal health and welfare and, therefore, farmer behaviour can be difficult, regardless of how good the farmer-advisor relationship is, how involved farmers are and how participatory the planning process has been. There are a variety of general and farm-specific barriers which need to be addressed in the first instance. Specific barriers are summarised in Table 1 as a series of questions following the ADKAR model. This provides a basic framework to consider when reflecting on how AHWP can be improved and acts as a useful checklist for advisors to consider when planning their approach. Another resource to refer to is the RESET model developed in the Netherlands (Lam et al., 2017) which categorises the five key levers for influencing farmer behaviour as:

- **R**ules and regulations
- **E**ducation and information
- **S**ocial pressure
- **E**conomics
- **T**ools

RESET has been recently applied in Wales in the UK (Rose et al., 2024). It is also worth noting that any desire to change or transform somebody (or something) comes with an inherent bias and imbalance in power, which many scholars encourage actors, such as advisors, to recognise and explore in a process of self-reflection (van Dijk et al., 2017; Darnhofer, 2020).

This section is not intended to be a review of the many behaviour change models and farming transition frameworks that exist in the literature. These are reviewed, e.g. in Glanville et al. (2020). The chapter is more of an overview of current challenges in implementing AHWP herd health improvement strategies in organic dairy farming. Individual farmer attitudes, beliefs, perceptions, motivations and personal goals will all need consideration in any successful AHWP process. This must be alongside consideration of wider socioeconomic and cultural factors and the specific farm context, which includes social dynamics within the farm team. Approaches that address all these are more likely to deliver results on the ground.

Table 1 The ADKAR change management model and its relationship to AHWP.

Element of ADKAR	Relevance to AHWP
Awareness of the need to change	• Is the issue/challenge perceived in the same way by all actors? The perception of issue(s) will and should be viewed from different angles and this will be shaped by different farming identities and backgrounds. • Who creates and provides the necessary information on the issue? • What is the accessibility and availability of relevant information/data about the issue? • Are different learning styles and communication styles accommodated? • Are data being turned into information that is being discussed, appraised and turned into knowledge based on what is already known?
Desire to participate in change	• Poor experiences in the past, or poor relationships between farmers/advisors can lead to apathy – has this been discussed and reflected upon? The farmer-advisor relationship and trust in one another are important to inspire change and may need time to develop. • How aligned are the farmers-advisors? Misaligned values about the necessary change/actions can be a barrier to change (e.g., on the efficacy of homeopathy). Alignment of life and professional goals with necessary changes (e.g. work-life balance, seeing friends/family, making a profit, etc.) can help drive change. • Is everyone speaking the same language/on the same page? Miscommunication/poor communication about necessary changes and actions can be a barrier to change. Paternalistic communication style, lack of shared decision-making and overly directive communication can lead to apathy/distrust.
Knowledge to change	• Whose knowledge matters – is everyone given a chance to share their knowledge and are they listened to? Where are the knowledge gaps? What can be done to overcome a hierarchy of knowledge and the science-practice gap? • Is there adequate knowledge mobilisation between different actors using participatory approaches and enhanced communication and facilitation strategies? Learning can be inherently social, so could peer-to-peer approaches enhance knowledge sharing? • Knowledge alone does not equate to action "Necessary but not sufficient"
Ability to make change	• How equipped and able to tackle the issue(s) is the farmer/farm team? What are the economic/financial restraints and limitations? Would a Cost-benefit analysis of the issue(s) help address differences in the perceived value of the change? • Is there a perceived lack of time, high workload, and/or lack of skilled farm labour hindering efforts? Is there a lack of skills/confidence in skills where training could help? • Are there external factors leading to paralysis/dis-empowerment (e.g. supply chain requirements/market demands/economics/debt/trade, etc.)? Can these be discussed and evaluated? • There is not one size fits all (e.g. Individual family businesses vs large Agri-businesses) so advice and solutions should not be one size fits all.

Element of ADKAR	Relevance to AHWP
Reinforcement of change	• Is there enough of the right sort of external support to sustain changes (e.g. policy incentives, advisors, peers, market demand/reward)? • What role could research and innovation play, if any? • What is needed now, later and in the future to support continual improvements?
Further contextual issues to consider	• Relevant food & farming policy – is this incentivising or hindering action on the issue? • Are there cultural/social/environmental challenges preventing action (e.g. weather, social stigma)? • Farmer identity and a sense of place are important factors that impact decision-making and changes on the farm (e.g. family ties to the land and producing food, see Kliebl et al., 2024)

Farmers and the wider farm team need to be central to all on-farm planning processes but often are not (Rose et al., 2018). This poses a challenge if pressure for improving dairy cow welfare comes from regulation or from wider society because that could impair "ownership" of the improvement process and build distrust and apathy. This risk was discussed by Vaarst and Fisker (2013) in the context of the Danish organic dairy sector which pioneered voluntary 'stable schools' bringing together small local groups of farmers to discuss common issues such as ways to reduce antibiotic use (discussed later in the chapter). The success of this initiative led to its being included in 2010 Danish legislation requiring larger dairy farms to put in place an obligatory health advisory service. As Vaarst and Fisker (2013) argue, there are potential contradictions between these voluntary, bottom-up and regulatory, top-down approaches.

Nevertheless, there are examples where policy instruments can support a multi-actor farm health team approach and enable the AHWP process to be led by those with on-the-ground experience. German legislation, e.g. requires farmers to undertake a self-assessment of animal welfare using a series of animal-based welfare indicators which enables farmers to contribute to the development of their own AHWP (discussed later in the chapter). Another example is the UK Government's Animal Health & Welfare Pathway.

7 Improving Animal Health and Welfare Planning: taking a holistic, systems approach

To manage a dairy cattle farm involves very diverse skillsets, knowledge and multiple processes that influence each other. Advisors supporting farms in their AHWP need to view farmers' decision-making regarding animal health and welfare in light of this larger system and consider the broader goals and priorities in farm management. Systems thinking is an approach that assesses problems by looking at whole systems and the relationships within them (see, e.g. Midgley et al. (2003)). Such thinking is part of the framework of organic principles which should always guide the planning on an organic farm. Advisors need to understand their own role as, most likely, just one of a group of several advisors to a farm with different perspectives and competencies. A systems approach is needed for AHWP.

Not only is animal husbandry just one piece in the herd health management puzzle, but the herd comprises animals of different ages and with different welfare needs that are both related to and yet independent of one another, making up a highly dynamic animal population that also provides a livelihood for the farmer(s). In addition, animals in integrated agroecological farming systems are managed within a wider framework of natural ecosystem processes in order to render farming systems more sustainable. Examples are agropastoral systems, crop-livestock farms or silvopastoral systems. A holistic approach is required to understand and mobilise the interrelationships and synergies between different elements of the system (Gliessman, 2007).

The different disease complexes challenging animal health are generally multifactorial (e.g. bovine respiratory disease) and are likely to affect each other. Many diseases of dairy cattle stem from the way the group or individuals are housed, cared for and managed. Decision-making in herd management is indeed very complex and requires knowledge about biological systems, such as physiology, medicine, animal behaviour and welfare, but also about a wide range of technical disciplines like economics, business management, legislation/regulation, marketing as well as advanced practical skills, such as calving, grazing and engineering, which further add to the complexity of herd management. Moreover, agroecology acknowledges the relevance of local knowledge and know-how to be able to deal with specific contexts (FAO, 2020). All this points to the importance of addressing herd health management from a holistic perspective.

The European CORE-organic project ANIPLAN (Planning for better animal health and welfare in dairy herds) identified nine principles for animal health and welfare planning on organic dairy farms that can be used in different farming contexts (Vaarst et al., 2011):

- P1: A health planning process should aim at continuous development and improvement, and should incorporate health promotion and disease-handling measures, based on a strategy including, current status and risks (animal-based + resource-based parameters), evaluation, action and review
- P2: Farm specific
- P3: Farmer ownership
- P4: External person(s) should be involved

- P5: External knowledge
- P6: Organic principles framework (systems approach)
- P7: Written
- P8: Acknowledge good aspects
- P9: Include all relevant people in the process

The ANIPLAN project recommended that the AHWP process should incorporate a learning cycle including the development of the health plan, continuous assessment of progress, evaluating and acting on recommendations. This is in alignment with results from a focus group study involving Swedish dairy farmers and farm managers where participants stressed the importance of follow-up as a central part of the AHWP process (Svensson et al., 2018). It also fits with the ADKAR framework discussed earlier in this chapter where reinforcement of any change is critical to its success.

Several studies have highlighted the importance of the process being guided by the farmer's own motivations, goals and priorities, suggesting the importance of investigating these aspects as part of a well-conducted AHWP. The people working with the animals on a day-to-day basis know the most about them and their current management, their knowledge and perspectives need to be included to ensure a well-informed evaluation of the status of the herd.

Since the 1990s Swedish veterinarians have offered strategic and systematic AHWP programs developed by the dairy sector and applying a holistic perspective to both organic and conventional dairy farms (Svensson et al., 2018, 2019). Sjöström et al. (2019) reported high implementation rates when a structured participatory approach to AHWP using systematic impact matrix analysis was used. Impact matrix analysis served to capture the complexity of the individual farm, assess the relationship between system variables as well as identify farm-specific drivers and focus areas (Krieger et al., 2017).

8 Improving AHWP: recognising and addressing power dynamics in decision-making and AHWP

8.1 Top-down versus bottom-up approaches to change

Taking a bottom-up approach to changing behaviours and practices on farms involves working from the ground up with the people who know the farm situation intimately. It starts with recognising that farmers have a wealth of experience and knowledge that is relevant and valuable for understanding and improving on-farm practices (Sumane et al., 2018). To quote a well-known anonymous saying: 'No one knows a farm better than a farmer'. It is this knowledge and experience which should drive and lead any change. From a global perspective, a bottom-up approach to food production (that recognises and prioritises local community knowledge, needs and livelihoods) not only leads to enhanced environmental, social and economic outcomes, but is necessary to avoid unintended consequences and environmental/social degradation (Sandhu, 2021). Bottom-up approaches are circular and iterative, involve many feedback loops, and include processes such as co-design and co-development (van Dijk et al., 2019). They aim to share the power involved in decision-making and pass control over what happens to those most affected or impacted by the change (Cornwall, 2008).

In contrast, top-down approaches to farming are very much led by either scientists, policymakers or advisors whom pass on their knowledge and expertise to 'end users' to influence change. It is often a very linear and one-way flow of knowledge with little opportunity for feedback from 'end users' or for collaborative processes such as co-design or co-innovation (Ingram et al., 2020). Top-down processes include much policy-making and designing of legislation, many advisor-led interventions around disease control, and has been one of the most common approaches to knowledge transfer in many countries around the world (Jones and Garforth, 1997). A top-down ethos does not prioritise the lived experiences of those it is aimed at changing, but rather the knowledge of perceived or traditionally-recognised 'experts' (Ivemeyer et al., 2015). Nevertheless, top-down approaches can be effective and often necessary to achieve wide-scale change; it is likely a combination of approaches that is needed when trying to shift complex behaviours.

8.2 Participatory approaches

Participatory approaches are common when following a bottom-up ethos as they are focused on democratising the decision-making process with all actors viewed as equal participants with varied skills, knowledge and backgrounds that come together to collaborate, learn from each other and solve shared issues (Cornwall, 2008). Participatory approaches often involve multiple actors working collaboratively but they can also encompass the farmer-advisor relationship as long as the relationship aims to share decision-making responsibilities and powers with all those involved (see e.g. https://ec.europa.eu/eip/agriculture/sites/default/files/eip-agri_brochure_participatory_approaches_2015_en_web.pdf). Figure 1 shows differing levels of participation by citizens and where they are on a continuum, from very little participation and therefore agency indecision-makingg (the bottom of the ladder) to full engagement in decision-making and full participation in the process (top of the ladder).

8.3 Participatory herd health planning

Participatory herd health planning involves co-developing health plans with those that are to implement them (i.e. farmers and farm workers) and including various sources of knowledge and expertise in a more bottom-up, equitable way. This shifts the focus of health planning from being top-down and advisor-led to an approach that is much more collaborative in style, structure and content. Those who are to implement any health plan have valid and insightful knowledge to share. Several studies have shown the importance of farmers' attitudes towards their animals as well as their goals and motivations in relation to productivity, health and welfare (e.g. Breuer et al., 2000; Waiblinger et al., 2002) in ensuring the success of health interventions (Ivemeyer et al., 2008). Taking a participatory approach can contribute to participants feeling more empowered and more likely to buy in to any agreed changes as well as sustain them in the long term.

Participatory herd health planning tools have been tested extensively. A recent example is the EU Horizon Disseminating Innovative Solutions for Antibiotic Resistance Management (DISARM) Project which was an international collaboration between farmers, veterinarians and other advisors, academics and industry to co-develop and implement innovative solutions for antibiotic resistance management (https://disarmproject.eu/what-we-do/farm-health-teams/). A key outcome was a practical tool kit for farmers and their advisors to use to establish a multi-actor farm health team, where diverse skills and knowledge come together to optimise the AHWP process.

In another example, a participatory approach was tested within the EU IMPRO (Impact Matrix Analysis and Cost-Benefit Calculations) Project to create an environment in which farmers could, in discussion with their herd health advisor, adapt indicators proposed by scientists for monitoring the five main production diseases on dairy cattle farms (Duval et al., 2016a). The study was conducted in France and Sweden, which have specific conditions in terms of national organic regulations and existing advisory services. In both countries, 20 certified organic dairy farmers and their advisors participated in the study. All of the farmers adapted the initial monitoring plan proposed by scientists to the specific production and animal health situations on their farms. This led to the co-construction of forty unique and farm-specific combinations of indicators for herd health monitoring. Advisors can benefit from this participatory approach because it transforms monitoring tools provided by scientists into farm-specific tools, it allows farmers ownership of the co-design process and favours dialogue between farmers and advisors (Duval et al., 2016a).

The co-designed tools were later used in a pilot study to test a herd health and production management (HHPM) programme. Farmers and their animal health advisors met 3 to 4 times in a period of 12 months. During each visit, they were expected to monitor herd health using the co-designed monitoring tool, to identify and implement necessary disease preventive practices and to evaluate their effectiveness. Evaluation of the programme showed that compliance with monitoring activities was relatively high. The monitoring tool also successfully triggered health alerts to identify disease issues, suggesting farmers did choose pertinent indicators and threshold levels. Moreover, the HHPM programme served broader purposes, since the meetings were considered, by both farmers and advisors, as an opportunity to discuss the animal health situation on the farm and assess animal health management practices. In addition, both French and Swedish advisors acknowledged that they discussed topics which they had not discussed during their usual collaboration with farmers. Both French and Swedish advisors reported to have learned more about farmers' objectives and their animal health management practices and, in the case of Swedish advisors, more about certain herd health problems (Duval et al., 2017a).

9 Improving AHWP: harnessing peer-to-peer learning

As pointed out previously, central to animal health and welfare planning is a farmer-owned process of continuous development and improvement.

Some ways of delivering such a process and based on the principles highlighted in the ANIPLAN process (see section 7) are discussed below.

9.1 Stable schools

One well-established approach that implements the mentioned principles is the 'farmer field school' concept for farmer learning, knowledge exchange and empowerment, which has been pioneered and used in primarily developing countries (Sones, 2003). The approach has more recently been modified by developing 'Stable schools' to support Danish farmers in achieving specific health and welfare goals (especially avoiding the use of antibiotics; Vaarst et al., 2007). Farmers take responsibility for health and welfare planning by identifying issues, setting goals and acting to improve the current situation facing the herd based on farm-specific data (e.g. animal-based welfare indicators or existing data like milk recording data). Stable schools are led by an external person taking on the role of facilitator, providing and pre-processing available farm data but not giving specific advice apart from when requested by participating farmers (Vaarst et al., 2011). Stable schools incorporate health promotion and disease management, based on a strategy of risk assessment which forms the basis for evaluation, action and review (Vaarst et al., 2010).

Stable schools comprise a group of about 5 to 10 farms (one or more participants per farm) which meet as a group on one another's farm in turn. The first 30 min are used for a farm walkabout, where the host farmer introduces his/her farm and herd and share one 'success case' in their herd management. Afterwards, two problem areas are discussed. The host farmer can explain their problem area and raise questions without being interrupted by the others. Then all other farmers can report their experiences and offer their suggestions

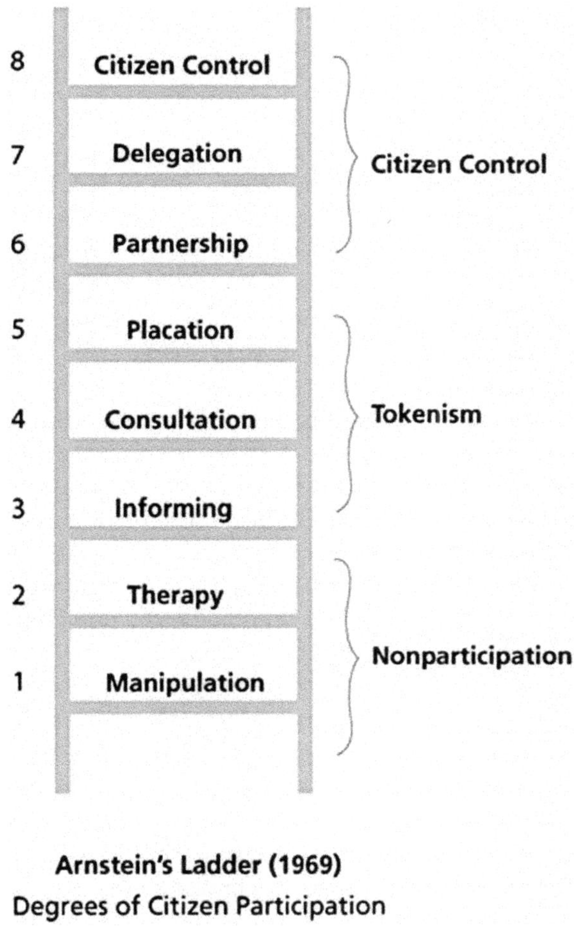

Figure 1 Ladder of Participation (Source: Arnstein, 1969).

to solve the problem. The host farmer chooses both the success case and the particular problem area with which he or she wants to work. A facilitator is part of the process by invitation, including keeping the agenda, directing the discussion where needed and writing the minutes (Vaarst et al., 2007).

Initiatives such as the EU ANIPLAN project (https://www.coreorganic.org/core1/research/projects/aniplan/index.html) have used stable schools successfully as an intervention tool and have shown improvements on the majority of the participating farms (e.g. Ivemeyer et al., 2012). Farmers and facilitators have been convinced by the approach and benefits for dairy herds. The degree of farmers' attention towards their herds and their ownership of the process appear to be crucial success factors for herd health and welfare improvement. In some European countries, this method has been implemented in advisory practice (reviewed by Ivemeyer et al., 2015).

9.2 Farmer Action Groups

Farmer Action Groups (FAGs) are modelled on the Stable School approach of common experiential learning and goal-orientated action using a peer-to-peer learning framework. The use of FAGs was studied as part of a participatory action research PhD project in the UK between 2016 and 2018 aimed at reducing reliance on the use of antibiotics in dairy farming (Morgans et al., 2021). The study identified the importance of mobilisation of different forms of knowledge from the participatory nature of the groups and the facilitation style adopted by the project facilitators. The participants also reported that confidence in their knowledge and understanding of antimicrobial stewardship increased. The groups fostered a sense of solidarity as the participants were equals going through the same process as a peer group, which helped them tackle often risky changes to their antibiotic use practices. Many farms reduced their usage dramatically in just 2 short years and, through a participatory action planning process, more than 80% of participant farms implemented more than a third of their Action Plan within a year (Morgans et al., 2021).

Based on these examples, AHWP with organic dairy farmers could and should adopt this kind of peer support model to inspire and galvanise farmers into action on AHW issues. This could be in addition to trusted advisory support and could also help farmers find out how they could implement changes in a practical way by learning from those that have similar challenges. The action planning process could also be a tool to build into health planning to allow the farmer to have the final say about what they implement as well as adapting and improving recommendations from their peers or veterinarian advisors.

9.3 Facilitation

Key to the implementation of Farmer Field Schools, Stable Schools and Farmer Action Groups were facilitators. Facilitation is a communication skill, a process and a profession used and valued by many industries and sectors worldwide. Facilitators aim to catalyse, coordinate and coach a group to achieve its goals. The word facilitation means to make something easier. There are several definitions of facilitation, depending on the area it is applied to but, in summary, it is a creative process where people are supported to work cooperatively and effectively together and involve participants equally and in a meaningful way (Hogan, 2003). In the context of AHWP, facilitation can offer another beneficial way to improve collaborative working between the farmer and their advisor. In successful peer-to-peer learning approaches, the facilitator is critical in catalysing a group and enabling them to work more effectively than if a facilitator was not present (van Dijk et al., 2017). An example of the use of facilitation is the EU Hennovation Project focussed on improving laying hen welfare (https://hennovation.eu/onewebmedia/Hennovation%20Final%20Report%20vs3.pdf).

10 Improving AHWP: farmer-veterinarian communication

The importance of communication skills in AHWP has received increasing attention in recent years (Bard et al., 2017; Ritter et al., 2018, 2019; Svensson et al., 2019; DeGroot et al., 2021). Communication skills are central tools that directly influence the farmer-veterinarian relationship, the success of the consultation and the ability of that relationship to support implementation of actions promoting animal health and welfare. In human and small animal medicine, where communication research has a longer tradition, attention and respect for clients' experiences and options, empathy and shared decision-making are considered key attributes of effective clinical communication (Silverman et al., 2013; Adams and Kurtz, 2017). Recent findings indicate the benefits of such client-centred communication, as applied in AHWP on dairy farms. Farmers reported higher satisfaction with herd health visits where the veterinarian provided less information and was more attentive to farmer viewpoints (lower ratio veterinary to farmer talk; Ritter et al., 2019) or where farmers' scores of the veterinarian's producer-centeredness were higher (DeGroot et al., 2023). Higher farmer satisfaction as well as a lower rating of the veterinarian's dominance during visits was associated with higher preparedness to adopt recommendations (Ritter et al., 2019).

However, in AHWP, the context can be more complex than in human or small animal medicine, requiring a more complex and longer-term interaction (Kleen et al., 2011). Identifying the farmer's goals and mindset is a central part of such an approach according to Lam et al. (2011), who have suggested this is best done using open-ended questions and awareness of farmers' perceptions about underlying barriers to performing a certain behaviour (see Table 1). Bard et al. (2019) identified that, although the information contained in the veterinarian's advice was valued, it was the trusting relationship, the shared understanding and meaningful interpretation of the advice, resulting from good veterinarian communication, that most likely influenced farmer behaviour change. Their findings suggest there are benefits from a collaborative communication approach to investigate and build a shared understanding of the farmer's immediate and long-term priorities, motivations and goals. Such a communication style aligns well with a bottom-up approach.

Motivational interviewing (MI) is a client-oriented and collaborative communication methodology that aims to facilitate a person's internal motivation to change (Miller and Rollnick, 2012). It was originally developed in alcohol abuse counselling, has been largely adopted in medical and psychological science and is increasingly being implemented across a number of other sectors. There is strong research evidence to support its efficacy in facilitating different lifestyle changes (Lundahl et al., 2010). Recent evidence has supported its usefulness and efficacy in veterinary AHWP (Svensson et al., 2020a, 2020b; Bard et al., 2022; Svensson et al., 2022).

MI methodology combines:

- relational skills that aim to build a working relationship between the advisor and the client
- technical skills that serve to elicit the client's expressions of own motives for change and to strengthen these and support plans for taking action

Investigating the client's goals and motivations and emphasizing the client's autonomy are central to MI, meaning that the responsibility for the action or change is left with the client (Miller and Rollnick, 2021).

To use persuasion or give advice without emphasizing the person's autonomy has been shown to be negatively associated with behaviour change (Apodaca and Longabaugh, 2009). MI methodology instead aims to provide information in dialogue with the client by awaiting for the client to request information, asking for permission to give the information, or emphasizing the client's autonomy in deciding whether to take advice (Miller and Rollnick, 2012).

Clients' statements expressing consideration of, motivation for, or commitment to a behaviour change – Change Talk – has been demonstrated to be strongly correlated with clients later adopting that behaviour change (Apodaca and Longabaugh 2009). In MI research, the amount of Change Talk expressed by the client has therefore been used as an indirect measure of implementation. Using this approach, it was found that veterinarians achieving moderate MI skills achieved 1.5 times more Change Talk with clients compared to clients who engaged in consultations with non-trained veterinarians (Svensson et al., 2020a). Such skills can be delivered through appropriate training, e.g. 6 one-day workshops in a stretched-out design over a 6-7-month period including coaching and specific feedback (Svensson et al., 2020b). The influence of veterinary communication on farmer Change Talk was also demonstrated after brief MI training (Bard et al., 2022). MI was rated as highly relevant to their profession by cattle veterinarians (Svensson et al., 2020b).

MI has the potential to increase clarity around farmer goals and priorities, improve the farmer-veterinarian relationship and level of farmer participation in AHWP, help veterinarians customize their advice to the farm and support effective implementation of preventive measures by the farmer and farm staff. Research also suggests that learning to master MI may be one means to increase veterinarians' engagement in AHWP (Svensson et al 2022).

11 Improving AHWP: on-farm self-assessment of animal-based welfare indicators

Since 2014 the Animal Welfare Act (§ 11 Section 8 Animal Welfare Act) in Germany has required livestock keepers to carry out an on-farm self-assessment mainly using animal-based welfare indicators, in addition to daily routine inspections. On-farm self-assessments aim to sharpen the perceptions of stockpersons and to overcome a potential 'blindness' towards aspects of their own herd and farm situation (Zapf et al., 2017). It facilitates early identification of specific animal welfare risks and thus helps to continually improve management, providing the on-farm self-assessment is conducted and documented regularly (e.g. once or twice a year). Additionally, on-farm self-assessment offers the opportunity of giving greater objectivity to broader discussions in society about farm animal welfare (Zapf et al., 2017). During the last decade, several tools and guidelines for on-farm self-assessment by farmers (or advisors) have been developed, such as Tierwohl Check (see https://tierwohl-check-sh.de/).

Tools include animal-based indicators derived from existing data like milk recording data, i.e. different indicators based on somatic cell count, metabolic risk indicators based on fat-protein ratio, as well as indicators based on longevity, mortality (https://q-check.org/). The following guidelines and tools have been provided:

- 1. a range of defined animal-based indicators to assess on-farm health and behaviour indicators (e.g. lameness, dirtiness, integument/skin alterations, body condition, rising behaviour, avoidance distance; Brinkmann et al., 2020; Zapf et al., 2015)
- 2. online training courses to ensure indicators are used properly and reliably
- 3. reference values, including target and alarm values, per indicator to more accurately evaluate the current welfare situation of the herd (e.g. Cimer et al., 2021).

Michaelis et al. (2022) showed that online and live training are feasible methods to enable farmers to self-assess animal welfare in a reliable way. In this study, 73 livestock farmers (keeping cattle, pigs or poultry) trained online and 69 farmers trained in person, and then rated the impact of the training on a scale from 0 (not true) to 100 (true). They gave a score of 75 (median) in improving their animal observation skills a score of 70 in encouragement to better assess indicators on the own farm, and a score of 77 in perception that self-assessment supports animal management (Michaelis et al., 2022).

Reference values to evaluate farm animal welfare can be based on normative target and alarm values or on the status quo in the sector as a whole using a benchmark system (Schultheiß et al., 2020). Benchmarking has been found to motivate and encourage farmers to make changes in their calf management (Sumner et al., 2018) and has been used within a health and welfare planning process that led to reduced medicine use (Morgans et al., 2021) without impairment of production and herd health, as demosntrated in 128 organic dairy farms in Europe (ANIPLAN project, Ivemeyer et al., 2012).

A challenge with the on-farm self-assessment method is that it is based on regulatory requirements and therefore contradicts the principle of farmer 'ownership'/self-motivation to improve the welfare status of the herd. Currently no one set of indicators or particular tool is required to fulfil regulatory requirements. However, it might be important for famers' own decision-making and motivation to use these kinds of instruments, as well as developing a herd-specific set of indicators and thresholds for monitoring animal health (Duval et al., 2016a).

12 Improving AHWP: the role of farmer associations and the wider supply chain

The role of farming associations, networks and the various accreditations that exist for organic farmers can contribute to positive attitudes and action towards optimising animal health. Being part of a professional peer group or network can galvanise farmers to act on specific health or husbandry challenges (e.g., being affiliated to a grazing dairy group). These group structures are not only sources of knowledge for farmers, above and beyond the local veterinarian or advisor, but provide wellbeing and social support, help with normalisation of new and innovative farm practices (Skaalsveen et al., 2020), and create or reinforce links between different farmers (Cofre-Bravo, 2019). In fact, many advisors contribute to and participate in these networks too. The benefits for farmers of being part of a network, especially if trialling new ideas or practices, are numerous for renormalising practices that may be at the vanguard of improving sustainability (Darnhofer, 2020). An example of such a network is Pasture for Life in the UK (https://www.pastureforlife.org/).

Food cooperatives and food supply chain companies have varying influence on herd health planning and animal health and welfare outcomes in organic dairy farming. This can range from being a source of information on food and hygiene standards and regulations (Casali et al., 2020) to offering a source of peer support since farmers are connected to one another by a common milk buyer (van Dijk et al., 2017). In some places, food supply chain companies carry a lot of power when it comes to influencing farmer behaviour around husbandry and herd health (Robinson et al., 2020). If a milk buyer/retailer requires a farmer to implement a change as part of their contract, there can be very little negotiation space available in order to keep a contract, which aligns more with a top-down ethos and little sharing of decision-making power, rather than a bottom-up and participatory approach where shared decision making occurs.

In the UK in 2021, the milk cooperative Arla introduced a ban on the elective euthanasia of male dairy calves under the age of 6 weeks (Compassion in World Farming, 2020). Most of the industry already agreed with banning this unpopular and unethical practice and the ban was initially a voluntary requirement before being made compulsory. Nevertheless, it demonstrates the influence a cooperative can have over on-farm practices relating to animal health and welfare. Various UK supermarket retailers regularly introduce rules for their producer groups on herd health matters, such as Johnes disease screening (Robinson et al., 2020) and medicine auditing (van Dijk et al., 2017; Begemann et al., 2020), often before it becomes an industry-wide requirement. Indeed, organic dairy farming and the relevant certification requirements are often ahead of the curve regarding the aforementioned examples (Barkema et al., 2015).

13 Future trends

Organic dairy herd health planning is an area of ongoing development. In recent years, there has been a focus on the farmer-advisor relationship and different communication strategies to improve this. The role of industry-led change and co-developing policy and animal health initiatives, using multi-actor approaches is gaining traction, especially in the context of antimicrobial stewardship and environmental sustainability. As more scholarly debate develops into the value of a farming systems perspective and the role of social learning and multi-actor networks as a way of empowering change within agriculture, the AHWP process will grow into much more than a regulatory requirement and one-way exchange of information, and into a continual and iterative learning process with many perspectives shared and many approaches used to foster action and change for the benefit of animal health and welfare.

Animal health is inextricably linked to human and planetary health, as highlighted at the start of this chapter. As we expand our understanding and application of One Health, One Welfare and Planetary Health concepts, the relevance and need to apply the multi-actor and bottom-up approaches discussed in this chapter becomes even more imperative. Nevertheless, the current global socioeconomic and environmental context poses fresh and rapidly evolving challenges within our food system as we face a poly-crisis, from climate and biodiversity to health and wellbeing. Organic dairy health and welfare planning has a unique role to play in ensuring dairy production is as environmentally sustainable as possible, farming in harmony with nature and minimising harmful inputs for cow, human and planetary health but, at the same time, producing safe and nutritious food to tackle inequalities within and beyond our food systems, not least for the farmers whose livelihoods depend on it.

14 Where to find further information

Principles of organic production for animal welfare as applied to the EU:

Animal welfare:

- Personnel keeping animals must possess the necessary basic knowledge and skills regarding the health and welfare needs of the animals.
- Particular attention should be paid to housing conditions, husbandry practices, respect for set stocking densities and minimum surfaces for indoor and outdoor areas.
- The number of livestock must be limited to minimise overgrazing, erosion, or pollution caused by animals or by the spreading of their manure.
- Animals should have, whenever possible, access to open air or grazing areas.
- Tethering or isolating livestock is prohibited aside from individual animals for a limited period of time and only for welfare, safety or veterinary reasons.
- Hormones or similar substances are not permitted, unless as a form of veterinary therapeutic treatment for an individual animal.
- When the animals are ill, allopathic veterinary medicinal products including antibiotics may be used where necessary and under strict conditions. This is only allowed when the use of phytotherapeutic, homeopathic and other products is inappropriate.
- The use of immunological veterinary medicines is permitted.

Taken from: https://agriculture.ec.europa.eu/farming/organic-farming/organic-production-and-products_en#:~:text=EU%20regulations%20on%20organic%20production,%2C%20yeast%2C%20aquaculture%20and%20more.

European case study projects related to animal health, disease and antibiotic reduction

DISARM (https://disarmproject.eu/):

A thematic network that aimed to develop, collate and spread practical and scientific knowledge and skills to reduce reliance on antibiotics across European livestock production. The development of a guide to establish and facilitate a Multi Actor Farm Health Team was a key outcome of this collaboration.

ROADMAP (https://www.roadmap-h2020.eu/):

The overall aim of ROADMAP is to foster transitions towards prudent antimicrobial use (AMU) in animal production in a large variety of contexts. This change will be achieved through an improvement of antimicrobial decision systems all along the food and drug supply chains.

Avant (https://avant-project.eu/):

The strategic objective of AVANT is to significantly reduce antimicrobial consumption by demonstrating three field interventions that target the animal species and disease condition accounting for the majority of antimicrobials used in livestock. AVANT is a multi-actor inter-sectorial project aimed at developing alternatives to antimicrobials for the management of bacterial infections in pigs, especially diarrhoea during the weaning period, as the major indication for antimicrobial use in livestock in Europe. During pre-clinical studies, the efficacy, toxicity, and mode of action of these interventions are tested, and their dosage and formulation are optimised. The results and a survey for veterinarian-, farmer-in and consumers perceptions of antimicrobial alternatives, will be used together with legal and economic considerations to select three interventions for large-scale farm trials, assessing clinical efficacy and impact on antimicrobial use.

Healthy Livestock was an EU-China collaborative research programme to study the contributions of enhanced animal health and welfare on reducing the need to use antimicrobials in pigs and poultry.

ANIPLAN (https://www.coreorganic.org/core1/research/projects/aniplan/index.html):

The aim of this project was to minimise the use of antibiotics in organic dairy herds through animal health and welfare planning. The objectives were to:

1) Develop animal health and welfare planning principles for organic dairy farms under diverse conditions based on an evaluation of current experiences.
2) Application of animal health and welfare assessment based on the Welfare Quality® parameters in different types of organic dairy herds across Europe. This will result in an overview of the herds and allow for potential adaptations for the organic situation (e.g., pasture systems, longer cow/calf contact). For calves, a special system will be developed by the Norwegian partners, and combined and tested together with the Welfare Quality® assessment system.
3) Develop guidelines for communication about animal health and welfare promotion in different settings. This can be part of existing animal health advisory services or farmer groups such as the Danish Stable School system and the Dutch network program.

Examples of multi-actor and stakeholder approaches from Europe

Case study: Liaison https://liaison2020.eu/optimizing-group-work/.

Case study: DISARM's Multi-Actor Farm Health Teams https://disarmproject.eu/what-we-do/farm-health-teams/.

Examples of farmer-led innovation and research

Case study: Innovative farmers field labs https://www.innovativefarmers.org/.

15 References

Adams, C. L. and Kurtz, S. (2017). *Skills for Communicating in Veterinary Medicine*. 1st ed. MHRI Publishing, Parsippany, NJ.

Arcari, P. (2017). Normalised, human-centric discourses of meat and animals in climate change, sustainability and food security literature. *Agriculture and Human Values* 34, 69–86. https://doi.org/10.1007/s10460-016-9697-0.

Arnstein, S. R. (1969). A ladder of citizen participation. *Journal of the American Institute of Planners* 35(4), 216–224. https://doi.org/10.1080/01944366908977225.

Aplin, K., Morgans, L., Palczynski, L., Main, D., Debbaut. C., Hepworth, L. and Reed, J. (2023). Calf health veterinary services: Making them work for calves, farmers and veterinarians. *Veterinary Record*. https://doi.org/10.1002/vetr.3051.

Alonso, M., González-Montaña, J. and Lomillos, J. (2020). Consumers' concerns and perceptions of farm animal welfare. *Animals* 10(3). https://doi.org/10.3390/ani10030385.

Apodaca, T. R. and Longabaugh, O. R. (2009). Mechanism of change in motivational interviewing: A review and preliminary evaluation of the evidence. *Addiction* 104, 705–715.

Bard, A. M., Main, D. C. J., Haase, A. M., Whay, H. R., Roe, E. J. and Reyher, K. K. (2017). The future of veterinary communication: Partnership or persuasion? A qualitative investigation of veterinary communication in the pursuit of client behaviour change. *PLoS One* 12(3). https://doi.org/10.1371/journal.pone.0171380.

Bard, A. M., Main, D., Roe, E., Haase, A., Whay, H. R. and Reyher, K. K. (2019). To change or not to change? Factors influencing farmers' enactment of veterinary advice on UK dairy farms. *Journal of Dairy Science* 102(11), 10379–10394. https://doi.org/10.3168/jds.2019-16364.

Bard, A. M., Main, D. C. J., Haase, A. M., Whay, H. R. and Reyher, K. K. (2022). Veterinary communication can influence farmer change talk and can be modified following brief motivational interviewing training. *PLoS One* 17(9). https://doi.org/10.1371/journal.pone.0265586.

Begemann, S., Watkins, F., Van, H. I., Vivancos, R., Christley, R. and Perkins, E. (2020). The governance of UK dairy antibiotic use: Industry-Led policy in action. *Frontiers in Veterinary Science Sec Veterinary Humanities and Social Sciences* 7. https://doi.org/10.3389/fvets.2020.00557.

Bellet, C. (2018). Change it or perish? Drug resistance and the dynamics of livestock farm practices. *Journal of Rural Studies* 63, 57–64. ISSN 0743-0167. https://doi.org/10.1016/j.jrurstud.2018.08.016.

Bennedsgaard, T. W., IKlaas, C. and Vaarst, M. (2010). Reducing use of antimicrobials – Experiences from an intervention study in organic dairy herds in Denmark. *Livestock Science* 131(2–3), 183–192. ISSN 1871-1413. https://doi.org/10.1016/j.livsci.2010.03.018.

Beynon, S. A., Wainwright, W. A. and Christie, M. (2015). The application of an ecosystem services framework to estimate the economic value of dung beetles to the U.K. cattle industry. *Ecological Entomology* 40, 124–135. https://doi.org/10.1111/een.12240.

Blanco-Penedo, I., Sjöström, K., Jones, P., Krieger, M., Duval, J., van Soest, F., Sundrum, A. and Emanuelson U. (2019). Structural characteristics of organic dairy farms in four European countries and their association with the implementation of animal health plans. *Agricultural Systems* 173, 244–253. https://doi.org/10.1016/j.agsy.2019.03.008.

Bloland, R. (2023). Amendments to the Standards, Edition: March 13, 2023. https://www.bioland.de/fileadmin/user_upload/Verband/Dokumente/Richtlinien_fuer_Erzeuger_und_Hersteller/Bioland-Richtlinien_englisch_2023-03_01.pdf.

Bruges, M. and Smith, W. (2008). Participatory approaches for sustainable agriculture: a contradiction in terms? *Agriculture and Human Values* 25, 13–23. https://doi.org/10.1007/s10460-007-9058-0.

Bracke, M. B. M., Spruijt, B. M. and Metz, J. H. M. (1999). Overall animal welfare assessment reviewed. Part 1: Is it possible? *Netherlands Journal of Agricultural Science* 47, 279–291.

Brand, A., Noordhuizen, J. P. T. M. and Schukken, Y. H. (2001). *Herd Health and Production Management in Dairy Practice.* 3rd ed. Wageningen Pers, Wageningen.

Brinkmann, J., Cimer, K., March, S., Ivemeyer, S., Pelzer, A., Schultheiß, U., Zapf. R. and Winckler, C. (2020). *Tierschutzindikatoren: Leitfaden für die Praxis - Rind: Vorschläge für die Produktionsrichtungen Milchkuh, Aufzuchtkalb, Mastrind [Animal welfare indicators – cattle: guidelines for dairy cows, rearing calves, beef cattle].* KTBL, Darmstadt.

Brock, C., Pempek, J. A., Jackson-Smith, D., Habing, G. G., da Costa, L. and Weaver, K. (2022). Managing organic dairy herd health: Current roles and possible future roles for veterinarians with organic dairy clientele. *Journal of Dairy Science* 105, 8328–8341. https://doi.org/10.3168/jds.2021-21720.

Bulut, E., Stout, A., Wemette, M., Llanos-Soto, S., Schell, R. C., Greiner, S. A., Shapiro, M. A., Moroni, P. and Ivanek, R. (2021). How does public perception of antibiotic use on dairy farms contribute to self-reported purchasing of organic? *Journal of Food Science* 86(5), 2045–2060. https://doi.org/10.1111/1750-3841.15720.

Casali, M., Mendonça, B., Brito, M., Santos, M., Lima, P., Siqueira, T., Damasceno, J. and Bánkuti, F. (2020). Information asymmetry among dairy producers in Paraná, Brazil. *Semina: Ciências Agrárias*. https://doi.org/10.5433/1679-0359.2020v41n1p295.

Chambers, R. (1994). The origins and practice of participatory rural appraisal. *World Development* 22(7), 953–969. https://doi.org/10.1016/0305-750X(94)90141-4.

Cheng, L., Zhang, X., Reis, S., Ren, C., Xu, J. and Gu, B. (2022). A 12% switch from monogastric to ruminant livestock production can reduce emissions and boost crop production for 525 million people. *Nature Food* 3, 1040–1051. https://doi.org/10.1038/s43016-022-00661-1.

Cimer, K., Ivemeyer, S., March, S. and Brinkmann, J. (2021). On-farm self-assessment of animal welfare. Project brief. https://literatur.thuenen.de/digbib_extern/dn064338.pdf.

Clifford, K., Desai, D., Prazeres da Costa, C., Meyer, H., Klohe, K., Winkler, A. S., Rahman, T., Islam, T. and Zaman, M. H. (2018). Antimicrobial resistance in livestock and poor quality veterinary medicines. *Bulletin of the World Health Organization* 96(9), 662–664. https://doi.org/10.2471/BLT.18.209585.

Cofre-Bravo, G., Klerkx, L. and Engler, A. (2019). Combinations of bonding, bridging, and linking social capital for farm innovation: how farmers configure different support networks. *Journal of Rural Studies*. https://doi.org/10.1016/J.JRURSTUD.2019.04.004. https://www.ciwf.org.uk/news/2020/12/progress-for-male-dairy-calves#:~:text=A%20cross%2Dindustry%20coalition%20is,helping%20farmers%20to%20achieve%20this. Accessed June 2023.

Cornwall, A. (2008, July). Unpacking 'participation': models, meanings and practices. *Community Development Journal* 43(3), 269–283. https://doi.org/10.1093/cdj/bsn010.

Curry, N. and Winter, M. European Briefing. (2000). The transition to environmental agriculture in Europe: Learning processes and knowledge networks. *European Planning Studies* 8(1), 107–121. https://doi.org/10.1080/096543100110956.

Darnhofer, I. (2020). Farming from a process-relational perspective: making openings for change visible. *Sociologia Ruralis* 60(2). https://doi.org/10.1111/soru.12294.

DeGroot, A., Coe, J., Kelton, D., Miltenburg, C., Wichtel, J. and Duffield, T. (2021). Factors associated with food-animal producer visit-specific satisfaction following on-farm interaction with a veterinarian. *Veterinary Record* 188(6), e15. https://doi.org/10.1002/vetr.15.

Demeter International. (2022). Production, Processing and Labelling International Standard for the use and certification of Demeter, Biodynamic and related trademarks (Oct 2022). https://demeter.net/wp-content/uploads/2022/10/20220929_BFDI_Standard_englVersion_final_fs.pdf.

Dumont, B., Fortun-Lamothe, L., Jouven, M., Thomas, M. and Tichit M. (2013). Prospects from agroecology and industrial ecology for animal production in the 21st century. *Animal* 7(6), 1028–1043. https://doi.org/10.1017/S1751731112002418.

Dumont, B. (ed.), Dupraz, P. (ed.), Aubin, J., Benoit, M., Bouamra-Mechemache, Z., Chatellier, V., Delaby, L., Delfosse, C., Dourmad, J. Y., Duru, M., Frappier, L., Friant-Perrot, M., Gaigné, C., Girard, A., Guichet, J. L., Havlik, P., Hostiou, N., Huguenin-Elie, O., Klumpp, K., Langlais, A., Lemauviel-Lavenant, S., Le Perchec, S., Lepiller, O., Méda, B., Ryschawy, J., Sabatier, R., Veissier, I., Verrier, E., Vollet, D., Savini, I., Hercule, J. and Donnars, C. (2016). *Rôles, impacts et services issus des élevages en Europe. Synopsis of the collective scientific assessment*. INRAE, France.

Doherty, B., Sidhu, Y., Heron, T., West, C. D., Seaton, A., Gulec, J., Prado, P. and Flores Martínez, P. (2020). Citizen participation in food systems policy making: a case study of a citizens' assembly. *Emerald Open Research* 2, 22. https://doi.org/10.35241/emeraldopenres.13609.1.

Enticott, G., Donaldson, A., Lowe, P., Power, M., Proctor, A. and Wilkinson, K. (2011). The changing role of veterinary expertise in the food chain. *Philosophical Transactions of the Royal Society of London. Series B, Biological Sciences* 366(1573), 1995–1965. https://doi.org/10.1098/rstb.2010.0408. PMID: 21624916; PMCID: PMC3130392.

Derks, M., van Werven, T., Hogeveen, H. and Kremer, W. D. J. (2013). Veterinary herd health management programs on dairy farms in the Netherlands: Use, execution, and relations to farmer characteristics. *Journal of Dairy Science* 96(3), 1623–1637. https://doi.org/10.3168/jds.2012-6106.

De Schutter, O. (2010). *Agroecology and the right to food*. Report presented at the 16th Session of the United Nations Human Rights Council.

Döring, T.F., Vieweger, A., Pautasso, M., Vaarst, M., Finckh, M.R. and Wolfe, M.S. (2015). Resilience as a universal criterion of health. *Journal of the Science of Food and Agriculture* 95, 455–465. https://doi.org/10.1002/jsfa.6539.

Duval, J. E., Bareille, N., Madouasse, A., De Joybert, M., Sjöström, K., Emanuelson, U., Bonnet-Beaugrand, F. and Fourichon, C. (2017a). Evaluation of the impact of a herd health and production management program in organic dairy cattle farms: a process evaluation approach. *Animal* 12(7), 1475–1483. https://doi.org/10.1017/S1751731117002841.

Duval, J. E., Bareille, N., Fourichon, C., Madouasse, A. and Vaarst, M. (2017b). How can veterinarians be interesting partners for organic dairy farmers? French farmers' point of views. *Preventive Veterinary Medicine* 146, 16–26. https://doi.org/10.1016/j.prevetmed.2017.07.013.

Duval, J. E., Fourichon, C., Madouasse, A., Sjöström, K., Emanuelson, U. and Bareille, N. (2016a). A participatory approach to design monitoring indicators of production diseases in organic dairy farms. *Preventive Veterinary Medicine* 128, 12–22. https://doi.org/10.1016/j.prevetmed.2016.04.001.

Duval, J. E., Bareille, N., Fourichon, C., Madouasse, A. and Vaarst, M. (2016b). Perceptions of French private veterinary practitioners on their role in organic dairy farms and opportunities to improve their advisory services for organic dairy farmers. *Preventive Veterinary Medicine* 133, 10–21. https://doi.org/10.1016/j.prevetmed.2016.09.008.

Emanuelson, U., Sjöström, K. and Fall, N. (2018). Biosecurity and animal disease management in organic and conventional Swedish dairy herds: A questionnaire study. *Acta Veterinaria Scandinavica* 60(1), 23. https://doi.org/10.1186/s13028-018-0376-6.

Enticott, G. (2012). Regulating animal health, gender and quality control: A study of veterinary surgeons in Great Britain. *Journal of Rural Studies* 28(4), 559–567. https://doi.org/10.1016/j.jrurstud.2012.05.004.

Enticott, G. (2023). 'Good farmers' and 'real vets': social identities, behaviour change and the future of bovine tuberculosis eradication. *Irish Veterinary Journal* 76(S1), 17. https://doi.org/10.1186/s13620-023-00245-w.

European Council. (2018). REGULATION (EU) 2018/848 of the European Parliament and the council on Organic production and labelling of organic products and repealing Council Regulation (EC) No 834/2007.

FAO. (2017). *Report of the "Regional Symposium on Agroecology for Sustainable Agriculture and Food Systems in Europe and Central Asia".* 23–25 November 2016, Budapest, Hungary. ISBN 978-92-5-109851-6.

Food and AgriculturEOrganisation of the United Nations (FAO). (2020). The 10 elements of agroecolog - Guiding the transition to sustainable food and agricultural systems. https://openknowledge.fao.org/server/api/core/bitstreams/3d7778b3-8fba-4a32-8d13-f21dd5ef31cf/content

Fraser, D. (2008). Understanding animal welfare. *Acta Veterinaria Scandinavica* 50, S1.

Glanville, C., Abraham, C. and Coleman, G. (2020). Human behaviour change interventions in animal care and interactive settings: A review and framework for design and evaluation. *Animals* 10(12), 2333. https://doi.org/10.3390/ani10122333.

Gliessman, S. (2007). *Agroecology: The Ecology of Sustainable Food Systems*. 2nd ed. Taylor and Francis Group, Boca Raton, FL.

Graham, D. W., Bergeron, G., Bourassa, M. W., Dickson, J., Gomes, F., Howe, A., Kahn, L. H., Morley, P. S., Scott, H. M., Simjee, S., Singer, R. S., Smith, T. C., Storrs, C. and Wittum, T. E. (2019). Complexities in understanding antimicrobial resistance across domesticated animal, human, and environmental systems. *Annals of the New York Academy of Sciences* 1441(1), 17–30. https://doi.org/10.1111/nyas.14036.

Green, M., Green, L., Huxley, J., Statham, J. and Statham, S. (2012). Concepts in dairy herd health. In M. Green (eds.), *Dairy Herd Health,* 1–9. CABI, Wallingford, UK.

Hall, J. and Wapenaar, W. (2012). Opinions and practices of veterinarians and dairy farmers towards herd health management in the UK. *Veterinary Record* 170(17), 441https://doi.org/10.1136/vr.100318.

Grodkowski, G., Gołębiewski, M., Slósarz, J., Grodkowska, K., Kostusiak, P., Sakowski, T. and Puppel, K. (2023). Organic milk production and dairy farming constraints and prospects under the laws of the European Union. *Animals* 13(9), 1457. https://doi.org/10.3390/ani13091457.

Hansen, B. G. and Østerås, O. (2019). Farmer welfare and animal welfare: exploring the relationship between farmer's occupational well-being and stress, farm expansion and animal welfare. *Preventive Veterinary Medicine* 170, 104741. https://doi.org/10.1016/j.prevetmed.2019.104741.

Hartcher, K. M., and Lum, H. K. (2019). Genetic selection of broilers and welfare consequences: A review. *World's Poultry Science Journal* 76(1), 154–167. https://doi.org/10.1080/00439339.2019.1680025. https://healthforanimals.org/wp-content/uploads/2021/06/achieving-the-sustainable-development-goals-the-value-of-healthier-animal.pdf. Accessed June 2024.

Hellec, F., Manoli, C. and Joybert, M. D. (2021). Alternative medicines on the farm: a study of dairy farmers' experiences in France. *Frontiers in Veterinary Science* 8, 563957. https://doi.org/10.3389/fvets.2021.563957.

Herzing, M., Wickström, H., Jacobsson, A., Källmen, H. and Forsberg, L. (2023). Enhancing compliance with waste sorting regulations through inspections and motivational interviewing. *Waste Management and Research* 41(8), 1342–1348. https://doi.org/10.1177/0734242X231154145.

HLPE. (2019). *Agroecological and Other Innovative Approaches for Sustainable Agriculture and Food Systems That Enhance Food Security and Nutrition*. A report by the High Level Panel of Experts on Food Security and Nutrition of the Committee on World Food Security, Rome, Italy.

Hoelzer, K., Bielke, L., Blake, D. P., Cox, E., Cutting, S. M., Devriendt, B., Erlacher-Vindel, E., Goossens, E., Karaca, K., Lemiere, S., Metzner, M., Raicek, M., Suriñach, M. C., Wong, N. M., Gay, C. and Immerseel F. V. (2018). Vaccines as alternatives to antibiotics for food producing animals. Part 1: challenges and needs. *Veterinary Research* 49, 64. https://doi.org/10.1186/s13567-018-0560-8.

Hogan, C. (2003). *Practical Facilitation: A Toolkit of Techniques*. Kogan Page. ISBN 0749438274.

Ingram, J., Gaskell, P., Mills, J. and Dwyer, J. (2020). How do we enact co-innovation with stakeholders in agricultural research projects? Managing the complex interplay between contextual and facilitation processes. *Journal of Rural Studies* 78, 65–77. https://doi.org/10.1016/j.jrurstud.2020.06.003.

Ivemeyer, S., Bell, N., Brinkmann, J., Cimer, K., Gratzer, E., Leeb, C., March, S., Mejdell, C., Roderick, S., Smolders, G., Walkenhorst, M., Winckler, C. and Vaarst, M. (2015). Farmers taking responsibility for herd health development: stable schools in research and advisory activities as a tool for dairy health and welfare planning in Europe. *Organic Agriculture*, 135–141. https://doi.org/10.1007/s13165-015-0101-y.

Ivemeyer, S., Smolders, G., Brinkmann, J., Gratzer, E., Hansen, B., Henriksen, B., Huber, J., Leeb, C., March, S., Mejdell, C., Nicholas, P., Roderick, S., Stöger, E., Vaarst, M., Whistance, L., Winckler, C. and Walkenhorst, M. (2012). Impact of animal health and welfare planning on medicine use, herd health and production in European organic dairy farms. *Livestock Science* 145, 63–72.

Jones, G. and Garforth, C. (1997). The history, development, and future of agricultural extension. In B. Swanson, R. Bentz and A. Sofranko (eds.), *Improving Agricultural Extension: A Reference Manual*. FAO, Rome.

Kleen, J. L., Atkinson, O. and Noordhuizen, J. P. T. M. (2011). Communication in production animal medicine: Modelling a complex interaction with the example of dairy herd health medicine. *Irish Veterinary Journal* 64, 8. https://doi.org/10.1186/2046-0481-64-8.

Klebl, F., Feindt, P. H. and Piorr, A. (2024). Farmers' behavioural determinants of on-farm biodiversity management in Europe: A systematic review. *Agriculture and Human Values* 41, 831–861. https://doi.org/10.1007/s10460-023-10505-8.

Klonek, F. E. and Kauffeld, S. (2012). Sustainability goes change talk: can motivational interviewing be used to increase pro-environmental behavior? In A. J. Spink, F. Grieco, O. E. Krips, et al. (eds.), *Proceedings of Measuring Behavior*. Utrecht, The Netherlands.

Knierim, U. (2002). Grundsätzliche ethologische Überlegungen zur Beurteilung der Tiergerechtheit bei Nutztieren [Basic ethological considerations concerning the evaluation of the welfare of farm animals]. *Dtsch Tierarztl Wochenschr* 109(6), 261–266.

Krieger, M., Sjöström, K., Blanco-Penedo, I., Madouasse, A., Duval, J. E., Bareille, N., Fourichon, C., Sundrum, A. and Emanuelson, U. (2017a). Prevalence of production disease related indicators in organic dairy herds in four European countries. *Livestock Science* 198, 104–108.

Krieger, M., Hoischen-Taubner, S., Emanuelson, U., Blanco-Penedo, I., de Joybert, M., Duval, J. E., Sjöström, K., Jones P. J. and Sundrum, A. (2017b). Capturing systemic interrelationships by an impact analysis to help reduce production diseases in dairy farms. *Agricultural Systems* 153, 43–52. https://doi.org/10.1016/j.agsy.2017.01.022.

Lam, T. J. G. M., Jansen, J., van den Borne, B. H. P., Renes, R. J. and Hogeveen, H. (2011). What veterinarians need to know about communication to optimise their role as advisors on udder health in dairy herds. *New Zealand Veterinary Journal* 59(1), 8–15. https://doi.org/10.1080/00480169.2011.547163.

Lam, T. J. G. M., Jansen, J. and Wessels, R. J. (2017). The RESET mindset model applied on decreasing antibiotic usage in dairy cattle in the Netherlands. *Irish Veterinary Journal* **70**, 5. https://doi.org/10.1186/s13620-017-0085-x.

Lundahl, B., Moleni, T., Burke, B. L., Butters, R., Tollefson, D., Butler, C. and Rollnick, S. (2013). Motivational interviewing in medical care settings: a systematic review and metaanalysis of randomized controlled trials. *Patient Education and Counseling* 93(2), 157–168.

Lundahl, B. W., Kunz, C., Brownell, C., Tollefson, D. and Burke, D. L. (2010). A meta-analysis of motivational interviewing: Twenty-five years of empirical studies. *Research on Social Work Practice* 20, 137–160. https://doi.org/10.1177/1049731509347850.

Mahon, N., Clark, B., Proctor, A. and Holloway, L. (2021). Exploring farmers' understanding of and responses to endemic animal health and welfare issues in the UK. *Veterinary Record*, 00e941. https://doi.org/10.1002/vetr.941.

Manyi-Loh, C., Mamphweli, S., Meyer, E. and Okoh, A. (2018). Antibiotic use in agriculture and its consequential resistance in environmental sources: Potential public health implications. *Molecules* 23(4), 795. https://doi.org/10.3390/molecules23040795.

Marley, C. K., Weller, R. F., Neale, M., Main, D. C. J., Roderick, S. and Keatinge, R. (2010). Aligning health and welfare principles and practices in organic dairy systems: a review. *Animal* 4, 259–71.

Michaelis, S., Schubbert, A., Gieseke, D., Cimer, K., Zapf, R., Lühken, S., March, S., Brinkmann, J., Schultheiß, U. and Knierim, U. (2022). A comparison of online and live training of livestock farmers for an on-farm self-assessment of animal welfare. *Frontiers in Animal Science,* 3, 915708. https://doi.org/10.3389/fanim.2022.915708.

Michie, S., van Stralen, M. M. and West, R. (2011). The behaviour change wheel: A new method for characterising and designing behaviour change interventions. *Implementation Science* 6, 42. https://doi.org/10.1186/1748-5908-6-42.

Midgely, G. (ed.). (2003). *Systems Thinking*. Sage Publications Ltd, London.

Migliorini, P. and Wezel, A. (2017). Converging and diverging principles and practices of organic agriculture regulations and agroecology: a review. *Agronomy for Sustainable Development* 37, 63. https://doi.org/10.1007/s13593-017-0472-4.

Miftari, I., Haas, R., Meixner, O., Imami, D. and Gjokaj, E. (2022). Factors influencing consumer attitudes towards organic food products in a transition economy–insights from Kosovo. *Sustainability* 14, 5873. https://doi.org/10.3390/su14105873.

Miller, W. R. and Rollnick, S. (2012). *Motivational Interviewing. Helping People Change*. 3rd ed. Guilford Press, New York. ISBN 9781609182274.

Morgans, L. C., Bolt, S., Bruno-McClung, E., van Dijk, L., Escobar, M. O., Buller, H. J., Main, D. C. J. and Reyher, K. A. (2021). A participatory, farmer-led approach to changing practices around antimicrobial use on UK farms. *Journal of Dairy Science* 104(2), 2212–2230. https://doi.org/10.3168/jds.2020-18874.

National Office for Animal Health. (2002). https://www.noah.co.uk/wp-content/uploads/2022/08/NOAH-Livestock-Vaccination-Guideline-August-2022.pdf. Accessed March 31st 2023.

Phillips, H. N. and Heins, B. J. (2022). Alternative practices in organic dairy production and effects on animal behavior, health, and welfare. *Animals* 12(14), 1785. https://doi.org/10.3390/ani12141785.

Poore, J. and Nemecek, T. (2018) Reducing food's environmental impacts through producers and consumers. *Science* 360(6392), 987–992. https://doi.org/10.1126/science.aaq0216. Erratum in: (2019). *Science*. 363(6429).

Pretty, J. N. (1995). Participatory learning for sustainable agriculture. *World Development* 23(8), 1247–1263. https://doi.org/10.1016/0305-750X(95)00046-F.

Richens, I. F., Hobson-West, P., Brennan, M. L., Lowton, R., Kaler, J. and Wapenaar, W. (2015). Farmers' perception of the role of veterinary surgeons in vaccination strategies on British dairy farms. *Veterinary Record* 177(18), 465. https://doi.org/10.1136/vr.103415.

Ries, J., Jensen, K. C., Müller, K. E., Thöne-Reineke, C. and Merle, R. (2022). Impact of veterinary herd health management on German dairy farms: Effect of participation on farm performance. *Frontiers in Veterinary Science* 9, 841405. https://doi.org/10.3389/fvets.2022.841405.

Ritter, C., Adams, C. L., Kelton, D. F. and Barkema, H. W. (2018). Clinical communication patterns of veterinary practitioners during dairy herd health and production management farm visits. *Journal of Dairy Science* 101, 10337–10350. https://doi.org/10.3168/jds.2018-14741.

Ritter, C., Adams, C. L., Kelton, D. F. and Barkema, H. W. (2019). Factors associated with dairy farmers' satisfaction and preparedness to adopt recommendations after veterinary herd health visits. *Journal of Dairy Science* 102, 4280–4293. https://doi.org/10.3168/jds.2018-15825.

Robinson, P. A. (2020). "They've got to be testing and doing something about it": farmer and veterinarian views on drivers for Johne's disease control in dairy herds in England. *Preventive Veterinary Medicine* 182, 105094. https://doi.org/10.1016/j.prevetmed.2020.105094.

Rose, V. H., Morgan, E. R., Hertzberg, H., Bartley, D. J., Bosco, A., Charlier, J., Chartier, C., Claerebout, E., de Waal, T., Hendrickx, G., Hinney, B., Höglund, J., Ježek, J., Kašný, M., Keane, O. M., Martínez-Valladares, M., Mateus, T. L., McIntyre, J., Mickiewicz, M., Munoz, A. M., Phythian, C. J., Ploeger, H. W., Rataj, A. V., Skuce, P. J., Simin, S., Sotiraki, S., Spinu, M., Stuen, S., Thamsborg, S. M., Vadlejch, J., Varady, M., von Samson-Himmelstjerna, G. and Rinaldi, L. (2020). Increasing importance of anthelmintic resistance in European livestock: creation and meta-analysis of an open database. *Parasite* 27, 69. https://doi.org/10.1051/parasite/2020062.

Rose, D. C., Keating, C. and Morris, C. (2018). Understand how to influence farmers' decision-making behaviour a social science literature review on behalf of AHDB. https://projectblue.blob.core.windows.net/media/Default/Imported%20Publication%20Docs/FarmersDecisionMaking_2018_09_18.pdf.

Rose, D. C., Schillings, J., Breen, J. and Morrison, R. (2024). A RESET of dairy farmer animal health planning behaviour: symbiotic advisory relationships and knowledge brokering in HerdAdvance. *The Journal of Agricultural Education and Extension*, 1–24. https://doi.org/10.1080/1389224X.2024.2371295.

Ruston, A., Shortall, O., Green, M., Brennan, M., Wapenaar, W. and Kaler, J. (2016). Challenges facing the farm animal veterinary profession in England: A qualitative study of veterinarians' perceptions and responses. *Preventive Veterinary Medicine* 127, 84–93. https://doi.org/10.1016/j.prevetmed.2016.03.008.

Sanders, J. and Heß, J. (eds.). (2019). *Leistungen des ökologischen Landbaus für Umwelt und Gesellschaft. 2. überarbeitete und ergänzte Auflage*. Johann Heinrich von Thünen-Institut, Braunschweig, 398 p, Thünen Rep 65. https://www.thuenen.de/media/publikationen/thuenen-report/Thuenen_Report_65.pdf.

Schröder, M. J. A. and McEachern, M. G. (2004). Consumer value conflicts surrounding ethical food purchase decisions: a focus on animal welfare. *International Journal of Consumer Studies* 28, 168–177.

Schultheiß, U., Zapf, R., Schubbert, A., Rauterberg, S., Gieseke, D., Brinkmann, J., March, S., Cimer, K. and Knierim, U. (2020). Erarbeitung eines Orientierungsrahmens zur Beurteilung des Tierwohls bei der Eigenkontrolle landwirtschaftlicher Betriebe [Development of reference values for the evaluation of results of on-farm animal welfare self-assessments]. In *Aktuelle Arbeiten zur artgemäßen Tierhaltung, conference proceedings DVG-Tagung* November 26th–27th 2020, KTBL -Schrift 520, 30–37. https://www.ktbl.de/fileadmin/user_upload/Allgemeines/Download/DVG-Tagung/P_11520.pdf.

Silverman, J., Kurtz, S. and Draper, J. (2013). *Skills for Communicating with Patients*. 3rd ed. Radcliffe Publishing, London.

Sjöström, K., Sternberg-Lewerin, S., Blanco-Penedo, I., Duval, J. E., Krieger, M., Emanuelson, U. and Fall, N. (2019). Effects of a participatory approach, with systematic impact matrix analysis in herd health planning in organic dairy cattle herds. *Animal* 13, 358–366. https://doi.org/10.1017/S1751731118002008.

Skaalsveen, K., Ingram, J. and Urquhart, J. (2020). The role of farmers' social networks in the implementation of no-till farming practices,. *Agricultural Systems* 181, 102824. https://doi.org/10.1016/j.agsy.2020.102824.

Skjølstrup, Nanna K. et al. (2021). The antimicrobial landscape as outlined by Danish dairy farmers. *Journal of Dairy Science*, 104, 10, 11147 –11164

Sones, K. R., Duveskog, D. and Minjauw, B. (eds.). (2003). *Farmer Field Schools: The Kenyan experience: Report of the farmer field school stakeholder" forum held 27th March 2003*. FAO/KARI/ILRI, Nairobi, Kenya, 65 pp.

Sumner, C. L., von Keyserlingk, M. A. G. and Weary, D. M. (2018). How benchmarking motivates farmers to improve dairy calf management. *Journal of Dairy Science* 101(4), 3323–3333. https://doi.org/10.3168/jds.2017-13596.

Sandhu, H. (2021). Bottom-up transformation of agriculture and food systems. *Sustainability* 13, 2171. https://doi.org/ 10.3390/su13042171.

Sundrum, A. (2001). Organic livestock farming: A critical review. *Livestock Production Science* 67(3), 207–215. https://doi.org/10.1016/S0301-6226(00)00188-3.

Speksnijder, D. C., Graveland, H., Eijck, I. A. J. M., Schepers, R. W. M., Heederik, D. J. J., Verheij, T. J. M. and Wagenaar, J. A. (2017). Effect of structural animal health planning on antimicrobial use and animal health variables in conventional dairy farming in the Netherlands. *Journal of Dairy Science* 100(6), 4903–4913. ISSN 0022-0302. https://doi.org/10.3168/jds.2016-11924.

Speksnijder, D. Reduction of veterinary antimicrobial use in the Netherlands. The Dutch success model. https://doi.org/10.1111/zph.12167.

Steinfeld, H., Gerber, P., Wassenaar, T., Castel, V., Rosales, M. and de Haan C. (2006). *Livestock's Long Shadow: Environmental Issues and Options*. Food and Agriculture Organization of the United Nations (FAO), Rome, Italy.

Sjöström, K., Sternberg-Lewerin, S., Blanco-Penedo, I., Duval, J. E., Krieger, M., Emanuelson, U. and Fall, N. (2019). Effects of a participatory approach, with systematic impact matrix analysis in herd health planning in organic dairy cattle herds. *Animal* 13(2), 358–366. https://doi.org/10.1017/S1751731118002008.

Sandra, S., Kunda, I., Knickel, K., Strauss, A., Tisenkopfs, T., des los Rios, I., Rivera, M., Chebach, T. and Ashkenazy, A. (2018). Local and farmers' knowledge matters! How integrating informal and formal knowledge enhances sustainable and resilient agriculture. *Journal of Rural Studies* 59, 232–241. ISSN 0743-0167. https://doi.org/10.1016/j.jrurstud.2017.01.020.

Stokes, J. E., Rowe, E., Mullan, S., Pritchard, J. C., Horler, R., Haskell, M. J., Dwyer, C. M. and Main, D. C. J. (2022). A "Good Life" for dairy cattle: Developing and piloting a framework for assessing positive welfare opportunities based on scientificeEvidence and farmer expertise. *Animals* 12(19), 2540. https://doi.org/10.3390/ani12192540.

Svensson, C., Alvåsen, K., Eldh, A. C., Frössling, J. and Lomander, H. (2018). Veterinary herd health management - experience among farmers and farm managers in Swedish dairy production. *Preventive Veterinary Medicine* 155, 45–52. https://doi.org/10.1016/j.prevetmed.2018.04.012.

Svensson, C., Lind, N., Reyher, K. K., Bard, A. M. and Emanuelson, U. (2019a). Trust, feasibility and priorities influence Swedish dairy farmers' adherence and non-adherence to veterinary advice. *Jouernal of Dairy Science* 102, 10360–10368. https//doi.org/10.3168/jds.2019-16470.

Svensson, C., Emanuelson, U., Bard, A. M., Forsberg, L., Wickström, H. and Reyher, K. K. (2019b). Communication styles of Swedish veterinarians involved in dairy herd health management: a motivational Interviewing perspective. *Journal of Dairy Science* 102, 10173–10185. https//doi.org/10.3168/jds.2018-15731.

Svensson, C., Forsberg, L., Emanuelson, U., Reyher, K. K., Bard, A. M., Betnér, S., von Brömssen, C. and Wickström, H. (2020a). Dairy veterinarians' skills in motivational interviewing are linked to client verbal behavior. *Animal* 14, 2167–2177. https//doi.org/10.1017/S175173112000107X.

Svensson, C., Wickström, H., Emanuelson, U., Bard, A. M., Reyher, K. K. and Forsberg, L. (2020b). Training in motivational interviewing improves cattle veterinarians′ communication skills in veterinary herd health management. *Veterinary Record* 187, 191. https//doi.org/10.1136/vr.105646.

Svensson, C., Lomander, H. and Kokko, S. (2022a). Veterinary herd health management – Experiences and perceptions among Swedish dairy cattle veterinarians. *Journal of Dairy Science* 105, 6820–6832. https://doi.org/10.3168/jds.2021-21184.

Svensson, C., Wickström, H., Forsberg, L., Betnér, S., von Brömssen, C., Reyher, K. K., Bard, A. M. and Emanuelson, U. (2022b). Dairy herd health management activities in relation to training of veterinarians in motivational interviewing. *Preventive Veterinary Medicine* 204, 105679. https://doi.org/1016./j.prevvetmed.2022.105679.

Thomsen, P. T., Shearer, J. K. and Houe, H. (2023). Prevalence of lameness in dairy cows: A literature review. *The Veterinary Journal* 295, 105975. https://doi.org/10.1016/j.tvjl.2023.105975.

Vaarst, M. and Fisker, I. (2013). Potential contradictions connected to the inclusion of stable schools in the legislation for Danish organic dairy farms. *The Open Agriculture Journal* 7, 118–124.

Vaarst M., Bennedsgaard T.W., Klaas I., Nissen B., Thamsborg S.M., Ostergaard S., (2006). Development and daily management of an explicit strategy of nonuse of antimicrobial drugs in twelve Danish organic dairy herds. *Journal of Dairy Science* 89, 1842–1853.

Vaarst, M., Nissen, T. B., Østergaard, S., Klaas, I. C., Bennedsgaard, T. W. and Christensen, J. (2007). Danish stable schools for experiential common learning in groups of organic dairy farmers. *Journal of Dairy Science* 90(5), 2543–2554. https://doi.org/10.3168/jds.2006-607.

Vaarst, M. (2010). Fjøsskoler - når grupper av bønder [Stable schools–when groups of farmers work together towards a goal]. *Buskap* 7, 58–59. http://viewer.zmags.com/publication/c5b5d97d#/c5b5d97d/58.

Vaarst, M., Roderick, S., Smolders, G., Leeb, C., Walkenhorst, M., Winckler, C., Gratzer, E., Stöger, E. and Whistance, L. K. Brinkmann, J., March, S., Ivemeyer, S., Mejdell, C. M., Henriksen, B. I. F. and Nicholas, P. (2011). The dialogue with farmers. ANIPLAN Final Report, 64–79. http://orgprints.org/18406/.

Dijk, L., Hayton, A., Main, D. C. J., Booth, A., King, A., Barrett, D. C., Buller, H. J. and Reyher, K. K. (2017). Participatory policy making by dairy producers to reduce anti-microbial use on farms. *Zoonoses and Public Health* 64(6), 476–484.

van Dijk, L., Buller, H. J., Blokhuis, H. J., van Niekerk, T., Voslarova, E., Manteca, X., Weeks, C. A. and Main, D. C. J. (2019). HENNOVATION: learnings from promoting practice-led multi-actor innovation networks to address complex animal welfare challenges within the laying hen industry. *Animals* 9, 24. https://doi.org/10.3390/ani9010024.

van Dijk, L., Buller, H., MacAllister, L. and Main, D. (2017). Facilitating practice-led co-innovation for the improvement in animal welfare. *Outlook on Agriculture* 46(2), 131–137. https://doi.org/10.1177/0030727017707408.

van Soest, F. J. S., Mourits, M. C. M., Blanco-Penedo, I., Duval, J., Fall, N., Krieger, M., Sjöstrom, K. and Hogeveen, H. (2019). Farm-specific failure costs of production disorders in European organic dairy herds. *Preventive Veterinary Medicine* 168, 19–29. https://doi.org/10.1016/j.prevetmed.2019.03.029.

Wallenbeck, A., Rousing, T., Sørensen, J. T., Bieber, A., Spengler, f. A., Fuerst-Waltl, B., Winckler, C., Peiffer, C., Steiniger, F., Simantke, C., March, S., Brinkmann, J., Walczak, J., Woijek, P., Ribikaukas, V., Wilhelmsson, S., Skjerve, T. and Ivemeyer, S. (2018). Characteristics of organic dairy major farm types in seven European countries. *Organic Agriculture* 9, 293.

Yeates, J. W. and Main, D. C. (2008). Assessment of positive welfare: a review. *The veterinary journal* 175(3), 293–300. https://doi.org/10.1016/j.tvjl.2007.05.009.

Zapf, R., Schultheiß, U., Achilles, W., Schrader, L., Knierim, U., Herrmann, H. J., Brinkmann, J., Winckler, C. (2015). Indicators for on-farm self-assessment of animal welfare - Example: dairy cows. *Landtechnik* 70(6), 221–230. https://doi.org/10.15150/lt.2015.2678.

Zapf, R., Schultheiß, U., Knierim, U., Brinkmann, J. and Schrader, L. (2017). Assessing farm animal welfare - guedelines for on.farm self-assessment. *Landtechnik* 72(4), 214–220. https://doi.org/10.15150/lt.2017.3166.

Chapter 8

The use of plant bio-active compounds and other alternative therapeutic strategies in organic dairy farming

Maria Groot, Wageningen University & Research, The Netherlands; Ariane Maeschli and Michael Walkenhorst, Research Institute for Organic Agriculture (FiBL), Switzerland

1 Introduction

Organic farmers are encouraged to use natural methods such as minerals, homeopathy and phytotherapy to keep animals healthy (definitions of key terms in the chapter are shown in Table 1). In the case of herbs and medicinal plants, this might be, at least partly, based on animal self-medication, which is known for several animal species including ruminants (Provenza, 2018; Villalba and Provenza, 2007; Villalba et al., 2009). This behaviour is possible as long as ruminants have access to meadows with a high biodiversity of phytochemical rich herbs from which to choose in supporting their health. However, this subject is not covered by this chapter.

Although there has been an increasing interest of veterinarians in herbal medicine worldwide in recent years (Walkenhorst, 2017; Abo-El-Sooud, 2018; Kupper et al., 2018; Shin and Park, 2018; Melzig et al., 2022), herbal veterinary medicinal products are scarce and most products are sold as complementary feed or feed additives. Herbs and medicinal plants have been used in European veterinary medicine for centuries (Delafond and Lassaigne, 1853;

http://dx.doi.org/10.19103/AS.2024.0138.09

Table 1 Definitions

Complementary medicine: Complementary and alternative medicine (CAM) is the term for products or methods that are used next to regular medical products (complementary) and practices that are used instead of registered veterinary medicines.	
Feed additives: products used in animal nutrition for purposes of improving the quality of feed and the quality of food from animal origin, or to improve the animals' performance and health, e.g. providing enhanced digestibility of the feed materials. Feed additives may not be put on the market unless authorisation has been given following a scientific evaluation demonstrating that the additive has no harmful effects, on human and animal health and on the environment.	
Homeopathy: alternative medicine based on the theory of treating 'like with like'. It stimulates healing responses to diseases by administering substances that mimic the symptoms of those diseases in healthy people/animals.	
Monograph: document whose purpose is to provide a scientific summary of all data available on the safety and efficacy of a herbal substance/preparation intended for medicinal use. There are herbal monographs from ESCOP, EMA and WHO and others available	
Natural remedies: substances of health value derived directly from plants or other natural sources.	
Phytotherapy: the practice of using medicines derived from plants or herbs to treat or prevent health conditions.	
Sensory additives: products improving taste and smell of the feed.	
Zootechnical: favourably affecting the performance of animals in good health or used to favourably affect the environment.	

Fröhner, 1900; Steiner et al., 2022). Ethnoveterinary medicine is still used in many rural areas in Europe (Mayer et al., 2014; Vogl et al., 2016; Stucki et al., 2019; Mertenat et al., 2020; Schlittenlacher et al., 2022) and could help in finding medical solutions for the worldwide antibiotic resistance crisis. Most of the herbal products are used as complementary feed or feed additives without medical indication. However, there are still some registered herbal veterinary medicinal products on the European market (Händler et al., 2005).

The use of homeopathy in farm animals is also becoming increasingly popular among farmers in different countries (Hellec et al., 2021; Hektoen, 2004; Maeschli et al., 2019; Keller et al., 2019; Orjales et al., 2015). Homeopathic remedies are easy to apply: contact with a mucous membrane (oral, nasal or vaginal) is usually sufficient. They are also typically cheap and free of residues. But homeopathy is also very complex and a good knowledge of the remedies, as well as good observational skill on the part of the farmer, are essential. If these aspects are taken into account, homeopathic products can potentially be used on their own for some, clearly identified conditions, though their use remains controversial, as will be discussed.

2 Types of alternative therapeutic products

2.1 Complementary feeds

A complementary feed is defined as a compound feed which has a high content of certain substances but which, by reason of its composition, is sufficient for a daily ration only if used in combination with other feed (REGULATION (EC) No 767/2009). Many of these products are used to support animal health, particularly in dairy farming, because individual animals can be given a tailored diet using complementary feed on top of a standard ration.

2.2 Feed additives

Feed additives are products used in animal nutrition to achieve an effect on the feed itself, on the animals, on food products obtained from the animals consuming the feed additive, or on the environment (EFSA). They are regulated via Regulation (EC) No 1831/2003. There are different categories of feed additives:

- Technological additives (e.g. preservatives, antioxidants, emulsifiers, stabilizing agents, acidity regulators, silage additives)
- Sensory additives (e.g. flavors, colorants)
- Nutritional additives (e.g. vitamins, minerals, amino acids, trace elements)
- Zootechnical additives (e.g. digestibility enhancers, gut flora stabilizers, to promote animal welfare)
- Coccidiostats and histomonostats (medicines against coccidiosis and histomoniasis)

Most herbal products fall under the category of plant-based sensory additives, which are allowed for organic production if they consist only of extracts of agricultural products, including extract of sweet chestnut (*Castanea sativa* Mill) (Regulation 2021/1165). Non-organic herbal products may be used in accordance with Article 24(3)(e)(iv) of Regulation (EU) 2018/848 and, if not available in organic form, manufactured or prepared without chemical solvents and as maximum of 1% of the feed ration.

2.3 Natural/household remedies

About 200 different European plant species have been documented in recent ethnoveterinary research (Mayer et al., 2014; Vogl et al., 2016; Stucki et al., 2019; Mertenat et al., 2020; Schlittenlacher et al., 2022). Farmers typically use these phytochemical-rich plant species either in pure or processed form to treat or prevent animal diseases. Farmers mostly treat dermatological disorders (including wounds) and gastrointestinal disorders, but also use remedies infertility and urogenital disorders as well as parasitosis. These treatments are reported to be most used for ruminants, particularly dairy cows and calves.

2.4 Herbal products/phytotherapeutics

Herbal products or phytotherapeutics are used on a large scale in Europe both in organic and conventional farming. Since the word phyto'therapeutics' may lead to the suggestion that these products are an established part of veterinary medicine, phytogenics (plant derived) or herbal products may be a more appropriate name. In some countries, such as the Netherlands and Switzerland, the government is actively promoting the use of natural products to reduce the use of antibiotics (Groot et al., 2021; StAR, 2022). The Dutch government has invested in providing information for farmers and veterinarians concerning alternatives in disease management and the use of alternative, health-supporting products. In a project funded by the Dutch Ministry of Agriculture, Nature and Food Quality, Wageningen Food Safety Research (WFSR) has developed a number of guides aimed at organic farmers covering dairy cattle, poultry and pigs (Groot, n.d.). These guides include (in the English translation) (Groot et al., 2010, 2011a, 2011b) the following:

- Natural dairy cow health;
- Natural swine health;
- Natural poultry health; and
- Guide to keeping your animals healthy with herbs and other natural products.

These books are freely available via the internet. There are now updated guides for conventional farmers, with guides currently covering dairy cows, veal calves, pigs, poultry, turkeys, sheep, dairy goats and rabbits.

There is a recent Swiss publication on therapeutic options using complementary medicine to treat mild animal diseases, as part of a series of antibiotic therapy guides for cattle and pigs published to reduce antimicrobial use in animal production (StAR, 2022). This is linked to the 'Arzneipflanzen' (Medicinal Plants) Section of the official information system for veterinary drugs 'CliniPharm/CliniTox' (Kupper et al., 2018; CliniPharm/CliniTox, 2023). This information system about medicinal plants is only available in German.

The use of herbal products alone is not enough to reduce the use of antibiotics in both organic and conventional animal production. Most antibiotics are used in young animals (pigs, broilers, turkeys and veal calves) for disorders of the intestinal or respiratory tract. The immune system in young animals is not yet fully developed in the first weeks of life. In calves, a combination of insufficient intake of colostrum, poor management and exposure to pathogens often gives rise to a high incidence of disease and mortality. In intensive production systems, long-distance transport (e.g. pigs, veal calves), fasting, mingling animals from different sources, high stock densities, changes in environmental conditions, suboptimal hygiene and inappropriate diets contribute to stress and susceptibility to disease. To reduce the need for antibiotics, other methods are needed such as breeding (e.g. slower-growing breeds, more resilient animals, optimal instead of maximum production, dual-purpose breeds), better management (e.g. housing, hygiene, environmental management, stock density, feeding), improved quality control and supply chain management. This combined approach could even lead to the development of antibiotic-free production chains. Organic farming already uses some of these approaches. Organic production is generally less intensive than in conventional farms, e.g. with lower stock densities. In organic production, the use of antibiotics is also restricted by European regulations and natural healing methods are preferred for animal health management.

3 Research supporting the use of alternative therapeutic strategies: ethnoveterinary research and systematic reviews of the scientific literature

Besides scientific data about animal self-medication, which is not an issue of this chapter, there are three further sources of knowledge about the use of medicinal plants:

- Ethnoveterinary research including historical information and surveys of recent farmer knowledge;
- Systematic reviews of existing scientific evidence about medicinal plants and their veterinary use; and
- Clinical research.

This section looks at the first two sources, with the following section looking at clinical research.

3.1 Ethnoveterinary research

Herbal products for human and animal health have been used all over the world for thousands of years and, for many developing countries, these are often the only available or affordable medication. The earliest comprehensive European veterinary text was the *Hippiatrica*, a Byzantine collection of Greek texts written about 1500 years ago (McCabe, 2007). It includes more than 250 medicinal plants, mainly for horses. Since the first veterinary school was founded in Lyon in 1764, veterinary pharmacology was mainly based on medicinal plants. Veterinary pharmacology textbooks from the last 150 years, e.g. German texts by Röll (1866, Austria) or Fröhner (1900, Germany) as well from France (Delafond and Lassaigne, 1853), were full of recipes including medicinal plants. The same was true of the lectures of the Swiss veterinary pharmacologist (Stucki et al. 2019). Antibiotics appeared in veterinary medicine from 1946 (Anonymous, 1946) with a transformative effect: 'for many years, progress in veterinary medicine has been oriented towards the use of highly effective synthetic mono-component drugs in the analogy to human medicine and negated the use of supposedly ineffective medicinal plants in the context of orthodox medical therapy' (Melzig et al., 2022).

European ethnoveterinary research has been undertaken mainly in Mediterranean countries (Gonz´alez et al., 2020; Groot et al. 2011b; Benítez et al., 2012; Mayer et al., 2014; Viegi and Ghedira, 2014; Bullitta et al., 2018; Oliveira et al., 2021), in Austria (Vogl et al., 2016) and some in Eastern European countries, notably Romania (Bartha et al., 2015), Serbia (Subarevic et al., 2015), Belarus (Soukand et al., 2017; Shrubok, 2020) and Estonia (Kalle and Kass, 2020). Some European research had an explicit focus on ethnoveterinary use of plants, while others studies were broader ethnobiological or ethnomedicinal studies which incidentally covered ethnoveterinary issues (Mayer et al., 2014).

The systematic evaluation of farmers' knowledge and experience has the potential to provide valuable information about how medicinal plants have been used in practice with farm animals. Systematic research has been done in Switzerland and south eastern Germany during the past decade (Stucki et al., 2019; Schmid et al., 2012; Disler et al., 2014; Bischoff et al., 2016; Mayer et al., 2017; Mertenat et al., 2020; Schlittenlacher et al., 2022). These collected detailed information about homemade herbal remedies (plant species, plant part, manufacturing process) and their use (target animal species, type of use, route of administration, dosage, source of knowledge, frequency of use, last time of use and farmers' satisfaction with outcomes). A total of 183 interviews were conducted with 253 persons leading to 918 separate herbal remedy reports covering individual plant species (HSHR) and 1041 use reports (UR).

In about two thirds of the use reports, a daily dosage for oral preparations or a concentration for external use could be estimated in gram dry plant equivalent. Estimated oral daily dosages were calculated per kg metabolic body weight based on the animal species and the age category to be able to compare the results with common human dosages of medicinal plants (ESCOP, 2003, 2009, 2022; Wichtl, 2009). The use of more than 100 medicinal plants was mentioned in the interviews. About half of them were medicinal plants already mentioned in historical veterinary pharmacology textbooks. The calculated median oral dosages per kg metabolic body weight lies mainly between 0.2 and 0.8 g. Interestingly, in line with historical (Fröhner, 1900; Mangold and Reichert, 1907;

Zipperlen, 1959) or recent (Reichling et al., 2016; Brendieck-Worm and Melzig, 2021) veterinary textbooks as well as dose recommendations for human use (ESCOP, 2003, 2009, 2022; Wichtl, 2009). Ethnoveterinay-based dosages might be therefore an interesting focus for future clinical research.

More than half of the recent use reports were intended to treat gastrointestinal disorders and metabolic dysfunction (ATCvet-Code QA) or dermatological disorders like skin alterations and sores (QD). Other conditions included infertility and female genital diseases including mastitis (QG), diseases of the musculoskeletal system (QM), parasitological (QP) and respiratory diseases (QR). Frequently mentioned-plants were as follows:

- *Camellia sinensis* (L.) Kuntze, *Cinnamomum verum* J.Presl., Coffea L., *Linum usitatissimum* L., *Matricaria chamomilla* L. and *Quercus robur* L. for gastrointestinal diseases (QA); and
- *Arnica montana* L., *Calendula officinalis* L., *Malva neglecta* Wallr., *Matricaria chamomilla* L., *Hypericum perforatum* L. and *Sanicula europaea* L. for skin afflictions (OD).

Typical farmers descriptions of the preparation and use of herbal remedies are for example as follows:

- 'Prepare one litre of an infusion based on one table spoon of dried chamomile flowers. Let it steep for about 10 minutes and strain it through a fine sieve afterwards. Give calves with diarrhoea one litre of this infusion two times daily for three days' (NB: Beside diarrhoea, chamomile infusion is used to treat wounds, conjunctivitis, ulcers and coughs).
- 'Take a 1 litre jar and fill it with fresh flowers of marigold (about 100 g fresh matter or 20 g dry matter respectively) and fill it with Schnaps (60%). To prepare the final preparation take 15 drops of the essence per 200ml of water and wash out fresh or poor healing wounds 2 times daily' (NB: Follow-up treatments of sores, as well as rough skin or teats, can be treated with a (heated or non-heated) oil or swine fat extract of marigold flowers as an ointment).

3.2 Systematic reviews of existing scientific evidence about medicinal plants and their veterinary use

There is a significant body of scientific knowledge about phytochemical constituents, receptor-binding capacity of single phytochemical substances and their potential modes of action, as well as their *in vitro* and *in vivo* activity. Phytochemical substances demonstrate antioxidative, anti-inflammatory, antiviral, antifungal, antimicrobial, antiparasitic, anticancer, immune-modulative and tissue-protective activities. Phytochemical substances are also able to influence various metabolic processes, though, depending on dosage, some are toxic. Most of the available research work has been done during recent decades focussed on human medicine. There have been a number of major reviews of medicinal plants in Europe (ESCOP, 2003, 2022; HMPC, 2022) and worldwide (WHO Monographs, 2022), though similar veterinary monographs are missing.

It is important to review systematically the available *in vitro* and *in vivo* and (human and animal) clinical data both as a basis for clinical studies and as a context for assessing the results of ethnoveterinary research. Some recent reviews have been undertaken relating to diseases of young cattle and pigs (Ayrle et al., 2016). Other reviews have assessed studies of *in vitro*, *in vivo* and clinical data in the field of veterinary medicine (Latek et al., 2022; Farinacci et al., 2022; Kuralkar and Kuralkar, 2021; Li et al., 2021).

4 Research supporting the use of alternative therapeutic strategies: clinical research

Clinical data about the use of medicinal plants to treat or prevent animal diseases is still rare (Ayrle et al., 2016), particularly in Europe (Farinacci et al., 2021). Unfortunately, the available research is characterised by a high variety of methods used and how accurately these methods were reported. It is often not possible to infer the applied dosage in a general comparable unit (i.e. dry plant equivalent) or the phytochemical constituents of the plant material used; sometimes, not even the plant species and plant part are given. One factor is that it is commercial herbal products, in particular feed or feed additives, that are being tested. They are not patentable so producers do not want to disclose commercially sensitive information. Moreover, many publications focus more on performance than on health parameters. Natural products used as feed additives have shown to improve zootechnical parameters such as growth performance, feed conversion and gut health in farm animals, and so reduce the need for antibiotics (Huang et al., 2018; Jouany and Morgavi, 2007; Mahfuz et al., 2021; Nielsen, 2008; Rossi et al., 2020; Yang et al., 2015).

Infectious diseases of farm animal diseases are typically infectious diseases of young animals mainly affecting the respiratory and gastrointestinal tracts. Other diseases are those of nearly permanently lactating and reproducing female animals affecting the genitourinary system, particularly the udder and uterus. Metabolic diseases affecting females are also seen as one cause of udder and uterine diseases, claw and leg diseases and overall infertility (Duffield et al., 2009; Chapinal et al., 2012; Raboisson et al., 2014). These animal disease complexes are the reason why farm animals still receive the highest amount of antimicrobials in Switzerland and all over Europe (Bundesamt für Lebensmittelsicherheit und Veterinärwesen, 2022a, 2022b; European Medicines Agency, 2022).

Two good examples are two clinical trials investigating the treatment of subclinical ketosis of dairy cows with a multicomponent herbal extract. The herbal extract contains aqueous extracts of *Camellia sinensis* (L.) Kuntze, *Cichorium intybus* L., *Gentiana lutea* L., *Glycyrrhiza glabra* L., *Taraxacum officinale* F.H. Wick, *Trigonella foenum-graecum* L. and *Zingiber officinale* Roscoe in a total daily amount of 27 g dry plant equivalent. Dosages of 20–50 g per cow and day are the usual dose set out in historical veterinary pharmacology textbooks (Fröhner 1900) as well as in recent veterinary herbal medicine texts (Reichling et al., 2016; Brendieck-Worm and Melzig, 2018). Subclinical ketosis is seen as an important health issue affecting dairy cows, elevating the risk of a range of other diseases like mastitis, lameness, displaced abomasum or fertility disorders. About one-fifth to one-quarter of dairy cows develop subclinical ketosis during early lactation (Suthar et al., 2013, Brunner et al., 2019; Loiklung et al., 2022).

However, besides these two studies involving a herbal extract (Dorn et al., 2016, Durrer et al., 2020), there have only been four other studies with diseased animals and a comparison of herbal treatment with a placebo or untreated control group since 1986 (Ruegsegger and Schultz, 1986; Jeong et al., 2018; Nuber et al., 2016; Mann et al., 2017). In these four studies, the therapeutic agent was given for 3–7 days, and effects were monitored from 4 (Dorn et al., 2016) to 5 (Durrer et al., 2020) days of administering the herbal product.

The combined application of the herbal extract with sodium propionate (28 cows treated, Dorn et al., 2016) led to a significant reduction of milk acetone content during the first week of observation (including application) compared to placebo (28 cows treated) and to the application of pure sodium propionate (28 cows treated). While the application of pure sodium propionate led to an immediate reduction of milk acetone, the pure application of the herbal extract (41 cows treated; Durrer et al., 2020) led to a decrease of milk acetone from day 7 after the start of the treatment and reached a significantly lower milk acetone content on day 14 after the start of the treatment. By this point, the difference between pure sodium propionate (41 cows treated) and placebo (41 cows treated) in terms of milk acetone was small (Durrer et al., 2020).

Although the mode of action was not the main focus of both clinical trials, there is a broad knowledge based on *in vitro* and *in vivo* research about the mode of action of the seven medicinal plants. One approach is using bitter substances from *Cichorium intybus* L. or *Taraxacum officinale* F.H. Wick, e.g. (Schutz et al., 2006) or the bitter glycoside gentiopicrin from *Gentiana lutea* L. (Mirzaee et al., 2017). Another approach is the stimulation of gastric secretion and the anti-emetic effect of *Zingiber officinale* Roscoe (Marx et al., 2015). Nausea is known to be a frequent symptom of human diabetic ketoacidosis. Impaired fat metabolism and fat deposition in the liver may be stabilized by

- glycyrrhizin contained in *Glycyrrhiza glabra* L. (Eu et al., 2010);
- *Taraxacum officinale* F.H. Wick (Davaatseren et al., 2013); and
- *Trigonella foenum-graecum* L. (Nagulapalli Venkata et al., 2017).

As the liver is the main organ affected, subclinically ketotic cows may benefit also from the hepatoprotective effects of *Zingiber officinale* Roscoe (Carnuta et al., 2018), *Cichorium intybus* L. (Elgengaihi et al., 2016), *Glycyrrhica glabra* L. (Hong et al., 2015; Li et al,. 2014), and *Taraxacum officinale* F.H. Wick. However, no difference compared to placebo and a lower performance compared to sodium propionate was observed for the pure herbal extract with regard to gamma-glutamyl transferase (GGT) and glutamate dehydrogenase (GLDH). Average blood concentrations of GGT and GLDH were far beyond the threshold of liver impairment in placebo as well as in both treatment groups in the study by Durrer et al. (2020). In conclusion, the herbal extract seems to have an equal therapeutic outcome compared to the standard therapy with glucoplastic substances. The combination of the herbal extract and sodium propionate may even be superior to sodium propionate alone.

4.1 Homeopathic research

The effectiveness of homeopathy is a controversial topic. However, there have been some placebo-controlled clinical studies on humans and animals, as well as basic research using physico-chemical methods (Tournier et al., 2019) or bio-assays (Ücker et al., 2018), which have been able to prove specific effects of some homeopathic medicines.

There are several studies of homeopathic treatments for poultry, pigs and cattle (Doehring and Sundrum, 2016). In an RCT single-blinded study, Werner et al. (2010) found a positive effect of homeopathic treatment in mild and moderate mastitis compared to a placebo group and no difference to antibiotic treatment. Klocke et al. (2010) were able to show a positive effect in mastitis prevention at drying off, with a lower somatic cell count (SCC) after calving in the homeopathically treated group compared to a placebo group. In subclinical mastitis, Kiarazm et al. (2011) found a lower SCC as well as lower levels of bacteria in the homeopathic group compared to the placebo. In pigs, several RCT single or double-blind studies have shown a significant effect of homeopathic treatment, e.g. in the prevention of *E. coli* diarrhoea (Coelho et al., 2009; Camerlink et al., 2010). At farm level, lower intramammary antibiotic use was observed when farmers used homeopathic mastitis treatments (Stevens et al., 2016) and a significant reduction of antibiotic treatments in dairy cattle, in general, was achieved on farms that participated in a programme of homeopathic treatment (Maeschli et al., 2019).

4.2 Research on herbal remedies

Most research on herbal remedies relates their effects on performance and gut health. Three examples are:

- Digextra;
- Sangrovit; and
- Yucca Plus.

Research was carried out by the Veterinary Faculty of Utrecht on Digextra BB (Oudendorp, 2014). This is a product for young calves immediately after colostrum. It is a mix of fibre hydrolyzate from wheat straw and a fibre hydrolyzate from *Saccharum munja*. The product also contains glycerol and electrolytes (magnesium sulphate, potassium chloride and sodium chloride). Two groups of 70 bull calves were housed individually for the first 22 days and then grouped on the basis of equivalent weight. The animals received milk powder solution twice a day for the first 22 days and from day 22 also roughage, straw and maize. The test group received 10–12 g of Digextra in the milk for 65 days. Parameters investigated were manure quality, antibiotic treatment days and number of animals treated.

From day 3 till day 13, there were a large number of calves with diarrhoea. Animals became ill in both the control and treatment groups. Significantly fewer animals were ill in the treatment group. In the control group, 32 animals were treated with Biopect (a prebiotic product), five with Duoprim (an antibiotic), one with Baytril (a veterinary medicine) and two with Halocur (another veterinary medicine). Thirteen animals in the treatment group were treated with Biopect and one with Duoprim. This study showed that the manure quality was better in the animals that received Digextra; the consistency was thicker, the dung was darker in colour and contained no blood. The animals from this group showed significantly less diarrhoea than the control animals. In terms of treatments for pneumonia, diarrhoea and cryptosporidia, the animals supplemented with Digextra were treated for less time and fewer animals from this group required treatment. What was striking is that, as soon as there were changes in the feed given to the calves, the control animals reacted much more strongly. Digextra appears to show a stabilizing effect on gut health.

Research has also been done with Digextra in heifer calves receiving 10 g per day through the milk from birth until day 91 (Braun und Hellmann, 2014). On days 0, 14 and 91, 17 animals per group were weighed. The Digextra group consisted of 18 animals and the control group of 19 animals. In both periods (day 0–14 and day 14–91), Digextra group animals were significantly heavier than the control group. The daily gain was 721 g for the Digextra group and 641 g for the control group. The palatability of the milk was not influenced by the product.

Sangrovit consists of herbal ingredients and a natural extract of Papaveraceae *Macleaya cordata*. It contains mainly alkaloids as the active substance, particularly sanguinarine. Sangrovit contains at least 1.5% sanguinarine. Several field trials have shown positive effects of Sangrovit which improves the roughage intake of dairy cows, resulting in higher milk production, fat and protein contents and a decrease in somatic cell count (data from producer). Milk production increased by an average of 1.2 l per animal per day, while the fat content increased by 7.4% and the protein content by 6.4%. In a trial in Germany with heifers, milk production increased by 10.5%, fat content by 12.5% and protein content by 6.5%. Production remained at the same level during heat stress. In addition, an average reduction of the somatic cell count of 24% was measured and the feed intake during lactation increased by 5%. (Groot et al., 2016).

Yucca Plus consists of 100% *Yucca schidigera*, which contains many active compounds such as saponins and glycol-components. The glycol components in Yucca Plus directly bind with ammonia to form a non-toxic complex (Headon et al., 1990). Studies in monogastric animals in particular have shown that ammonia emissions to air can be reduced by up to 20–50% with the help of *Yucca schidigera*. The binding of *Yucca schidigera* to ammonia in cattle has been shown to result in lower ammonia emissions and less urea in both plasma and blood. A large meta-analysis (33 studies in ruminants) showed that rumen ammonia is on average 12.5% lower when *Yucca schidigera* is used (data Jadis Additiva). This is due to a reduction in the conversion of urea to ammonia in the liver, which also saves energy for the cow (Wallace et al., 1994).

5 Legislation

5.1 Veterinary medicinal products

The number of registered herbal veterinary medicines is limited in most European countries, in part due to high registration demands and costs. Remedies are mostly old registrations based on pre-2019 legislation. Thirteen pure herbal products are registered as veterinary medicines in the Netherlands 13, with around 20 in Germany and only 2 in Switzerland (CBG, 2022; CliniPharm/CliniTox, 2023). Most of these products are registered for oral administration. To the knowledge of the authors, there is at least one registered herbal veterinary medicinal product on the European market for (a) intrauterine endometritis therapy and (b) external application in the case of skin afflictions and treatment of wounds.

According to the EU Regulation on Veterinary Medicinal Products, applicable from 28 January 2022 (Regulation (EU) 2019/6), homeopathic single remedies that do not have a veterinary authorisation may no longer be used on animals by non-veterinarians. This has led to some homeopathic single remedies as well as combination preparations being approved specifically for use in veterinary medicine. In Switzerland, on the other hand, all homeopathic single remedies available with a potency of D6 or higher can be used by farmers.

5.2 Legislation for organic farming

European legislation on health management in organic farming states that phytotherapeutic and homeopathic products should be used in preference to treatment with chemically synthesised allopathic veterinary medicinal products, including antibiotics, provided that their therapeutic effect is effective for the species of animal and for the condition for which the treatment is intended (Regulation 2018/848). However, EU Regulation on veterinary medicinal products (Regulation (EU) 2019/6, article 86) does not mention phytotherapeutic agents. This is in contrast to human medicine where there is specific legislation for traditional herbal medicine (Directive 2001/83/EC, art. 16). Homeopathy is also mentioned in the directive, with a simplified registration procedure, but Art. 14 states that no specific therapeutic indication may appear on the labelling of the medicinal product.

5.3 Future potential legislation

New European veterinary legislation (Regulation (EU) 2019/6) does allow for the option of using traditional herbal medicines for animals. In EU-Regulation 2019/6, it is stated that '[t]here is insufficient information to date on traditional herbal products used to treat animals in order to allow the establishment of a simplified system. Therefore, the possibility of introducing such a system should be examined by the Commission based on the information provided by the Member States on the use of such products on their territory.' Member states need to gather this information by 2027. Recently (2023) a COST action (MedPlants4Vet, 'Medicinal plants for animal health care: Translating tradition into modern veterinary medicine') had been agreed to gather this information to support a clearer legal framework for traditional veterinary herbal medicine.

6 Conclusion and future trends

Some key points from this chapter can be summarised as follows:

- Ethnoveterinary herbal remedies are still used by many farmers in Europe.
- There are scientific monographs substantiating the health effects of herbs and their constituents, but these are lacking for veterinary use.

- Since most herbal products are used as feed additives or complementary feed, clinical studies are scarce.
- There is scientific evidence about the efficacy of homeopathic remedies in animals. In some studies missing effects may be due to a lack of individual adaptation.
- In contrast to Swiss legislation, EU Legislation prevents the use of therapeutic indications on homeopathic remedies.

The future use of plant bio-active compounds and other alternative therapeutic strategies in organic dairy farming depends on a number of factors. Broad acceptance by society, governments and regulators is needed to keep therapeutic diversity for individual treatments in the (veterinary) medicinal system. From the authors perspective, it is a misunderstanding of evidence-based medicine (EBM) to accept only therapies that are proven by systematic reviews. If so, evidence-based veterinary medicine will be nearly impossible for most species and indications. In contrast, David Sacket, the 'father of EBM' clearly stated that this 'external evidence' not only includes systematic reviews but also all kinds of clinical studies, case reports or even expert opinions. Beside the 'external evidence', the individual clinical experience of the clinicians as well as the 'more thoughtful identification and compassionate use of individual patients' predicaments, rights and preferences in making clinical decisions about their care' are two further equal pillars of EBM (Sacket et al., 1996).

However, research in veterinary homeopathy and veterinary herbal medicine is still very necessary and important. This needs the acceptance of the scientific community that complementary medicine is part of scientific discourse. Research field also needs public investment, as many such products are too simple and too cheap to be of interest to the industry which cannot see a return on investment (in research). The MedPlants4Vet website (https://www.medplants4vet.eu/) provides a good foundation for future research, as it aims to provide concise, easily accessible and properly validated information about herbal remedies and their traditional and current use for animals. This Cost action aims to generate basic information for a new legal framework that includes traditional herbal medicines for animals in Regulation (EU) 2019/6.

Veterinary homeopathy and veterinary herbal medicine should also be part of veterinary education, since it has been shown that veterinarians, based on their diagnostic capacity and their experience, are able to efficiently use complementary medicine as part of their therapeutic tools of common daily practice. Finally, given increasing concern about antimicrobial and antiparasitic resistance, interest in sustainable farming, circular agriculture and global sustainability goals, herbal and homeopathic remedies with proven efficacy or, at least, clearly based on clinical experience, offers a solution for future farming.

7 References

Abo-El-Sooud, K. (2018). Ethnoveterinary perspectives and promising future. *Int J Vet Sci Med* 6: 1–7. https://doi.org/10.1016/j.ijvsm.2018.04.001

Anonymous-. (1946). Penicillin in der Veterinärmedizin. In *Tierarztl Umsch (Tierärztliche Umschau)*1: 11–12.

Ayrle, H., Mevissen, M., Kaske, M., Nathues, H., Gruetzner, N., Melzig, M. and Walkenhorst, M. (2016). Medicinal plants–prophylactic and therapeutic options for gastrointestinal and respiratory diseases in calves and piglets? A systematic review. *BMC Vet Res* 12 (1): 1–31. https://doi.org/10.1186/s12917-016-0714-8

Bartha, S. G., Quave, C. L., Balogh, L. and Papp, N. (2015). Ethnoveterinary practices of Covasna County, Transylvania, Romania. *J Ethnobiol Ethnomed* 11: 35. https://doi.org/10.1186/s13002-015-0020-8

Benítez, G., Gonz´alez-Tejero, M. R. and Molero-Mesa, J. (2012). Knowledge of ethnoveterinary medicine in the province of Granada, Andalusia, Spain. *J Ethnopharmacol* 139 (2): 429–439. https://doi.org/10.1016/j.jep.2011.11.029

Bischoff, T., Vogl, C. R., Ivemeyer, S., Klarer, F., Meier, B., Hamburger, M. and Walkenhorst, M. (2016). Plant and natural product based homemade remedies manufactured and used by farmers of six central Swiss cantons to treat livestock. *Livest Sci* 189: 110–125.

Braun, G. und Hellmann, K. 2014. Untersuchung zur Verbesserung der Gewichtszunahme durch die Zugabe von Digextra BB zur Milchtränke bei Kälbern. Tierärztl. Umschau 69 (11): 477-480.

Brendieck-Worm, C. and Melzig, M. F. (Eds.). (2021). *Phytotherapie in der Tiermedizin*. Georg Thieme Verlag, Stuttgart.

Brunner, N., Groeger, S., Raposo, J. C., Bruckmaier, R. M. and Gross, J. J. (2019). Prevalence of subclinical ketosis and production diseases in dairy cows in Central and South America, Africa, Asia, Australia, New Zealand, and Eastern Europe. *Transl Anim Sci* 3: 84–92.

Bullitta, S., Re, G. A., Manunta, M. D. I. and Piluzza, G. (2018). Traditional knowledge about plant, animal, and mineral-based remedies to treat cattle, pigs, horses, and other domestic animals in the Mediterranean island of Sardinia. *J Ethnobiol Ethnomed* 14 (1): 1–26. https://doi.org/10.1186/s13002-018-0250-7

Bundesamt für Lebensmittelsicherheit und Veterinärwesen. (2022a). IS ABV Verschreibungen von Antibiotika für Tiere in der Schweiz 2021. https://www.blv.admin.ch/blv/de/home/tiere/tierarzneimittel/antibiotika/isabv.html

Bundesamt für Lebensmittelsicherheit und Veterinärwesen. (2022b). Strategie Antibiotikaresistenzen - Supplement Komplementärmedizin zum Therapieleitfaden Rinder und Schweine für Tierärztinnen und Tierärzte. https://www.blv.admin.ch/dam/blv/de/dokumente/tiere/tierkrankheiten-und-arzneimittel/tierarzneimittel/supplement-komplementaermedizin.pdf.download.pdf/Supplement_Komplementaermedizin_d.pdf

Camerlink, I., Ellinger, L., Bakker, E. J. and Lantinga, E. A. (2010). Homeopathy as replacement to antibiotics in the case of *Escherichia coli* diarrhoea in neonatal piglets. *Homeopathy* 99: 57–62.

Carnuta, M. G., Deleanu, M., Barbalata, T., Toma, L., Raileanu, M., Sima, A. V. and Stancu, C. S. (2018). *Zingiber officinale* extract administration diminishes steroyl-CoA desaturase gene expression and activity in hyperlipidemic hamster liver by reducing the oxidative and endoplasmic reticulum stress. *Phytomedicine* 48: 62–69.

CBG: https://english.cbg-meb.nl/

Chapinal, N., Leblanc, S. J., Carson, M. E., Leslie, K. E., Godden, S., Capel, M., Santos, J. E. P., Overton, M. W. and Duffield, T. F. (2012). Herd-level association of serum metabolites in the transition period with disease, milk production, and early lactation reproductive performance. *J Dairy Sci* 95 (10): 5676–5682. https://doi.org/10.3168/jds.2011-5132

CliniPharm/CliniTox. (2023). https://www.vetpharm.uzh.ch/tak/clinidoc.htm

Coelho, C., Soto, F., Vuaden, E., Melville, P., Oliveira, F. and Benites, N. (2009). Evaluation of preventive homeopathic treatment against colibacillosis in swine production. *Int J High Dilut Res* 8: 183–190.

Davaatseren, M., Hur, H. J., Yang, H. J., Hwang, J. T., Park, J. H., Kim, H. J., Kim, M. J., Kwon, D. Y. and Sung, M. J. (2013). *Taraxacum officinale* (dandelion) leaf extract alleviates high-fat diet-induced nonalcoholic fatty liver. *Food Chem Toxicol* 58: 30–36.

Delafond, O. and Lassaigne, J. (1853). *Traité de matière médicale et de pharmacie vétérinaires théorique et pratique*. 2 ed. Labé, Paris. 819.

Directive 2001/83/EC of the European Parliament and of the Council of 6 November 2001 on the Community code relating to medicinal products for human use. https://eur-lex.europa.eu/legal-content/en/ALL/?uri=CELEX%3A32001L0083

Disler, M., Ivemeyer, S., Hamburger, M., Vogl, C., Tesic, A., Klarer, F., Meier, B. and Walkenhorst, M. (2014). Ethnoveterinary herbal remedies used by farmers in four north-eastern Swiss cantons (St. Gallen, Thurgau, Appenzell Innerrhoden and Appenzell Ausserrhoden). *J Ethnobiol Ethnomed* 10: 32.

Doehring, C. and Sundrum, A. (2016). Efficacy of homeopathy in livestock according to peer-reviewed publications from 1981 to 2014. *Vet Rec* 179 (24): 628. https://doi.org/10.1136/vr.103779.

Dorn, K., Leiber, F., Sundrum, A., Holinger, M., Mayer, P. and Walkenhorst, M. (2016). A field trial on the effects of pure sodium propionate and a combination with herbal extracts on short term development of subclinical ketosis. *Livest Sci* 187: 87–95.

Duffield, T. F., Lissemore, K. D., McBride, B. W. and Leslie, K. E. (2009). Impact of hyperketonemia in early lactation dairy cows on health and production. *J Dairy Sci* 92 (2), 571–580. https://doi.org/10.3168/jds.2008-1507

Durrer, M., Mevissen, M., Holinger, M., Hamburger, M., Graf-Schiller, S., Mayer, P., Potterat, O., Bruckmaier, R., and Walkenhorst, M. (2020). Effects of a multicomponent herbal extract on the course of subclinical ketosis in dairy cows – a blinded placebo-controlled field-study. *Planta Med.* https://doi.org/10.1055/a-1260-3148

EFSA, topic feed additives. https://www.efsa.europa.eu/en/topics/topic/feed-additives

Elgengaihi, S., Mossa, A. T., Refaie, A. A. and Aboubaker, D. (2016). Hepatoprotective efficacy of *Cichorium intybus* L. extract against carbon tetrachloride-induced liver damage in rats. *J Diet Suppl* 13: 570–584.

ESCOP Monographs. (2003). *The Scientific Foundation for Herbal Medicinal Products*, 2nd edition. Georg Thieme Verlag, Stuttgart.

ESCOP Monographs online. (2022). United Kingdom: European scientific cooperative on phytotherapy (ESCOP). www.escop.com, 30.10.2022.

ESCOP Monographs, Supplement. (2009). *The Scientific Foundation for Herbal Medicinal Products*, 2nd edition. Georg Thieme Verlag, Stuttgart.

Eu, C. H. A., Lim, W. Y., Ton, S. H. and Abdul Kadir, K. (2010). Glycyrrhizic acid improved lipoprotein lipase expression, insulin sensitivity, serum lipid and lipid deposition in high-fat diet-induced obese rats. *Lipids Health Dis* 9: 81.

European Medicines Agency. (2022). Sales of veterinary antimicrobial agents in 31 European countries in 2021. https://www.ema.europa.eu/en/documents/report/sales-veterinary-antimicrobial-agents-31-european-countries-2021-trends-2010-2021-twelfth-esvac_en.pdf

Farinacci, P., Mevissen, M., Ayrle, H., Maurer, V., Sørensen Dalgaard, T., Melzig, M. F. and Walkenhorst, M. (2021). Medicinal plants for prophylaxis and therapy of common infectious diseases in poultry – A systematic review of *In Vivo* studies. *Planta Medica*. https://doi.org/10.1055/a-1543-5502

Fröhner, E. (Ed.). (1900). *Lehrbuch der Arzneimittellehre für Thierärzte 5*. Auflage. Verlag von Ferdinand Enke, Stuttgard.

Gonz´alez, J. A., Verde, A. and Pardo-de-Santayana, M. (2020). The use of plants for animal health care in the Spanish Inventory of Traditional Knowledge. In: McGaw L. J. and Abdalla, M. A. (Eds.), *Ethnoveterinary Medicine*. Springer, Cham, pp. 391–426.

Groot, M. J. Kruiden en diergezondheid. https://www.wur.nl/nl/landingspagina-redacteuren/nl/onderzoek-resultaten/onderzoeksinstituten/food-safety-research/wetenschappelijk-onderzoek/kruiden-en-diergezondheid.htm

Groot, M. J., Berendsen, B. J. A. and Cleton, N. B. (2021). The next step to further decrease veterinary antibiotic applications: Phytogenic alternatives and effective monitoring; the Dutch approach. *Front Vet Sci* 8: 709750. https://pubmed.ncbi.nlm.nih.gov/34540934/

Groot, M. J., Kleijer Ligtenberg, G. and van Asseldonk, A. (2010). Natural swine health: A guide to keeping your pigs healthy with herbs and other natural products. https://edepot.wur.nl/194290

Groot, M., van Asseldonk, A. and Puls-van der Kamp, I. (2011a). Natural poultry health: A guide to keeping your flock healthy with herbs and other natural products. https://edepot.wur.nl/194291

Groot, M. J., Kleijer-Ligtenberg, G., van Asseldonk, T. and Hansma, H. (2011b). Natural dairy cow health. https://www.naturallivestockfarming.com/wp-content/uploads/2015/09/Natural-Dairy-Cow-Health.pdf

Groot, M. J., Kleijer-Ligtenberg, G., van Asseldonk, T. and Hansma, H. (2016). Stalboekje melkvee: handboek voor natuurlijke melkveegezondheidszorg met kruiden en andere natuurproducten. https://www.wur.nl/nl/download/stalboekje-melkvee-2016.htm

Händler, J., Aslan, S., Findik, M., Kalender, H., Bastan, A., Kaymaz, M., Tomaschek, N. and Wesenauer, G. (2005). Wirksamkeit der intrauterinen Behandlung von puerperalen und post-puerperlaen Endometritiden mit Eucacomp bzw. Lotagen beim Rind. *Praktischer Tierarzt* 86 (3): 188–195.

Headon, D. and Dawson, K. (1990). Yucca extract controls atmospheric ammonia levels. *Feedstuffs* 62 (29): 15–16.

Hektoen, L. (2004). Investigations of the motivations underlying dairy farmers' use of homoeopathy. *Vet Rec* 155: 701–707. https://doi.org/10.1136/vr.155.22701

Hellec, F., Manoli, C. and Joybert, M. (2021). Alternative medicines on the farm: A study of dairy farmers' experiences in France. *Front Vet Sci* 8: 563957. https://doi.org/10.3389/fvets.2021.563957

HMPC: Committee on Herbal Medicinal Products of the European Medicines Agency. https://www.ema.europa.eu/en/committees/committee-herbal-medicinal-products-hmpc, 30.10.2022.

Hong, M., Li, S., Tan, H. Y., Wang, N., Tsao, S. W. and Feng, Y. (2015). Current status of herbal medicines in chronic liver disease therapy: the biological effects, molecular targets and future prospects. *Int J Mol Sci* 16: 28705–28745.

Huang, Q., Liu, X., Zhao, G., Hu, T. and Wang, Y. (2018). Potential and challenges of tannins as an alternative to in-feed antibiotics for farm animal production. *Anim Nutr* 4: 137–150.

Jeong, J. K., Choi, I. S., Moon, S. H., Lee, S. C., Kang, H. G., Jung, Y. H., Park, S. B. and Kim, I. H. (2018). Effect of two treatment protocols for ketosis on the resolution, postpartum health, milk yield, and reproductive outcomes of dairy cows. *Theriogenol* 106: 53–59.

Jouany, J. P. and Morgavi, D. P. (2007). Use of 'natural' products as alternatives to antibiotic feed additives in ruminant production. *Animal* 1 (10): 1443–1466. https://doi.org/10.1017/S1751731107000742

Kalle, R. and Kass, M. (2020). Local practice of cattle farming and ethnoveterinary medicine in Estonia: case study of Saaremaa and Muhumaa. In: McGaw, L. J. and Abdalla, M. A. (Eds.), *Ethnoveterinary Medicine*. Springer, Cham, 345–374.

Keller, D., Blanco-Penedo, I., De Joybert, M., & Sundrum, A. (2019). How target-orientated is the use of homeopathy in dairy farming?–A survey in France, Germany and Spain. *Acta Vet Scand* 61: 30. https://doi.org/10.1186/s13028-019-0463-3

Kiarazm, M., Talik, P. and Nava, H. G. (2011). Assessment of the effect of homoeopathic nosodes in subclinical bovine mastitis. *Ann Biol Res* 2: 552–562.

Klocke, P., Ivemeyer, S., Butler, G., Maeschli, A. and Heil, F. (2010). A randomized controlled trial to compare the use of homeopathy and internal teat sealers for the prevention of mastitis in organically farmed dairy cows during the dry period and 100 days post-calving. *Homeopathy* 99: 90–98.

Kupper, J., Walkenhorst, M., Ayrle, H., Mevissen, M., Demuth, D. and Naegeli, H. (2018). Online-Informationssystem für die Phytotherapie bei Tieren. *Schweizer Archiv für Tierheilkunde* 10: 589–595. https://doi.org/10.17236/sat00178

Kuralkar, P. and Kuralkar, S. V. (2021). Role of herbal products in animal production – an updated review. *J Ethnopharmacol* 278: 114246. https://doi.org/10.1016/j.jep.2021.114246

Latek, U., Chłopecka, M., Karlik, W. and Mendel, M. (2022). Phytogenic compounds for enhancing intestinal barrier function in poultry–A review. *Planta Med* 88: 218–236. https://doi.org/10.1055/a-1524-0358

Li, J. Y., Cao, H. Y., Liu, P., Cheng, G. H. and Sun, M. Y. (2014). Glycyrrhizic acid in the treatment of liver diseases: literature review. *Biomed Res Int* 2014: 872139.

Li, L., Sun, X., Zhao, D. and Dai, H. (2021). Pharmacological applications and action mechanisms of phytochemicals as alternatives to antibiotics in pig production. *Front Immunol* 12: 798553. https://doi.org/10.3389/fimmu.2021.798553

Loiklung, C., Sukon, P. and Thamrongyoswittayakul, C. (2022). Global prevalence of subclinical ketosis in dairy cows: A systematic review and meta-analysis. *Res Vet Sci* 144: 66–76.

Maeschli, A., Schmidt, A., Amman, W., Schurtenberger, P., Maurer, E. and Walkenhorst, M. (2019). Influence of a complementary and alternative medical advisory program by telephone on the use of antibiotics in Livestock in Switzerland. *Complement Med Res* 26 (3): 174–181. https://doi.org/10.1159/000496031

Mahfuz, S., Shang, Q. and Piao, X. (2021). Phenolic compounds as natural feed additives in poultry and swine diets: A review. *J Anim Sci Biotechnol* 12: 48. https://doi.org/10.1186/s40104-021-00565-3

Mangold, P. and Reicherter, E. (1907). *Neues Illustriertes Haustierarzneibuch* (10. Auflage (Erstauflage 1885). Enßlin & Laiblins Verlagsbuchhandlung.

Mann, S., Yepes, F. A. L., Behling-Kelly, E. and McArt, J. A. A. (2017). The effect of different treatments for early-lactation hyperketonemia on blood beta-hydroxybutyrate, plasma non-esterified fatty acids, glucose, insulin, and glucagon in dairy cattle. *J Dairy Sci* 100: 6470–6482.

Marx, W., Kiss, N. and Isenring, L. (2015). Is ginger beneficial for nausea and vomiting? An update of the literature. *Curr Opin Support Palliat* 9: 189–195.

Mayer, M., Vogl, C. R., Amorena, M., Hamburger, M. and Walkenhorst, M. (2014). Treatment of organic livestock with medicinal plants: A systematic review of European ethnoveterinary research. *Complement Med Res* 21 (6): 375–386. https://doi.org/10.1159/000370216

Mayer, M., Zbinden, M., Vogl, C. R., Ivemeyer, S., Meier, B., Morena, M., Maeschli, A., Hamburger, M. and Walkenhorst, M. (2017). Swiss ethnoveterinary knowledge on medicinal plants – a within-country comparison of Italian speaking regions with north-western German speaking regions. *J Ethnobiol Ethnomed* 13: 3.

McCabe, A. (2007). *A Byzantine Encyclopaedia of Horse Medicine: The Sources, Compilation, and Transmission of the Hippiatrica*. Oxford University Press, Oxford and New York. ISBN 978-0-19-153510-9.

Melzig, M. F., Mendel, M. and Walkenhorst, M. (2022). Editorial for the special issue «Veterinary Phytotherapy». *Planta Med* 88: 185–186. https://doi.org/10.1055/a-1739-6594

Mertenat, D., Dal Cero, M., Vogl, C. R., Ivemeyer, S., Meier, B., Maeschli, A., Hamburger, M. and Walkenhorst, M. (2020). Ethnoveterinary knowledge of farmers in bilingual regions of Switzerland - is there potential to extend veterinary options to reduce antimicrobial use? *J Ethnopharmacol* 246: 112184.

Mirzaee, F., Hosseini, A., Jouybari, H. B., Davoodi, A. and Azadbakht, M. (2017). Medicinal, biological and phytochemical properties of Gentiana species. *J Tradit Complement Med* 7: 400–408.

Nagulapalli Venkata, K. C., Swaroop, A., Bagchi, D. and Bishayee, A. (2017). A small plant with big benefits: Fenugreek (*Trigonella foenum-graecum* Linn.) for disease prevention and health promotion. *Mol Nutr Food Res* 61: 6.

Nielsen, B. K. (2008). Effects of botanicals rich in polyphenolics on animal productivity and health: A review (Book chapter). *New Res Live Sci Dairy Farm (P.Di Alberto (Editor), C. Costa (Editor),* 41–76.

Nuber, U., van Dorland, H. A. and Bruckmaier, R. M. (2016). Effects of butafosfan with or without cyanocobalamin on the metabolism of early lactating cows with subclinical ketosis. *J Anim Physiol and Anim Nutrit* 100: 146–155.

Oliveira, M., Hoste, H. and Custódio, L. (2021). A systematic review on the ethnoveterinary uses of mediterranean salt-tolerant plants: Exploring its potential use as fodder, nutraceuticals or phytotherapeutics in ruminant production. *J Ethnopharmacol* 267: 113464.

Orjales, I., López-Alonso, M., Rodríguez-Bermúdez, R., Rey-Crespo, F., Villar, A. and Miranda, M. (2016). Use of homeopathy in organic dairy farming in Spain. *Homeopathy* 105 (1): 102–108. https://doi.org/10.1016/j.homp.2015.08.005

Oudendorp, A. (2014). De invloed van vezelhydrolysaat uit tarwestro als natuurlijk middel op verschillende gezondheidsparameters en het geneesmiddelengebruik bij witvleeskalveren. Intern rapport.

Provenza, F. (2018). *Nourishment: What Animals Can Teach Us about Rediscovering Our Nutritional Wisdom*. Chelsea Green Publishing, Vermont. ISBN 9781603588027.

Raboisson, D., Mounié, M. and Maigné, E. (2014). Diseases, reproductive performance, and changes in milk production associated with subclinical ketosis in dairy cows: A meta-analysis and review. *J Dairy Sci* 97 (12): 7547–7563. https://pubmed.ncbi.nlm.nih.gov/25306269/

Regulation 2021/1165. https://eur-lex.europa.eu/legal-content/EN/TXT/?uri=CELEX%3A32021R1165

REGULATION (EC) No 1831/2003 OF THE EUROPEAN PARLIAMENT AND OF THE COUNCIL of 22 September 2003. https://eur-lex.europa.eu/legal-content/EN/TXT/HTML/?uri=CELEX:32003R1831&qid=1687942652986

REGULATION (EC) No 767/2009 OF THE EUROPEAN PARLIAMENT AND OF THE COUNCIL of 13 July 2009 on the placing on the market and use of feed, amending European Parliament and Council Regulation (EC) No 1831/2003 and repealing Council Directive 79/373/EEC, Commission Directive 80/511/EEC, Council Directives 82/471/EEC, 83/228/EEC, 93/74/EEC, 93/113/EC and 96/25/EC and Commission Decision 2004/217/EC

REGULATION (EU) 2018/848 OF THE EUROPEAN PARLIAMENT AND OF THE COUNCIL of 30 May 2018 on organic production and labelling of organic products and repealing Council Regulation (EC) No 834/2007, Article 1.5.2.3. https://eur-lex.europa.eu/legal-content/EN/TXT/?uri=CELEX%3A32018R0848

Regulation (EU) 2019/6 of the European Parliament and of the Council of 11 December 2018 on veterinary medicinal products and repealing Directive 2001/82/EC. https://eur-lex.europa.eu/eli/reg/2019/6/oj

Regulation (EU) 2021/1165 of 15 July 2021 authorising certain products and substances for use in organic production and establishing their lists (Text with EEA relevance). https://eur-lex.europa.eu/legal-content/EN/TXT/HTML/?uri=CELEX:32021R1165&from=EN

Reichling, J., Frater-Schröder, M., Saller, R., Fitzi-Rathgen, J. and Gachnian-Mirtscheva, R. (2016). *Heilpflanzenkunde für die Veterinärpraxis*. Springer-Verlag. https://link.springer.com/book/10.1007/978-3-662-48795-2

Röll, M. F. (1866). *Lehrbuch der Arzneimittellehre für Thierärzte*. 2 ed. Wilhelm Braunmüller, Vienna.

Rossi, B., Toschi, A., Piva, A. and Grilli, E. (2020). Single components of botanicals and nature-identical compounds as a non-antibiotic strategy to ameliorate health status and improve performance in poultry and pigs. *Nutr Res Rev* 33: 218–34. https://doi.org/10.1017/S0954422420000013

Ruegsegger, G. J. and Schultz, L. H. (1986). Use of a combination of propylene glycol and niacin for subclinical ketosis. *J Dairy Sci* 69: 1411–1415.

Sacket, D. L., Rosenbert, W. M. C., Gray, J. A. M., Haynes, R. B. and Richardson, W. S. (1996). Evidence based medicine: What it is and what it isn't. *BMJ* 32: 71–72.

Schlittenlacher, T., Knubben-Schweizer, G., Dal Cero, M., Vogl, C. R., Maeschli, A., Hamburger, M. and Walkenhorst, M. (2022). What can we learn from past and recent Bavarian knowledge for the future development of European veterinary herbal medicine? An ethnoveterinary study. *J Ethnopharmacol* 288: 114933. https://doi.org/10.1016/j.jep.2021.114933

Schmid, K., Ivemeyer, S., Hamburger, M., Vogl, C., Klarer, F., Meier, B. and Walkenhorst, M. (2012). Traditional use of herbal remedies in livestock by farmers in three Swiss cantons (Aargau, Zurich and Schaffhausen). Forschende Komplementärmedizin. *Forsch Komplement Med* 19: 125–136. https://doi.org/10.1159/000339336

Schutz, K., Carle, R. and Schieber, A. (2006). Taraxacum– a review on its phytochemical and pharmacological profile. *J Ethnopharmacol* 107: 313–323.

Shin, B. and Park, W. (2018). Zoonotic diseases and phytochemical medicines for microbial infections in veterinary science: Current state and future perspective. *Front Vet Sci* 5: 166. https://doi.org/10.3389/fvets.2018.00166

Shrubok, A. (2020). Belarusian ethnoveterinary medicine: Ritual practices and traditional remedies. In: McGaw, L. J. and Abdalla, M. A. (Eds.), *Ethnoveterinary Medicine*. Springer, Cham: 375–390.

Soukand, R., Hrynevich, Y., Prakofjewa, J., Valodzina, T., Vasilyeva, I., Paciupa, J., Shrubok, A., Hlushko, A., Knureva, Y., Litvinava, Y., Vyskvarka, S., Silivonchyk, H., Paulava, A., Koiva, M. and Kalle, R. (2017). Use of cultivated plants and non-plant remedies for human and animal home-medication in Liuba´n district, Belarus. *J Ethnobiol Ethnomed* 13 (1): 1–31. https://doi.org/10.1186/s13002-017-0183-6

Steiner, L., Jenny, U., Hirsbrunner, G. and Walkenhorst, M (2022). Phytotherapeutic treatments of gynecological diseases and fertility disorders in cattle – a veterinary historical analysis. *Schweizer Archiv für Tierheilkunde* 164 (9): 645–659. https://doi.org/10.17236/sat00367

Stevens, M., Piepers, S. and De Vliegher, S. (2016). Mastitis prevention and control practices and mastitis treatment strategies associated with the consumption of (critically important) antimicrobials on dairy herds in Flanders, Belgium. *J Dairy Sci* 99 (4): 2896—2903.

Strategie Antibiotikaresistenzen Schweiz (StAR). (2022). https://www.blv.admin.ch/blv/de/home/das-blv/strategien/nationale-strategie-antibiotikaresistenzen.html

Stucki, K., Dal Cero, M., Vogl, C. R., Ivemeyer, S., Meier, B., Maeschli, A., Hamburger, M. and Walkenhorst, M. (2019). Ethnoveterinary contemporary knowledge of farmers in pre-alpine and alpine regions of the Swiss cantons of Bern and Lucerne compared to ancient and recent literature – is there a tradition? *J Ethnopharmacol* 234: 225–244.

Subarevic, N., Stevanovic, O. and Petrujkic, B. (2015). Use of phytotherapy as a form of ethnoveterinary medicine in the area of Stara planina mountain in Serbia. *Acta Medico-Historica Adriat* 13 (1): 75–94.

Suthar, V. S., Canelas-Raposo, J., Deniz, A. and Heuwieser, W. (2013). Prevalence of subclinical ketosis and relationships with postpartum diseases in European dairy cows. *J Dairy Sci* 96: 2925–2938.

Tournier, A., Klein, S. D., Würtenberger, S., Wolf, U, and Baumgartner, S. (2019). Physicochemical investigations of homeopathic preparations: A systematic review and bibliometric analysis-part 2. *J Altern Complement Med* 25 (9): 890–901. https://www.liebertpub.com/doi/pdf/10.1089/acm.2019.0064

Ücker, A., Baumgartner, S., Sokol, A., Huber, R., Doesburg, P. and Jäger, T. (2018). Systematic review of plant-based homeopathic basic research: An update. *Homeopathy* 107 (2): 115–129. https://pubmed.ncbi.nlm.nih.gov/29621812/

Viegi, L. and Ghedira, K. (2014). Preliminary study of plants used in ethnoveterinary medicine in Tunisia and in Italy. *Afr J Tradit Complementary Altern Med* 11 (3): 189–199. https://doi.org/10.4314/ajtcam.v11i3.27

Villalba, J. J. and Provenza, F. D. (2007). Self-medication and homeostatic behaviour in herbivores: Learning about the benefits of nature's pharmacy. *Animal* 1(9): 1360–1370.

Villalba, J. J., Provenza, F. D. and Stott, R. (2009). Rumen distension and contraction influence feed preference by sheep. *J Anim Sci* 87 (1): 340–350.

Vogl, C. R., Vogl-Lukasser, B. and Walkenhorst, M. (2016). Local knowledge held by farmers in Eastern Tyrol (Austria) about the use of plants to maintain and improve animal health and welfare. *J Ethnobiol Ethnomed* 12: 40.

Walkenhorst, M. (2017). Tiermedizinische Publikationen in complementary medicine research. *Complement Med Res* 24: 338–341. https://doi.org/10.1159/000486074

Wallace, R. J., Arthaud, L., Newbold, C. J., et al. (1994). Influence of *Yucca schidigera* extract on ruminal ammonia concentrations and ruminal microorganisms. *Appl Environ Microbiol* 60 (6): 1762–1767.

Werner, C., Sobiraj, A. and Sundrum, A. (2010). Efficacy of homeopathic and antibiotic treatment strategies in cases of mild and moderate bovine clinical mastitis. J Dairy Res 77(4): 460-–467. https://doi.org/10.1017/S0022029910000543

WHO Monographs: World Health Organization, WHO Consultation on Selected Medicinal Plants, WHO Consultation on Selected Medicinal Plants (2nd: 1999 : Ravello-Salerno, Italy), WHO Consultation on Selected Medicinal Plants (3rd : 2001 : Ottawa, Ont.) & WHO Consultation on Selected Medicinal Plants (4th : 2005 : Salerno-Paestum, Italy). (2006). *WHO Monographs on Selected Medicinal Plants*. World Health Organization. https://apps.who.int/iris/handle/10665/42052, 30.10.2022

Wichtl, M. (Ed.). (2009). *Teedrogen und Phytopharmaka, Ein Handbuch für die Praxis auf wissenschaftlicher Grundlage*. 5. Auflage, Wissenschaftliche Verlagsgesellschaft mbH, Stuttgart.

Yang, C., Chowdhury, M. A. K., Hou, Y. and Gong, J. (2015). Phytogenic compounds as alternatives to in-feed antibiotics: potentials and challenges in application. *Pathogens* 4: 137–156. https://doi.org/10.3390/pathogens4010137

Zipperlen, W. (1959). *Praktischer, Illustrierter HAUSTIERARZT für Landwirte und Haustierbesitzer – zugleich ein Handbuch für Tierärzte und Hufschmiede* (14. Auflage (Erstauflage 1867). Verlag Max und Carl Ebner & Siebeneicher Verlag.

Chapter 9

Improving the management of parasites in organic dairy farming

Stig M. Thamsborg, Andrew Williams, and Helena Mejer, University of Copenhagen, Denmark; Nadine Ravinet, INRAE, France; Miguel Peña-Espinoza, University of Veterinary Medicine Vienna, Austria; and Philip J. Skuce, Moredun Research Institute, UK

1 Introduction: Natural life and parasites – and the ethos of organic farming

Parasites of animals, and plants as well, constitute an important part of any ecosystem. In fact, 50% of all living species have a lifestyle that can be regarded as parasitic (Poulin and Morand, 2004), and parasites comprise up to 3% of the total biomass in certain estuarine ecosystems (Kuris et al., 2008). Parasites are also an inevitable part of any livestock production system. Although organic farming as a principle aims for natural living (Hovi et al., 2003; Lund, 2005), there will be inherent restrictions imposed due to, e.g. limited space, age

http://dx.doi.org/10.19103/AS.2024.0138.10

structure of animal flock or herd and nutrition, which make it different from natural ecosystems. These factors will affect the balance between parasites and hosts, and in some cases result in disease. Some parasite infections affect only a few farms, and can be limited by biosecurity measures, whereas other infections are endemic and inherent to grazing systems and are therefore difficult, or impossible, to avoid. The management in organic dairy systems should aim for an ecological balance between parasite burden and animal welfare/ unavoidable production losses, and when needed, appropriate intervention (Thamsborg and Roepstorff, 2003). This chapter will briefly summarize the most important parasite infections and problems recognized in organic dairy farming – many of these are well known from conventional farming but levels and interventions may be different. Furthermore, we will discuss new directions in research within this area of organic dairy farming.

2 Cryptosporidiosis

The protozoan parasite, *Cryptosporidium parvum*, is frequently the cause of diarrhoea in neonatal calves. Clinically, profuse watery diarrhoea, dehydration and weight loss are observed, typically within the first 2–3 weeks of life. The infection is often self-limiting but can have a more serious course associated with early infection with a high infective dose and/or if the calf is co-infected with, e.g. rota-coronavirus. *C. parvum* has a broad host spectrum, including humans, while older calves (>1 month of age) are more often infected with cattle-specific species (*C. bovis* and, less commonly, *C. ryanae*) (Langkjær et al., 2006). *C. bovis* infection generally has a milder/non-clinical course (Åberg et al., 2019) and may predominate in some countries (Silverås and Blanco-Penedo, 2013).

It is most likely that *Cryptosporidium* spp. are present in all organic dairy herds, but even if infection is the rule, the associated disease is not observed everywhere. Predisposing factors for high infection levels are poor hygiene (e.g. lack of, or poor, cleaning of calf boxes), low colostrum allocation, high stocking density, and also some seasonal variation is seen, typically an increase in autumn. A Danish study showed higher shedding rates of *Cryptosporidium* spp. in calves from organic herds, where calves typically spend more time with the dam (minimum 48 h), compared to conventional herds and in calves from herds with short resting time between use of calving pens (Maddox-Hyttel et al., 2006). Moving calves after a maximum of 1 week from single pens to group pens is mandatory for welfare reasons in organic farming in the EU, with this practice resulting in more contacts between calves. A recent systematic review of 14 studies confirmed that organic production, large herd size, contact with several other calves and possibly co-infections were significant risk factors for *C. parvum* infection (Brainard et al., 2020). In summary, it is most likely that organic dairy herds may have more problems with cryptosporidiosis than conventional herds, but single responsible factors may be difficult to pinpoint. Whether the prolonged period of calf with dam is a decisive factor is still questionable, e.g. an Estonian study suggested that early removal of calves immediately after birth increases calves' excretion of oocysts (Lassen et al., 2009). Silverås and Blanco-Penedo (2013) emphasized the importance of farmers' attitude to biosecurity for disease occurrence, and interestingly, found no differences in prevalence of *Cryptosporidium* spp. (predominantly *C. bovis*) between organic and conventional calves (44.7% vs. 52.3%) or cows.

Cryptosporidiosis is transmitted between calves by means of oocysts, small egg-like structures passed in faeces of infected calves following an intestinal development cycle of 2–10 days. Unlike coccidia (below), oocysts of *Cryptosporidium* spp. are immediately infectious when they leave the calf. The oocysts survive for several months in cool, moist weather but are sensitive to desiccation. They are resistant to most disinfectants except gaseous ammonia and hydrogen peroxide (Tomazic et al., 2018). Still, the occurrence of cryptosporidiosis in an organic herd can best be limited by hygienic and other preventative management measures such as: clean, dry and disinfect calving pens, clean pens and utensils for new-born calves, avoid spread between calves, secure sufficient colostrum, clean group pens, stable groups in pens (all-in-all-out) and good quality hay in racks. Lastly, quick and appropriate intervention in case of diarrhoea may limit oocyst excretion (Tomazic et al., 2018).

Rearing dairy calves with nurse cows (not milked lactating cows of the herd) is a practice which has been increasingly adopted in organic farming. A lower frequency of non-specific neonatal diarrhoea cases has been reported in calves reared with nurse cows (or their dams) compared to classically reared calves (Michaud et al., 2018; Wagenaar and Langhout, 2007). In France, a cow-calf rearing system on organic farms has been described, consisting of a first phase of 4–5 days with the dam, followed by an optional phase of artificial feeding on whole milk (15 days on average), and a final phase of fostering by a nurse cow (1–5 calves per nurse cow; average age at fostering: 8 days) (Constancis et al., 2021). This study also focused on *Cryptosporidium* and found an oocyst shedding prevalence similar to classically reared calves but a lower intensity of shedding and less diarrhea in calves with nurse cows (Constancis et al., 2021). However, the study also revealed some specific risks for *Cryptosporidium* infection, e.g. practicing the transitional artificial milk feeding phase, and early fostering (≤3 days) when a high number of calves (>3) were fostered by the same nurse. This suggested that early changes in feeding and housing, together with the stress of multiple fostering (Johnsen et al., 2016), could increase the susceptibility and/or the early intake of oocysts.

Options for medical treatment are limited to halofuginone, primarily active on the extracellular stages (sporozoites and merozoites). The treatment should be initiated no later than one day after the onset of the diarrhoea, but even then, the curative efficacy is poor. In fact, the use of halofuginone is better indicated for prevention of diarrhoea caused by *C. parvum* when applied metaphylactic in all calves within 1–2 days of birth during a period of disease problems. The approach is questionable in an organic context, the administration protocol is cumbersome and difficult to apply rigorously and further, it is difficult to determine when treatments can be terminated without reappearance of disease. This reinforces the importance of hygienic measures aimed at lowering the infection pressure from the environment.

3 Coccidiosis

3.1 Coccidiosis in stabled calves

Coccidiosis typically affects calves under 6 months of age and is caused by different species of the protozoan parasites of the genus *Eimeria*, most frequently by *E. bovis* and *E. zuernii*. Practically all cattle herds are infected at some level. Within herds, there will be great variation in the manifestation of the disease among the calves, but in several studies, infection has been detected among the majority of stabled calves (50%–90%), independent of disease (Autzen et al., 2002; Enemark et al., 2013). Transmission of coccidiosis is by oocysts excreted in faeces of calves starting 18–21 days after infection. As long as it is possible to detect and count oocysts in faeces, the infection is classified as being a patent infection. Oocysts sporulate in the environment and become infective within 2–7 days. Clinical diagnosis of coccidiosis of calves is made by detecting high oocyst excretion (>1000 oocysts per g faeces, OPG) (Enemark et al., 2013), together with case history and clinical findings (often bloody diarrhoea). Clinical signs are roughly correlated with shedding of oocysts (Autzen et al., 2002). However, the vast majority of infections are subclinical and result mainly in reduced animal growth. Clinical disease is linked to a high level of infection, which in turn is related to environmental factors favoring development and survival of oocysts, high oocyst ingestion and/or host stress factors inducing high multiplication of the parasite within the hosts. The latter is fundamentally different from helminths (see next section), which have no multiplication in the host. Clinical disease often occurs 3–4 weeks after bringing calves together, e.g. in group pens. However, even in stables with individual pens in a row, the infection can run from one animal to another. If deep bedding is not removed in a pen from one group to the next, there is often a problem, especially if the last calves brought into the pen have been very young. In contrast, for calves older than 6 months, deep litter bedding can often be reused without problems (pers. communication). A number of factors predispose to high oocyst excretion, e.g. poor colostrum allocation, poor and moist feed and absence of hay racks (Autzen et al., 2002). Other factors include change of feed, buying animals from other herds, overcrowding (too few boxes), general poor hygiene and concurrent viral infections.

Improved management can, in many cases, solve the problem, but not in all. Management measures include proper colostrum allocation, thorough cleaning of boxes and utensils (e.g. by heating, possibly flame-burning, washing, disinfection and drying), avoiding stress and mixing, and appropriate coating and feeding. Furthermore, sick animals must be isolated. For treatment, triazine-derivatives (toltrazuril and diclazuril) are currently the drugs of choice. It is noted that toltrazuril is phytotoxic (see later section). Toltrazuril is registered for the prevention of clinical symptoms of coccidiosis and for reduction of oocyst shedding in calves in dairy herds where coccidiosis has been confirmed. Often a routine treatment to avoid expected recurring problems is applied (e.g. specific number of days after moving to a group pen), alternatively treatment is initiated after illness in the first individual in a herd (metaphylactic use). Since there are few other medicinal alternatives, there is a significant risk of developing toltrazuril resistance, as seen in sheep (Odden et al., 2018).

3.2 Coccidiosis on pasture

In early season, young dairy calves are often pastured in small permanent paddocks close to the farm buildings. This makes adaptation to grazing and welfare observations easier for the farmer, as well as facilitating supplementary feeding. However, the coccidia oocysts may survive on pasture from one grazing season to the next (and further), and highly infected animals turned-out may also rapidly contaminate paddocks. Although much less common than stable coccidiosis, this may result in clinical coccidiosis, similar to that mentioned above, in calves after 1–3 months on pasture. Even in cases without clinical disease, turn-out onto contaminated pasture, resulting in moderate oocyst excretion levels in organically reared calves, may be associated with significantly reduced weight gains (Nielsen et al., 2003). Prevention can thus be achieved by the use of clean pasture at the turn-out, i.e. paddocks not grazed by calves the year before, and also through supplementary feeding. Furthermore, fixed or stable groups of animals, i.e. without unnecessary mixing or new introductions later in the season, are strongly recommended. Drug treatment may be considered in clinical cases, as mentioned earlier, but toltrazuril and related triazine drugs are phytotoxic and a ban on the use of toltrazuril for calves above 80 kg must be observed. Clinical coccidiosis on pasture appears in some countries as a specific syndrome closely related to turn-out (Svensson et al., 1997). This clinical syndrome is characterized by watery diarrhoea beginning 4–7 days after turn-out and is associated with a very high excretion of *E. alabamensis* oocysts (>100,000 OPG) a few days later. The significance of *E. alabamensis* in organic farming is to the best of our knowledge, not clearly elucidated.

4 Pasture-borne infections

On all organic farms, calves, heifers and cows need to have access to pastures for a large part of their productive life, typically 120–150 days annually (Hovi et al., 2003; Sorge et al., 2015). While grazing, there is no doubt that the animals get plenty of opportunities to exercise natural behaviour, for example, social behaviour in larger groups, free movement and natural food selection. Natural living by itself achieves a high degree of animal welfare (Philips & Heins, 2022). The flipside of this is often a low growth rate and a series of infections that are closely related to grazing and particularly permanent pastures. These pasture-borne infections include parasites like coccidia, gastrointestinal nematodes (GIN), ectoparasites, lungworms and liver flukes. Several of these infections are commonly found in higher levels in target age groups in organic farms, compared to conventional farms (Höglund et al., 2010; Sorge et al., 2015; Olsen et al., 2015). Many conventional farms will have entirely indoor rearing for some stock. While liver flukes (and rumen flukes) are, with very few exceptions, acquired on pasture, the impact is often seen later, e.g. during the stabling period.

5 Gastrointestinal nematodes

5.1 Introduction

Gastrointestinal nematodes (GIN) are inherent in all grazing systems, and organic farmers need to strike a balance so that infection levels are sufficiently low to avoid clinical disease and/or serious production losses and high enough for immunity to establish. In the short term, the foundation for reaching this balance is appropriate management of pastures and nutrition. The typical age group at risk is young calves during their first season on pasture. Heavily infected calves will have watery diarrhoea in the second half of the season, typically July–August in the northern hemisphere, but a subclinical course with reduced weight gains (often up to 50%) and poor condition is more common (Charlier et al., 2014). If sufficiently exposed, calves will become solidly immune and the following year, as second-year grazing heifers (or steers) and later as cows, they will have very low GIN infection levels and seldom experience clinical problems. However, we know from studies in milking cows treated with anthelmintics that GIN infections may, on average, result in a production penalty of around 0.35–1 kg/milk day (Charlier et al., 2014).

It is often argued that exposure to GIN (infection pressure) and consequently infection levels could be higher in organic dairy systems, but this assumption is not always confirmed in comparative surveys. Studies of GIN faecal egg counts in first season heifers in Swedish organic dairy farms revealed generally low levels of infections (Höglund et al., 2001). It was clearly indicated that organic farmers were more aware of parasites and applied more alternative approaches as compared to conventional (Svensson et al, 2000). A comparative study in Minnesota found generally low levels of GIN infection in breeding-age heifers, but infection levels in this age group in organic herds were still significantly higher than in conventional dairy herds (Sorge et al., 2015). While organic farmers applied no or very little anthelmintic, thus enhancing opportunities for build-up of immunity, 71%–92% of conventional herds applied routine treatments (Sorge et al., 2015). Studies in cows have indicated no major differences in infection levels between organic and conventional herds using individual or bulk-tank milk levels of *Ostertagia*-specific antibodies (Blanco-Penedo et al., 2012) or FEC (Chincarini et al., 2022), while an earlier study found significantly higher bulk tank antibody levels in organic farms (Höglund et al., 2010).

5.2 Life cycle and diagnostics

The mature GIN (mainly *Ostertagia ostertagi* and *Cooperia* spp.) residing in the stomach (abomasum) and small intestine of cattle, respectively, produce eggs that are passed out with faeces onto pasture (Fig. 1). Eggs develop in the faecal pat to the infective third larval stage (L_3) within 1–6 weeks, depending on temperature and humidity. Transmission of L_3 from the faecal pat to the herbage is a continuous process, facilitated primarily by raindrop dispersal, thereby making L_3 available for ingestion by the animals. Once ingested, larvae develop to the adult stage within 3 weeks. However, *Ostertagia* L_3 ingested in the late season (from October in the northern hemisphere) become inhibited as L_4 in the abomasal mucosa and reactivated the following spring. This inhibition phenomenon can be particularly important in organic dairy production after heavy exposure during autumn and if no efficacious treatments are applied at or during the housing period to remove inhibited larvae. This has, however, not been studied in detail. The usual diagnostic procedure in young stock is counting of GIN eggs in faeces but, due to increasing levels of immunity and inhibition in the late season, faecal egg counts (FEC) are poorly correlated with the number of worms in the gut (i.e. the worm burden) at the end of the grazing season (after September in the northern hemisphere). In autumn at housing, in first-season grazing heifers, an alternative could be the measurement of serum pepsinogen levels, indicating abomasal lesions by *Ostertagia* (Charlier et al., 2023). The pepsinogen level can be used for several purposes: (i) to assess the effectiveness of the infection control plan during the past grazing season, (ii) to assess the need for treatment at housing and (iii) to adapt, if necessary, the

Figure 1 Life cycle of gastrointestinal nematodes (GIN) in cattle. Parasitic adult GIN (1) in the abomasum (e.g. *Ostertagia ostertagi*) or small intestine (e.g. *Cooperia* spp.) produce eggs (2) that are passed with the faeces into the environment. In the faecal pat, eggs develop until infective larvae (L_3) (3), which migrate to the herbage. Cattle are infected when consuming L_3 in the pasture (4). Ingested L_3 develops in the gastrointestinal tract (5) into L_4 and L_5, until reaching the adult stage (1). Alternatively, ingested L_4 late in the grazing season becomes inhibited as L_4 in the gastrointestinal mucosa and resume their development the following spring.

control plan for the next grazing season. Although found negatively associated with weight gain in some studies (Merlin et al., 2016; 2017), a universal threshold indicating levels of subclinical infection associated with growth retardation are not clearly defined. Measurements of bulk-tank milk antibody levels against GIN (primarily *Ostertagia*) are used extensively in some countries to evaluate exposure levels on-farm, but this indicator is not useful for young stock.

5.3 Prevention

Prevention of GIN infections in line with the goals of organic farming may include management of grazing to avoid build-up of infection levels on pasture (the so-called grazing management), and dietary interventions. Other approaches, like breeding of resistant animals, development of vaccines and biological control are also available (see later) but they may need further development or their role in organic farming may be questioned. They are briefly mentioned at the end of this chapter as cross-cutting topics. The routine, preventive use of anthelmintics is not allowed in organic farming systems (as well as in conventional systems in several countries) but there may be situations where the use of drugs can hardly be avoided, e.g. liver flukes. However, it is our opinion that grazing management and the use of supplementary feeding and/or bioactive forages will, to a large extent, prevent disease and serious losses and should therefore be an integral part of GIN prevention in organic dairy farming. Interestingly, a European survey of organic farmers' attitudes to parasites also revealed a willingness and a wish to apply non-medical approaches to control (Takeuchi-Storm et al., 2019). In all cases, daily supervision is very important to secure rapid intervention. Likewise, we have observed that calves that have had pneumonia or other ailments several times during the stabling period have difficulties coping with changes at the turn-out, e.g. exposure to parasites. These animals should not graze on extensive pastures.

5.4 Grazing management

Firstly, it is important to define 'clean' pastures (in a parasitological sense) as pastures with no or very low contamination with larvae (or eggs) of GIN (Box 1). The listed recommendations are strict and lower levels of safety (e.g. 'safe' instead of 'clean') can also work satisfactorily, if, e.g. a pasture has only been used for old immune stock for a brief period the year before. In this context, a clear advantage of organic farming is the relatively large proportion of farmland with grass/clover to ensure sufficient N-fixation, and this may be available for grazing, depending on the location of fields. Grazing management options for GIN prevention are commonly sub-divided as follows (Barger, 1996; Thamsborg et al., 1999) (see Table 1 for details):

- preventative strategies: avoiding build-up of contamination by grazing parasite-free animals on initially clean pasture;
- evasive strategies: evading contamination already built up by a move (or more) to a new, clean paddock
- diluting strategies: avoiding build-up of contamination by lowering the density of highly susceptible animals

---

Box 1 Parasite "clean" pastures – what is that?

Parasite "clean" pastures are defined best as pastures with no or very low contamination with larvae (or eggs) of gastrointestinal nematodes. This can be achieved by:

- Clean pastures at turn-out (as of mid-April/May in the Northern Hemisphere) are areas that:
  - have not been grazed by cattle the previous year
  - a new ley or undersown crop from the year before
- Clean pastures at mid-season (as of mid-July in the Northern Hemisphere) are areas that:
  - have not been grazed by cattle in the same year (used once for hay or silage cut)
  - have been grazed by animals other than cattle in the same year (e.g. sheep, horse)

These recommendations are very strict and lower levels of safety can also work satisfactory if, e.g. a pasture has only been used for old immune stock in a brief period the year before.

A classical preventative strategy is the turn-out of naive first-season dairy calves/heifers onto clean pasture in spring. Although infection levels eventually may build-up with this strategy, there is also the risk of under-exposure and thus lack of immunity. However, this has not been demonstrated. Other preventative strategies include delayed turn-out, resulting in lower levels of infection with overwintering L_3 on pasture, or grazing calves for 1–2 months at the end of the season, in order to achieve a certain level of immunity. If clean pasture is not available at the turn-out and young stock go on contaminated pasture, it is advisable to move animals to clean paddocks around mid-July (northern temperate regions) (Foldager et al., 1981; Larsson et al., 2006). This evasive strategy will not remove infections already established in the animals and calves will excrete eggs on the new paddock, but it will lower the risk of high contamination levels later in the season, particularly if the date of housing is not too late in the autumn (October). Repeated moves to clean pasture during the season are advisable for first-season calves, and the effect is comparable to repeated anthelmintic treatments (Eysker et al., 1998). The correct timing will depend on local knowledge of when the increase in pasture contamination takes place during summer and the length of the grazing season. Rotational grazing can be regarded as repeated moves, and vets and farmers sometimes believe that once animals are on rotational grazing, the GIN risk will be controlled. However, not all rotational systems are (equally) effective, and returning animals to paddocks grazed (read: contaminated) in the same season can be detrimental (Kristensen et al., 2006)(Fig. 2). Therefore, the variation in risk of GIN infection between rotational systems is hard to predict as it depends on the number of plots, the time spent on each plot (the return time), the weather, and the density of animals on each plot.

On permanent pastures, like rangeland, prevention is best achieved by dividing the area into two (if possible, more), where one part (e.g. one-third of

Table 1 Grazing management strategies against gastrointestinal nematodes - what is that? Practices can easily be combined and please note that other authors may classify some practices differently (modified from Barger (1996) and Thamsborg et al. (1999)).

	Preventative	Evasive	Diluting
Goal	Avoiding build-up of contamination by grazing parasite-free animals on initially clean pasture	Evading contamination already built up by a move (or more) to a new, clean paddock	Avoiding build-up of contamination by lowering the the density of highly susceptible animals
Practices	• The turn-out of naive first-season calves on clean pasture in spring • Delayed turn-out	• Move animals to clean pasture before pasture contamination peaks in early July (northern temperate climates) • Repeated moves to clean pasture, depending on the length of the grazing season • (Drug treatment should be avoided at the time of moving due to the risk of anthelmintic resistance)	• Co-grazing or alternate grazing (shifting paddocks) of cattle with another species, e.g. sheep or horses • Co-grazing young calves with older resistant stock • Reducing the animal density/stocking rate per se • Use of nurse cows

the area) is grazed by first-season calves before mid-July (northern hemisphere) and the remainder afterwards (Kristensen et al., 2006). Areas that are not used should preferably be cut for hay or silage, and if this is not possible, they can be grazed by second-season cattle or other livestock species. If the amount of feed is insufficient at the end of the season, calves should be moved home to clean pastures within the crop rotation, e.g. aftermath. Dilution can be achieved by mixing young susceptible calves with another species, e.g. sheep or horses, or with older resistant stock (second-season heifers) (Nansen et al. 1990), or simply by reducing the animal density or stocking rate, i.e. the number of animals per unit area (Nansen et al., 1988; Shaw et al., 1998; Thamsborg et al., 1999; Andersen et al., 2002). The mixed species/age groups can be performed either as co-grazing or alternate grazing (shifting paddocks), the effect will depend on the ratio between susceptible and non-susceptible, i.e. inert stock, e.g. the sheep–cattle ratio on the farm, as discussed in detail by Barger (1996). The earlier mentioned use of nurse cows may be regarded as a diluting strategy, see later under *Case study*.

5.5 Dietary interventions to regulate GIN

Nutritional interventions can regulate infections with GIN based on three main mechanisms (Coop and Kyriazakis, 2001): (a) increasing the ability of the animals

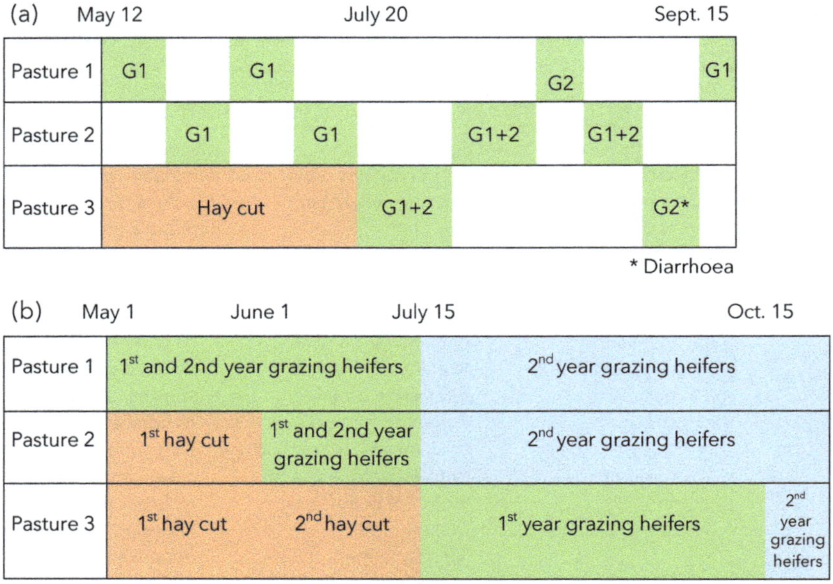

Figure 2 Schematic representation of grazing history on a farm with first and second-season calves/heifers in a system with three paddocks. Above show events for two groups (G1 & G2) of first season calves leading to clinical problems in calves in G2 in September (a). Below is a possible scenario where acceptable control has been achieved by co-grazing between age groups and moving first-season calves to clean pasture in July (b) (modified from Jensen (2002)).

to cope with the negative effects of GIN and sustain productive/reproductive functions despite infection ("resilience"); (b) enhancing the immune response of the animals against GIN that can control and potentially resolve the infection ("resistance") and; (c) directly disrupting the GIN in the animal by the intake of bioactive (anti-parasitic) compounds in the diets. Dietary interventions to control GIN in ruminants mainly include supplementary protein feeding and the use of bioactive forages with natural anthelmintic properties.

Supplementary feeding with protein-rich diets has been confirmed in several studies to increase the resilience and resistance of small ruminants against GIN (reviewed by Hoste et al., 2016), but surprisingly, very few trials have been conducted to verify these findings in cattle. In Denmark, Jørgensen et al. (1992) reported that grazing heifer calves with ad libitum access to lucerne (*Medicago sativa*) pellets in mid-season had a temporary reduction in FEC and less signs of clinical parasitism compared with control animals. However, whether these effects were due to the enhanced dietary protein or to a reduced pasture intake of the lucerne-fed group (leading to lower larval intake) are unclear. In Zimbabwe, Magaya et al. (2000) described that grazing steers of the indigenous Tuli breed supplemented with cottonseed meal (crude protein 33%) had significantly higher weight gains and a trend of lower GIN egg output compared with non-supplemented calves. In Sweden, supplementation of grazing calves with a commercial concentrate and hay for 1 month after turn-out did not prevent the development of subclinical and clinical parasitism, including reduced weight gain and need for anthelmintic treatment (Larsson et al., 2006). More recently, a large-scale study in Brazil reported a positive correlation between weight gain and FEC in grazing Nellore (*Bos taurus indicus*) calves supplemented with protein-rich diets, thus suggesting a positive effect on the animals' resilience, whereas a negative correlation between FEC and weight gain was observed in non-supplemented animals (Zapa et al., 2021). More experimental and on-farm studies are necessary to elucidate the effects and mechanisms of protein supplementation in cattle infected with GIN, including the evaluation of different protein sources, duration and timing of protein supplementation and responses in different cattle breeds.

Bioactive forages or crops are plants that, besides their nutritional value, contain bioactive phytochemicals with health benefits for the animal (Waller and Thamsborg, 2004). Due to their dual positive effects on animal nutrition and health, these bioactive plants are also described as 'nutraceuticals' (Hoste et al., 2015), as opposed to phytotherapy (see below in *Treatment*). In this regard, the inclusion of bioactive forages in cattle diets should be evaluated considering both their nutritional and health-promoting benefits, as well as their palatability for the animals. Further information on the main bioactive compounds identified in these forages and their anti-parasitic activity is available (e.g. for condensed tannins by Hoste et al., 2016; for sesquiterpene lactones by Peña-Espinoza et al., 2018; Valente et al., 2021). Different bioactive forages have been evaluated for their anthelmintic effects against GIN in cattle *in vivo*, namely chicory (*Cichorium intybus*), sainfoin (*Onobrychis viciifolia*), birdsfoot trefoil (*Lotus corniculatus*) and plantain (*Plantago lanceolata*). Current evidence supports the observation that the inclusion of some of these bioactive forages in cattle diets can help to selectively regulate infection levels of certain GIN via a reduction of adult worm burden and/or egg excretion. Forage chicory (Fig. 3) as silage or grazed fresh forage may lower *O. ostertagi* adult worm burdens by 60%–66% in calves (fed with ≥70% of chicory dry matter [DM] in daily intake) compared to calves fed a control (iso-proteic/iso-energetic) diet (Peña-Espinoza et al., 2016). Feeding chicory-based diets also resulted in reduced FEC excretion of *O. ostertagi* eggs by 65% within 4 weeks in calves grazed on forage chicory compared with animals grazing ryegrass, while *C. oncophora* infection levels were unaffected (Peña-Espinoza et al. 2016). An earlier study failed to show any difference in FEC and serum pepsinogen in naturally infected beef steers grazed on either a chicory/ryegrass pasture (~24% chicory DM in the field) or a pure stand of ryegrass (Marley et al. 2014). The available *in vivo* evidence thus indicates that a relatively high proportion of the forage intake needs to be chicory to elicit anthelmintic effects (≥70% of chicory DM intake), and the effect may be selective and only against abomasal worms. However, the lower effective intake of forage chicory to affect GIN *in vivo* and length of feeding period remain to be elucidated (Peña-Espinoza et al., 2018), but use of forage chicory at levels ≥50% DM intake is highly palatable for cattle and can increase milk production and/or polyunsaturated milk fatty acids in grazing dairy cows (Chapman et al., 2008; Muir et al., 2014; Mangwe et al., 2020). Furthermore, anti-parasitic effects of chicory by-products such as root pulp have been reported (Peña-Espinoza et al., 2020), and therefore, they could be further explored as alternative chicory-based strategies for cattle infected with *O. ostertagi*.

Feeding the leguminous forage sainfoin has also been demonstrated to significantly reduce GIN infection levels in small ruminants and cattle. Calves fed sainfoin-pellets (90% of the daily DM intake) had a 51% reduction of *O. ostertagi* burdens and lower pepsinogen in comparison with animals fed a control diet (Desrues et al., 2016). Moreover, structural damage to the cuticle of *O. ostertagi* isolated from the abomasum of sainfoin-fed calves was observed. No effect of sainfoin feeding was observed on *C. oncophora* in the small intestine, which may be related to the low activity of tannins in this gut compartment (Desrues et al., 2017). Further studies should clarify whether lower intake levels of sainfoin is having sufficient anti-parasitic activity. A study in Canada

Figure 3 Bioactive forage with anthelmintic activity in cattle: Chicory (*Cichorium intybus*).

explored the anthelmintic effects of birdsfoot trefoil, reporting that dairy heifers naturally infected with GIN and supplemented for 45 min daily with fresh birdsfoot trefoil (equivalent of 3.1% of DM intake) had significantly lower FEC, in comparison with heifers grazing the same pastures but supplemented with a control haylage (Shepley et al., 2015). In an earlier study from Chile, calves with access to forage plantain (67.3% plantain DM in the herbage) had significantly lower FEC of *O. ostertagi* and *Trichostrongylus* spp. in comparison with animals grazing only grass/clover and exposed to similar parasite larval challenge from pasture throughout the trial (Sievers and Nannig, 2006). Further research on the practical applicability of these bioactive forages as complementary parasite control in organic cattle may help their uptake in different farming settings.

5.6 Treatment

Daily supervision of all grazing livestock, including organic cattle, is imperative to secure a high level of welfare. Whenever needed, sick animals should be treated and cared for in consultation with the local vet or health advisor. Treatment with anthelmintics (mainly benzimidazoles, macrocyclic lactones or pyrantel/levamisole) will, in most cases, lead to rapid recovery, but does not remove the underlying problem if the paddock is heavily contaminated with GIN. Planning is needed to avoid this, and perhaps monitoring by FEC, e.g. a couple of months after turn-out to gauge the level of overwintering infection in first-season calves. In a few cases of recurring problems in certain areas, long-acting anthelmintics or repeated drug use may be justified. This may, however, compromise the development of natural immunity, affect invertebrate biodiversity (see section 8), and may increase the risk of anthelmintic resistance.

Use of medicinal plants or phytotherapy (i.e. dosing of parasitized animals with plant extracts or plant preparations) is part of a long ethno-veterinary tradition in many countries, including the EU (Waller et al., 2001; Calzetta et al., 2020). Further confirmation of the anecdotal anthelmintic effects (and potential toxicity) of these plant preparations remains to be conducted in cattle with *in vivo* controlled studies under farm conditions, which could validate their recommendation. For the interested readers, specific textbooks cover this topic in detail (e.g. Wynn and Fougère, 2007).

6 Lungworms

6.1 Introduction

Lungworm infections in cattle are caused by the nematode *Dictyocaulus viviparus*. Cattle lungworm infections may be subclinical, with reduced animal growth and milk production, or clinical with severe clinical respiratory disease known as bovine parasitic bronchitis (or 'husk') in both young and in adult animals (Ploeger, 2002; May et al., 2018; Forbes, 2021a). Infections with *D. viviparus* have been reported worldwide, although most clinical outbreaks and negative production impacts are observed in high rainfall temperate regions such as northern Europe but with large variation between and within countries and regions (Springer et al., 2021; Vanhecke et al., 2021). A study in Sweden evaluating antibody levels against *D. viviparus* in bulk-tank milk reported that organic farms had a higher lungworm prevalence (18% of organic farms positive) than conventional herds (9% of farms positive), with positive conventional farms located in the proximity of the infected organic herds (Höglund et al., 2010).

6.2 Epidemiology and diagnostics

Cattle get infected with *D. viviparus* by ingesting infective L_3 while grazing in contaminated pastures. The ingested larvae reach the small intestine, from where they invade the lymphatic system, moult to L_4, and migrate to the lungs to continue their development until the adult stage. Adult *D. viviparus* parasitizes trachea and major bronchi, and female worms begin producing eggs at 3–4 weeks after infection. Female *D. viviparus* are very prolific and produce in average 10,000 eggs per worm per day, for up to 6 weeks (Ploeger and Eysker, 2000), which has a significant impact on lungworm epidemiology (see below). Lungworm eggs hatch within the lungs and the L_1 is coughed up by the host, ingested and passed out with the dung onto pasture. Patent infections can be diagnosed by the detection of L_1 in faeces by the Baermann method. In the faeces when conditions are optimal (warm and humid), the larvae can complete their development to infective L_3 within 1 week. The infective L_3 reaches the pasture mainly by being propelled from the dung to the surrounding herbage during the sporangial discharge of the endocoprophilous fungus *Pilobolus* spp. (Robinson, 1962; Jørgensen et al., 1982). The high numbers of *D. viviparus* L_1 excreted into the environment by infected animals, and their fast development to L_3 in the dung during warm months, can lead to a very rapid build-up of pasture larval populations that can predispose to outbreaks of parasitic bronchitis, which in northern Europe are normally observed between July and October (Sargison, 2007; Forbes, 2018a). However, lungworm L_3 only survive ~4–6 weeks on pasture during summer (mainly due to desiccation), resulting in a highly dynamic rise and decline of larval contamination throughout the grazing season (Jørgensen, 1980; Forbes, 2021a). Overwintering of lungworm infections to the following year (in regions with winter stabling) occurs in two ways: (a) by the survival of infective L_3 on pasture and soil during winter, which are then picked up by animals at turn-out and (b) by carrier animals that harbour arrested larvae from the previous grazing season that will continue their development to adult worms and patent infections in spring (McLeonard and Dijk, 2017). Due to the short survival of infective L_3 on pasture, carrier animals are key as source of lungworm for both the maintenance of an immune memory in the herd (see below), as well as responsible for pasture contamination that can lead to lungworm disease. Cattle develop immunity to *D. viviparus* mainly by: (a) reducing the establishment of newly ingested L_3 from pasture and (b) lowering the number of adult worms in the lungs. However, without further parasite exposure, the duration of immunity against incoming L_3 is brief and only lasts for about 4 months, while protection towards adult worms lasts for up to 2 years (Michel et al., 1965). Some of the practical implications of this are: (a) first season grazing cattle are susceptible to both L_3 and adult lungworm at turn-out; (b) second season heifers and adult cows, exposed to lungworm during the previous grazing season, can still be protected against adult worms after winter stabling, but their resistance to larval challenge from pasture is expected to fade by turn-out; (c) young and adult cows in a previously lungworm-free herd have no immunity and are susceptible to infections brought to the herd by carrier animals (Forbes, 2021a).

Lung colonization by the developing and adult worms can lead to severe pulmonary damage, resulting on a sudden drop in milk yield and clinical respiratory signs such as frequent coughing and rapid and forced respiration, sometimes with fatal outcomes in heavily infected animals (Forbes, 2021a). Clinical respiratory signs in infected cattle (including reduced tolerance to exercise), although unspecific, can be seen from 15 days after infection, particularly when animals are moved around, whereas the post-mortem diagnosis of adult *D. viviparus* (3–8 cm of length) in the lung airways is straightforward (McLeonard and Dijk, 2017). In a farm with unknown presence of *D. viviparus*, the infection status of the herd can be established by faecal (Baermann method for detection of L_1) or serological (antibody levels in serum) diagnostic methods using bulk-tank milk, or preferably of individual cattle (6–9 heifers and 10–15 adult cows per farm) to increase the chance of detecting a positive animal (Ploeger et al., 2012). Regular monitoring of antibodies against lungworm in bulk-tank milk samples during the grazing season may in the future offer potential for prediction of clinical outbreaks and production losses in adult cows (Charlier et al., 2023).

6.3 Risk factors and prevention

Risk factors linked with outbreaks of lungworm disease involve the introduction of susceptible animals to massively contaminated pasture at turn-out or later in the grazing season, co-grazing of susceptible cattle and carrier animals, leading to a rapid build-up of pasture contamination, and introduction of carrier animals into a naïve herd and of naïve animals into an infected herd (McLeonard and Dijk, 2017). Factors associated with lungworm infections related to milk yield losses include the purchase of new animals (that could introduce infected animals into a naïve herd, or naïve animals to an endemically infected herd), routine anthelmintic treatment of first grazing season cattle (limiting natural infections and the development of herd immunity) and presence of (semi-) permanent water bodies in the grazing areas (Vanhecke et al., 2021).

Prevention of lungworm infections with grazing management is difficult to implement due to the rapid development of excreted L_1 into infective L_3, and it would require the rotation of animals to new pastures every week (McLeonard and Dijk, 2017). Low stocking rates could help to reduce the risk of lungworm disease by diluting the infection, while temporary housing of severely affected animals can be implemented to prevent further lung damage induced by continuous larval challenge from pasture (Younie et al., 2004). Controlled low-level infections with lungworm L_3 during grazing is beneficial for the development of acquired immunity in the herd (as is the case for GIN), but how to achieve this considering the rapidly changing epidemiology of *D. viviparus* is challenging. A live, irradiation-attenuated lungworm vaccine is available in some countries and provide a very good controlled exposure and protection in susceptible animals (≥8 weeks of age, with two doses) at turn-out, although it requires that vaccinated animals are exposed to low levels of natural booster infections on pasture to acquire solid immunity (Forbes, 2018b). Lungworm vaccine should only be used in herds confirmed positive to *D. viviparus* (and not in combination with anthelmintics) because it contains live larvae that could develop to adult worms in vaccinated cattle, that can then become carrier animals (Forbes, 2021a). Anthelmintic treatment to single or groups of animals may be necessary in outbreaks of lungworm disease, but routine deworming of the entire herd, besides being not allowed in organic systems, is considered counterproductive because it may hinder the development and maintenance of herd immunity. Nevertheless, quarantine anthelmintic treatment of recently purchased animals may be recommended to avoid introduction of the parasite to lungworm-free herds (Forbes, 2018a). Anthelmintics commonly used against GIN are also active against lungworm. Alternative control approaches to reduce the pasture contamination with infective lungworm L_3, such as the use of nematophagous fungi (Fernández et al., 1999) and dung beetles (Gormally, 1993), have been explored but practical use needs to be demonstrated.

7 Liver flukes

7.1 Introduction

Liver fluke infection is an all too common problem in grazing ruminants, and organic dairy cattle are no exception. Although most often sub-clinical in cattle, liver fluke disease (or fasciolosis) can result in significant production losses (Charlier et al., 2014). Due to its mud snail intermediate host, infections are invariably associated with heifers (and/or dry cows) grazing wet pastures

(Takeuchi-Storm et al., 2017). An abattoir study in Denmark (Olsen et al., 2015), based on liver condemnation data, found that organic herds had significantly higher prevalence compared to conventional herds when they were of medium or large size. This was suggested to be related to either more access to pasture or to lower treatment levels in organic herds. That said, this does not appear to be a universal finding. For example, in a study of the major helminth species affecting organic dairy cattle in Spain (Orjales et al., 2017), the herd- and individual-level prevalences of *F. hepatica* in organic systems (54.5% and 13.3%), respectively, as determined by coproantigen ELISA, were similar to that in conventional systems from the same highly exposed area (Mezo et al., 2008). In a similar study in Swedish dairy cattle, no significant differences in seropositivity between organic and conventional herds were noted (Novobilsky et al., 2015).

The causative agent of fasciolosis in temperate regions is typically, *Fasciola hepatica*, with *F. gigantica* more of a problem in the tropics. Like most trematode parasites, the liver fluke has a complicated life cycle, in this case involving a mud snail intermediate host (*Galba truncatula*), so the fluke is typically found on low-lying or boggy ground, i.e. ideal mud snail habitat. Snails become infected by ciliated larvae called miracidia, which hatch from the fluke eggs under favourable conditions of temperature and rainfall, once released from the dung. The infective stage, the metacercaria (or cyst stage), is released from the snails. Once the metacercariae are ingested by a suitable definitive host while grazing, the immature flukes escape from their cysts, penetrate the wall of the intestine, and move into the peritoneal cavity, where they go off in search of the liver. Once located, the growing flukes penetrate the liver capsule and migrate through the liver parenchyma, feeding on blood and liver tissue as they do so, ultimately ending up as blood-sucking adult (>10-week old) flukes in the bile duct system. From there, they begin shedding eggs, upwards of 10,000 eggs per fluke per day, which are released in the dung, starting the cycle over again. Cattle mount an immune response to flukes that form the basis of immunological diagnostic tests but it does not protect them from reinfection at a later stage. A longitudinal study deploying a panel of liver fluke diagnostic tests, showed that young stock was consistently seen as the main high-risk group, acquiring infection at approx. 1.5–2 years of age and carrying the infection into first lactation if untreated (Takeuchi-Storm et al., 2018). Re-infections in the milking herd were likely due to grazing dry cows on wet pastures or probably more rarely, transmission in paddocks used for milking cows. Transmission by freshly cut grass is possible, whereas silage or hay will not sustain survival of viable metacercariae for more than 2–3 weeks, if properly produced (John et al., 2019; 2020).

7.2 Impact and diagnostics

Liver fluke infection typically manifests in two main forms in livestock, namely acute fluke disease, caused by mass migration of immature flukes over a short time following ingestion of a large number of metacercarial cysts at once, or chronic fluke disease, associated with a gradual build-up of significant numbers of adult flukes in the bile ducts. The typical presentation of liver flukes in cattle would tend to be chronic infection, characterized by scarring and fibrosis of the liver tissue, leading to clinical signs such as reduced daily liveweight gain, reduced feed conversion efficiency and reduced reproductive success. Studies on Danish organic farms have indicated losses of 6% in mean 305 day energy corrected milk yield (580 kg) in infected herds and around 10% (919 kg) at individual level for infected cows in third or later lactation (Takeuchi-Storm et al., 2021). Liver flukes are typically detected through the presence of eggs in faeces (FEC), although this only applies to adult flukes >10 weeks of age and sensitivity is generally low with few eggs found. A coproantigen test is also available for use on faecal samples, as this is somewhat more sensitive than the FEC and can detect flukes as young as ~6–7 weeks of age. There is also a serological liver fluke test, which can detect anti-fluke antibodies in blood (or milk) as the earliest sign of infection and is a very useful test in young first season grazing animals. In addition to these tests, there is also a bulk tank anti-fluke antibody test, which is a very useful and convenient test for fluke in dairy cows and assessing level of transmission in cows (Takeuchi-Storm et al., 2017). In the latter case, it must be considered whether dry cows that graze on wet areas could later contribute antibodies to the bulk tank milk.

7.3 Prevention

The risk of liver fluke infection is dictated to a large extent by prevailing climatic and ground conditions, but also by grazing management in terms of where and when stock are grazing. Drainage of persistently wet fields to reduce mud snail habitats was advocated in the past but recent agri-environmental schemes encourage retention or rewetting of pasture to promote other biodiversity benefits. These changes have been suggested as possible risk factors for liver flukes in some areas (Pritchard et al., 2005), whereas others have failed to see an effect of rewetting (May et al., 2022). Eradication of the liver fluke is unlikely, given its multiple life stages and hosts in the environment, including definitive hosts such as red deer, rabbits, hares, coypu, etc., which share grazing land with livestock (French et al., 2019). Consequently, most practical on-farm liver fluke control aims to use animal and pasture management to limit exposure to infection and prevent clinical disease and/or production losses. This can be very challenging on persistently wet (or 'fluky') farms (Takeuchi-Storm et al., 2021). Potential options include fencing off persistently wet areas within fields, avoiding grazing certain fields altogether, or perhaps to move animals away from wet fields before the major increase in infective metacercariae on herbage, but this requires a sound understanding of fluke risk in time and space and the ability to recognize potential mud snail habitats. Recent advances in detection of environmental DNA (eDNA), released from mud snails and/or fluke stages, can help pinpoint such areas and contribute to on-farm fluke risk assessment (e.g. Jones et al., 2018; Davis et al., 2020). Options for biological control of liver flukes have also been explored, usually focusing on the mud snail intermediate host. For example, a number of wetland birds, e.g. ducks, have been shown to eat significant numbers of mud snails, and are actively encouraged in rewetting schemes, but they have also been shown to carry the snails to other locations, potentially spreading fluke risk (van Leeuwen, 2012). There is also a suggestion that the less pathogenic rumen fluke (see later) is out-competing the liver fluke in the *G. truncatula* snail population, in certain parts of the UK and Europe (Jones et al., 2015, 2017; Rondelaud et al., 2022). This may, however, also be related to improved control of liver fluke, providing rumen fluke with a niche to exploit. Finally, measures to ensure grazing livestock are on a good plane of nutrition will also help ensure animals cope better with any acquired infection.

By far the main approach to liver fluke control is through the routine use of flukicides. There is a limited number of active compounds available and these typically target different fluke stages in the final host. As a result of the age profile of the respective flukicides, it is imperative to use the right product at the right time of year to kill the right age of fluke. This is best informed by monitoring and diagnostic testing but most treatments are currently administered, based on the time of year, rather than the appropriate diagnostic indicators. Interestingly, none of the available flukicides has any persistence in the host animal, so livestock are prone to reinfection if grazing high risk pasture following treatment. This can often be mistaken for lack of efficacy or resistance. There is confirmed resistance to triclabendazole, most likely selected through routine treatment and over-use in sheep. There have also been reports of closantel inefficacy in cattle, which is a cause for concern (Novobilsky and Hoglund, 2015). One of the main issues with flukicide treatment, especially in dairy cattle, is the appearance of residues in milk and associated withhold periods (extended in organic dairy systems). This can preclude effective treatment at key times, e.g. lactation, and has led some dairy farmers to treat heifers and dry cows in winter (Novobilsky et al., 2020). Still, consistent use of triclabendazole in all infected dry cows is feasible and can be used to initiate control of infections (Takeuchi-Storm et al., 2018). Treatment with an appropriate flukicide product at or around housing represents an opportunity to reduce the impact of liver flukes on animals when the infection risk is low, but also a treatment with an appropriate adulticide before turn-out can limit fluke egg output and reduce the potential for subsequent infection of mud snails.

8 Rumen flukes

In recent years, there have also been numerous reports of rumen fluke infections in grazing livestock, most notably cattle. These are ubiquitous in more tropical regions but increasingly common in temperate regions, indeed, often more common than liver fluke infections. The rumen fluke life cycle is very similar to that of the liver fluke, and recent research confirms they share the same intermediate host, *G. truncatula*, in most of Europe (Jones et al., 2017). As a result, rumen flukes are often found on the same farms and in the same animals as liver flukes. Recent phylogenetic studies have revealed that the predominant, if not the only, rumen fluke species infecting European livestock is *Calicophoron daubneyi*, rather than *Paramphistomum cervi*, as initially suspected (Gordon et al., 2013). There is still some debate about the pathogenicity of rumen flukes, though adults appear to be relatively well tolerated in the rumen. In cases where clinical signs and even deaths have been reported, it is invariably associated with heavy infections of immature stages (e.g. O'Shaughnessy et al., 2018). Interestingly, most anecdotal reports of rumen flukes in cattle appear to involve dairy cattle, possibly reflecting how sensitive highly productive animals are to a relatively minor perturbation, and numerous accounts of how quickly animals respond when rumen fluke are successfully treated. Interestingly, in the few cases where it has been examined systematically, rumen flukes appear to be more prevalent in beef than dairy cattle (e.g. Delafosse, 2022). There is only one flukicide with activity against rumen fluke, namely oxyclozanide, but this is a liver fluke drug in its own right and there is no label claim for rumen flukes in most countries where oxyclozanide is available.

9 Environmental aspects of anti-parasitic drug use

As mentioned earlier, certain triazine drugs (e.g. ponazuril – the major metabolite of toltrazuril in faeces) are phytotoxic and after use, care is to be exercised when depositing faeces of treated animals. Likewise, a number of frontline dewormers, e.g. macrocyclic lactones (e.g. ivermectin) and triclabendazole, have the potential to impact negatively on important dung, soil and aquatic invertebrates, when excreted with faeces and/or urine of treated animals (Wall and Strong, 1987; Gilbert et al., 2019). This is perhaps to be expected, given their likely mode(s) of action, but there is increasing concern around the use and disposal of some of these compounds, especially in an organic farming context and with the current focus on 'regenerative' farming practices, biodiversity loss and environmental persistence of man-made compounds. For these reasons, macrocyclic lactones are banned in organic farming in some countries. Furthermore, anti-parasitic drug use, in general, needs to be carefully considered in nature conservation areas. That said, there is still some debate about the actual impact of anthelmintic treatments in practice under genuine field or farm conditions (e.g. Webb et al., 2010; Forbes, 1996, 2021b), and more on-farm studies are urgently needed.

10 Case study

The use of nurse cows for calves was mentioned earlier. Here we elaborate on a 2-year follow-up on GIN infections in 20 French organic dairy farms rearing calves (n = 405) with nurse cows (n = 199) (Constancis, 2021; Constancis et al., 2022). In the first year, calves around 1 month of age (or even younger) were turned out with nurse cows in early spring, summer or autumn, depending on the date of birth, the duration of the first grazing season (FGS) being thus variable. The duration of the second grazing season (SGS) with 312 heifers was long (mean: 278 days). No anthelmintic treatments were applied during the grazing seasons. Pepsinogen and antibody levels were universally low at the end of FGS, indicating a low exposure to GIN. This low GIN infection could be related to factors specific to the nurse cow rearing system. First, mixed grazing of resistant (nurse cows) and susceptible (calves) animals allowed a dilution effect, as immune nurse cows ingested a lot of infective larvae on pastures while excreting very low numbers of eggs, resulting in a cleansing effect of pastures. This effect will, of course, depend on the number of calves per nurse cow. Secondly, the predominantly milk diet of calves during the first months of grazing will have lowered the level of exposure compared to weaned calves grazing on their own pasture. However, this low level of GIN infections was also subjected to factors unrelated to the nurse cow rearing system, e.g. other grazing management practices (rotational grazing) and meteorological conditions (summer drought). During the SGS, the build-up of GIN infections was variable between groups and depended, in part, on the duration of the FGS. In heifers with a short FGS (<90 days), the development of GIN infection corresponded to the classical course in non-immunized animals, whereas in heifers with a long FGS (> 180 days), the low FEC reflected a prior development of immunity against GIN. We believe, despite diverse farming conditions, illustrated in Fig. 4, that in a system with calves and nurse cows, a long-sought balance between infectivity of the plots, ingestion of grass, establishment of immunity and acceptable levels of infection in calves is achievable.

11 Conclusion and future trends

The parasite infections in organic dairy farming selected for this review may have a major negative impact on productivity and animal health if preventative measures are neglected. For cryptosporidiosis and coccidiosis, a high level of hygiene, appropriate appliances and pens, and good management practices

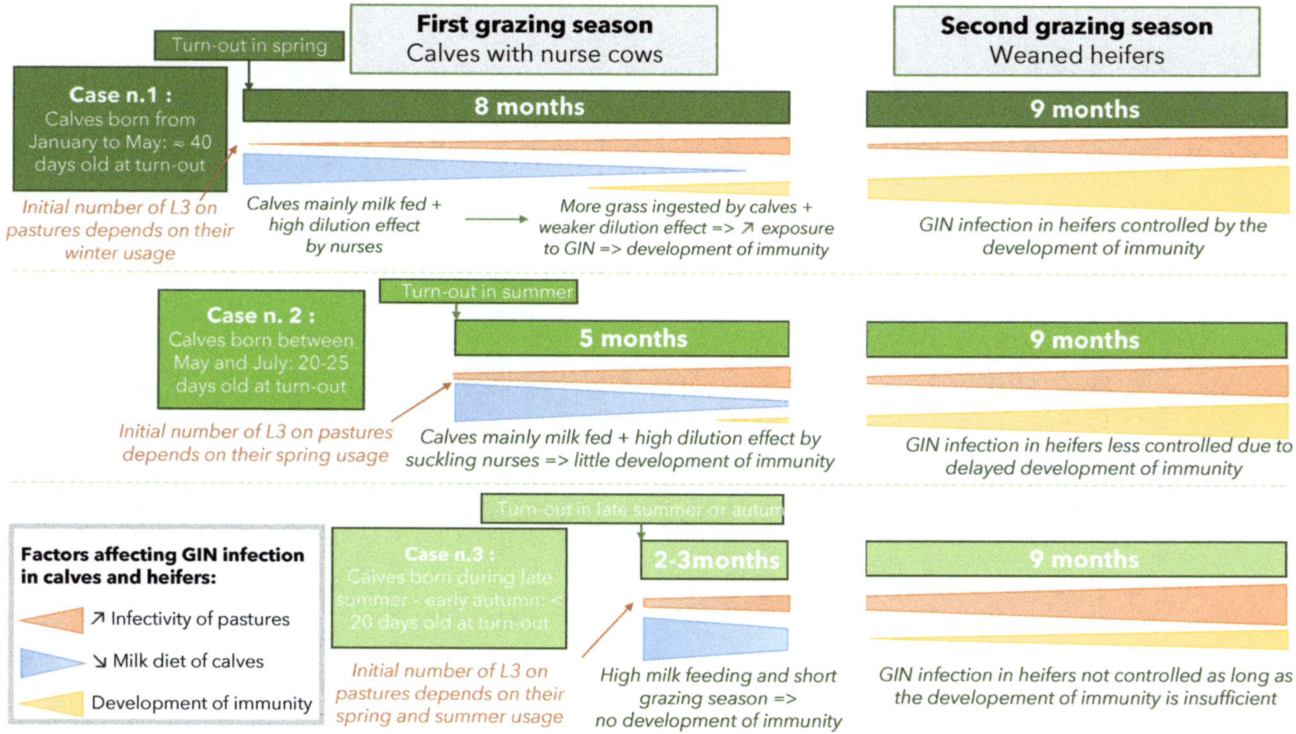

Figure 4 Description of the evolution of the three factors (pasture infectivity, proportion of milk in diet and development of immunity) affecting GIN infections in a nurse cow-calf rearing system, according to the date of birth and age at first turn-out of calves (first grazing season with suckling nurse cows and second grazing season as weaned heifers) (source: Constancis C., 2021).

around calving will in most instances prevent severe problems in stabled calves. At present, alternative approaches are limited in this age group but it is likely that a system with calves and nurse cows (or dams) may offer possibilities for better welfare and less disease problems. However, certain risks have been identified and need to be addressed. Drug treatments are likely unavoidable in certain situations and should be applied for animal welfare reasons for limited periods.

As outlined above, there are ample opportunities for coping with pasture-borne infections, particularly GIN, through grazing management. The limiting factors here are planning and availability of appropriate pastures, cost of fencing (+ perhaps diagnostics) and higher workload for the farmer. Drug use in such systems can be either therapeutic, based on clinical disease or pre-emptive, i.e. treatment performed only when a certain threshold is reached, the so-called targeted treatments (Charlier et al., 2023). The latter approach is explored nowadays in conventional farming, aiming to maintain a subpopulation of GIN unexposed to anthelmintics (i.e. *refugia*-based control) and thus slowing the development of anthelmintic resistance (Charlier et al., 2022). However, for organic farming, the ultimate goal should be no medicinal interventions. Whether this is even realistic, needs more applied research on integrated control on-farm to assess the efficacy of grazing management in combination with, e.g. supplementary feeding with protein-rich diets, bioactive forages, or targeted anthelmintics. This requires long, cumbersome study protocols and substantial funding, as also impacts on the environment and slowdown of anthelmintic resistance have to be considered.

Likely scenarios for future control of GIN in organic farming, may also include vaccination, biological control and breeding for resistance. Given sufficient exposure to sub-clinical infection with GIN, most cattle will become immune in time. This provides encouragement that it may be feasible to develop a vaccine(s) to protect cattle from infection, particularly against *O. ostertagi* as it is considerably more pathogenic than *C. oncophora*. Significant progress has been made using either purified 'native' extracts of worms or recombinant (engineered) worm proteins. Native extracts have provided significant protection in a number of vaccine trials in cattle, for example, *O. ostertagi* larval antigens reduced cumulative faecal egg counts by 70%–85% and worm counts by 64% (Halliday and Smith, 2010). These native antigen preparations offer encouraging proof of concept, but the main drawback is that these approaches cannot be scaled up readily for commercial vaccine production (Claerebout and Geldhof, 2020).

Biological control is, by definition, the use of one living microorganism introduced into the environment to obtain control or reduce effects of a pathogenic target microorganism. In fact, any natural enemy of free-living stages of GIN could be exploited in this way if feasible, e.g. earthworms, dung beetles or mites. However, to date, nematode-destroying fungi are the only realistic option in the short term and there is a commercial product available (Healey et al., 2018). These microfungi, found worldwide, are introduced to the pastures and feed on soil nematodes as well as L_1–L_3 of most species of GIN, or can in periods be saprophytic. The effect is variable and there a number of limitations, but the best results have actually been obtained in the uninfected first-season calves when they are introduced to pastures in spring (see Charlier et al., 2022 for a recent review). Its compatibility with organic principles needs to be evaluated.

It has been long known that resistance to GIN (and other helminths) is a heritable trait in livestock, including cattle (Zvinorova et al., 2016), and proof-of-concept selective breeding programs based on FEC and pepsinogen have been established in beef cattle (e.g. Gasbarre et al., 2001). Breeding for resistance to GIN is a practical option in sheep farming in several countries but to our knowledge, it is not in use in dairy cattle production. Major limitations are a lack of routine monitoring of infections and/or genomic tools, and often resistance is negatively correlated with production parameters. However, breeding for resistance (or resilience) is very much in line with the principles of organic farming and should be further investigated.

Liver fluke infections are complicated to control on certain farms where there is a general lack of dry pastures free of mud snails. In these cases, implementing satisfactory control will be difficult without the use of anthelmintics. An effective liver fluke vaccine to protect grazing livestock from infection and/or ameliorate the impact of fluke infection would be highly desirable and represents the Holy Grail of liver fluke research, globally. However, despite a considerable research effort and several promising candidates, a commercial liver fluke vaccine remains elusive (Molina-Hernández et al., 2015).

Despite the obstacles and research needs outlined above, grazing and the use of grasslands should still be an integral part of organic dairy production. Although greenhouse gas (GHG) emissions need to be limited, the use of feed sources that are inedible to humans and mono-gastric animals still makes cattle relevant in the context of improving biodiversity and nature conservation, particularly wet meadows, rangelands and mountain pastures. Further, the use of large herbivores in grasslands may increase the soil carbon pool in these areas (Naidu et al., 2022). If the risk of parasites is becoming critical, more robust second-season heifers (or steers) should be used. However, more research at the systems level is needed on how disadvantages like GHG emissions and poor feed efficiency are balanced with benefits related to the use of inedible resources, animal welfare, provision of meat for a growing (world) population, increased carbon-sequestration and enhanced biodiversity. Here lies an obligation for organic dairy farming to contribute.

12 Where to find further information

The following websites provide a good overview of the subject:

- COWS – Control of worms sustainably: https://www.cattleparasites.org.uk/.
- SCOPS – Sustainable control of parasites (this is a website for small ruminants but still, very useful): https://www.scops.org.uk/
- Weideparasiten (website with information on parasite control in grazing ruminants in German, including an interactive decision tree): https://www.weide-parasiten.de.

13 Acknowledgements

NR was supported by the project GrazyDaiZy within CORE Organic Cofund (co-supervising the PhD of Caroline Constancis in this project). HM, ARW and SMT were supported by the project ROAM-FREE within CORE Organic Cofund, an ERA-NET funded by the European Commission's Horizon 2020 Framework Programme for Research and Innovation Contract No. 727495.

14 References

Åberg, M., Emanuelson, U., Troell, K. and Björkman, C. (2019), Infection dynamics of *Cryptosporidium bovis* and *Cryptosporidium ryanae* in a Swedish dairy herd, *Vet Parasitol.* 276S:100010.

Andersen, H.R., Kristensen, T., Bligaard, H.B. and Thamsborg, S.M. (2002), Studeproduktion ved afgræsning af ferske enge [Production of Holstein Friesian steers on riparian meadows], *Report from Danish Institute of Agricultural Sciences (DJF)*, 40:87pp.

Autzen, S., Maddox-Hyttel, C. and Monrad, J. (2002). Infection with *Eimeria*-species in calves, *Danish Vet. J.* 85(18):6–10.

Barger, I. (1996), Grazing Management for the Control of Nematode Parasites. In: *Sustainable Parasite Control in Small Ruminants*, ACIARLF, eds. Le Jambre, L., Knox, M.R., 146–150.

Blanco-Penedo, I., Höglund, J., Fall, N. and Emanuelson, U. (2012). Exposure to pasture borne nematodes affects individual milk yield in Swedish dairy herds, *Vet. Parasitol.* 188(1–2):93–98.

Brainard, J., Hooper, L., McFarlane, S., Hammer, C. C., Hunter, P. R. and Tyler, K. (2020) Systematic review of modifiable risk factors shows little evidential support for most current practices in *Cryptosporidium* management in bovine calves, *Parasitol. Res.* 119:3571–3584.

Calzetta, L., Pistocchini, E., Leo, A., Roncada, P., Ritondo, B.L., Palma, E., di Cave, D. and Britti, D. (2020), Anthelminthic medicinal plants in veterinary ethnopharmacology: a network meta-analysis following the PRISMA-P and PROSPERO recommendations, *Heliyon* 6:e03256.

Chapman, D., Tharmaraj, J. and Nie, Z. (2008), Milk-production potential of different sward types in a temperate southern Australian environment, *Grass Forage Sci.* 63:221–233.

Charlier, J., van der Voort, M., Kenyon, F., Skuce, P. and Vercruysse, J. (2014) Chasing helminths and their economic impact on farmed ruminants, *Trends Parasitol.* 30(7):361–367. https://doi.org/10.1016/j.pt.2014.04.009.

Charlier, J., Bartley, D.J., Sotiraki, S., Martinez-Valladares, M., Claerebout, E., von Samson-Himmelstjerna, G., Thamsborg, S.M., Hoste, H., Morgan, E.R. and Rinaldi, L. (2022), Anthelmintic Resistance in Ruminants: challenges and Solutions. In *Advances in Parasitology,* Rollinson, D. and Stothard, R. (red.). Academic Press, 115:171–227.

Charlier, J., Williams, D.J., Ravinet, N. and Claerebout E. (2023), To treat or not to treat: diagnostic thresholds in subclinical helminth infections of cattle, *Trends Parasitol.* 39(2):139–151. https://doi.org/10.1016/j.pt.2022.11.014.

Chincarini, M., Lanzoni, L., Di Pasquale, J., Morelli, S., Vignola, G., Paoletti, B. and Di Cesare, A. (2022), Animal welfare and parasite infections in organic and conventional dairy farms: a comparative pilot study in Central Italy, *Animals* 12:351.

Claerebout, E. and Geldhof, P. (2020), Helminth vaccines in ruminants: from development to application, *Vet Clin N America Food Animal Practice* 36(1):159–171. https://doi.org/10.1016/j.cvfa.2019.10.001.

Constancis, C. (2021), *Rearing dairy calves with nurse cows in organix farming: description of practices and epidemiological studies on cryptosporidiosis and gastrointestinal and respiratory strongylosis*, PhD Thesis, 30th September 2021, Oniris, Nantes, 196 pp.

Constancis, C., Chartier, C., Leligois, M., Brisseau, N., Bareille, C., Strube, C. and Ravinet, N. (2022), Gastrointestinal nematode and lungworm infections in organic dairy calves reared with nurse cows during their first grazing season in western France. *Vet Parasitol.* 302:109659. https://doi.org/10.1016/j.vetpar.2022.109659.

Constancis, C., Ravinet, N., Bernard, M., Lehebel, A., Brisseau, N. and Chartier, C. (2021), Rearing system with nurse cows and risk factors for *Cryptosporidium* infection in organic dairy calves, *Prev. Vet. Med.* 190. https://doi.org/10.1016/j.prevetmed.2021.105321.

Coop, R.L. and Kyriazakis, I. (2001), Influence of host nutrition on the development and consequences of nematode parasitism in ruminants. *Trends Parasitol.* 17:325–330.

Davis, C.N., Tyson, F., Cutress, D., Davies, E., Jones, D.L., Brophy, P.M., Prescott, A., Rose, M.T., Williams, M., Williams, H.W. and Jones, R.A. (2020), Rapid detection of *Galba truncatula* in water sources on pasture-land using loop-mediated isothermal amplification for control of trematode infections, *Parasit Vectors* 13(1):496. https://doi.org/10.1186/s13071-020-04371-0.

Delafosse, A. (2022), Rumen fluke infections (Paramphistomidae) in diarrhoeal cattle in western France and association with production parameters, *Vet. Parasitol. Reg. Stud. Rep.* 29:100694. https://doi.org/10.1016/j.vprsr.2022.100694.

Desrues, O., Mueller-Harvey, I., Pellikaan, W., Enemark, H. and Thamsborg, S. (2017), Condensed tannins in the gastrointestinal tract of cattle after sainfoin (*Onobrychis viciifolia*) intake and their possible relationship with anthelmintic effects. *J. Agric. Food Chem.* 65:1420–1427.

Desrues, O., Peña-Espinoza, M., Hansen, T., Enemark, H.L. and Thamsborg, S.M. (2016), Anti-parasitic activity of pelleted sainfoin (*Onobrychis viciifolia*) against *Ostertagia ostertagi* and *Cooperia oncophora* in calves. *Parasites Vectors* 9:329.

Enemark, H.L., Dahl, J. and Enemark, J.M.D. (2013), Eimeriosis in Danish dairy calves - Correlation between species, oocyst excretion and diarrhoea. *Parasitol Res.* 112:169–176.

Eysker, M., van der Aar, W.M., Boersema, J.H., Githiori, J.B. and Koyman, F.N.J. (1998), The effect of repeated moves to clean pasture on the build up of gastrointestinal nematode infections in calves. *Vet Parasitol.* 76:81–94.

Fernández, A.S., Larsen, M., Nansen, P., Grønvold, J., Henriksen, S.A., Bjørn, H. and Wolstrup, J. (1999), The efficacy of two isolates of the nematode-trapping fungus *Duddingtonia flagrans* against *Dictyocaulus viviparus* larvae in faeces, *Vet Parasitol.*, 85:289–304. https://doi.org/10.1016/S0304-4017(99)00122-3.

Foldager, J., Sejrsen, K., Brolund Larsen, J., Nansen, P., Jørgensen, R.J. and Hansen J.W. (1981), Control of infection with gastrointestinal helminths in calves and heifers on pasture, 514. *Report from Danish Institute of Animal Sciences*, Frederiksberg, 325 pp.

Forbes, A. (1996), Environmental assessments in veterinary parasitology: a balanced perspective. *Int. J. Parasitol.* 26(5):567–569. https://doi.org/10.1016/0020-7519(96)00019-7.

Forbes, A. (2018a), Lungworm in cattle: epidemiology, pathology and immunobiology, *Livestock* 23:59–66. https://doi.org/10.12968/live.2018.23.2.59

Forbes, A. (2018b), Lungworm in cattle: treatment and control, *Livestock* 23:102–106. https://doi.org/10.12968/live.2018.23.3.102.

Forbes, A. (2021a), *Lungworm infections in cattle. In: Parasites of Cattle and Sheep: A Practical Guide to Their Biology and Control.* CABI International, Wallingford, Oxfordshire, UK; Boston, MA, USA, pp. 116–138.

Forbes, A. (2021b), Environmental risk assessment of veterinary parasiticides used in cattle, *Livestock* 26:15–25.

French, A.S., Zadoks, R.N., Skuce, P.J., Mitchell, G., Gordon-Gibbs, D.K. and Taggart, M.A. (2019), Habitat and host factors associated with liver fluke (*Fasciola hepatica*) diagnoses in wild red deer (*Cervus elaphus*) in the Scottish Highlands, *Parasit Vectors* 12(1):535. https://doi.org/10.1186/s13071-019-3782-3.

Gasbarre, L.C., Leighton, E.A. and Sonstegard, T. (2001), Role of the bovine immune system and genome in resistance to gastrointestinal nematodes, *Vet. Parasitol.* 98(1–3):51–64. https://doi.org/10.1016/s0304-4017(01)00423-x.

Gilbert, G., MacGillivray, F.S., Robertson, H.L. and Jonsson, N.N. (2019), Adverse effects of routine bovine health treatments containing triclabendazole and synthetic pyrethroids on the abundance of dipteran larvae in bovine faeces, *Sci. Rep.* 9(1):1–10.

Gordon D.K., Roberts, L.C.P., Lean, N., Zadoks, R.N., Sargison, N.D. and Skuce, P.J. (2013), Identification of the rumen fluke, *Calicophoron daubneyi*, in GB livestock: possible implications for liver fluke diagnosis, *Vet. Parasitol* 195(1–2):65–71. https://doi.org/10.1016/j.vetpar.2013.01.014.

Gormally, M.J., (1993), Laboratory investigations of Aphodius beetles (Scarabaeidae) as biological control agents of the Cattle Lungworm, Dictyocaulus viviparus (Nematoda: Dictyocaulidae), *Biocontrol Sci. Technol.* 3, 499–502. https://doi.org/10.1080/09583159309355305.

Halliday A.M. and Smith W.D. (2010), Protective immunization of calves against *Ostertagia ostertagi* using fourth stage larval extracts, *Parasite Immunol.* 32(9–10):656–663. https://doi.org/10.1111/j.1365-3024.2010.01221.x.

Healey, K., Lawlor, C., Knox, M.R., Chambers, M., Lamb, J. and Groves, P. (2018), Field evaluation of *Duddingtonia flagrans* IAH 1297 for the reduction of worm burden in grazing animals: pasture larval studies in horses, cattle and goats, *Vet. Parasitol.* 258:124–132.

Hoste, H., Torres-Acosta, J.F.J., Quijada, J., Chan-Perez, I., Dakheel, M.M., Kommuru, D.S., Mueller-Harvey, I. and Terrill, T.H. (2016), Interactions between nutrition and infections with *Haemonchus contortus* and related gastrointestinal nematodes in small ruminants, *Adv. Parasitol.* 93:239–351. https://doi.org/10.1016/bs.apar.2016.02.025.

Hoste, H., Torres-Acosta, J.F.J., Sandoval-Castro, C., Mueller-Harvey, I., Sotiraki, S., Louvandini, H., Thamsborg, S.M. and Terrill, T.H. (2015), Tannin containing legumes as a model for nutraceuticals against digestive parasites in livestock, *Vet. Parasitol*, 212:5–17. https://doi.org/10.1016/j.vetpar.2015.06.026.

Hovi, M., Sundrum, A. and Thamsborg, S.M. (2003), Animal health and welfare in organic livestock production in Europe: current state and future challenges. *Livest. Prod. Sci.* 80:41–53.

Höglund, J., Svensson, C. and Hessle, A. (2001), A field survey on the status of internal parasites in calves on organic dairy farms in South-Western Sweden, *Vet. Parasitol.* 99:113–128.

Höglund, J., Dahlström, F., Engström, A., Hessle, A., Jakubek, E.-B., Schnieder, T., Strube, C. and Sollenberg, S. (2010), Antibodies to major pasture borne helminth infections in bulk-tank milk samples from organic and nearby conventional dairy herds in South-Central Sweden, *Vet. Parasitol.* 171:293–299. https://doi.org/10.1016/j.vetpar.2010.04.002.

Jensen, A. M. (2002), [Strategies for prevention of pasture-borne parasites and Salmonella], *Seminar on Forage for cattle*, Danish Cattle Advisory Services, Aarhus.

John, B.C., Davies, D.R., Williams, D.J. and Hodgkinson, J.E. (2019), A review of our current understanding of parasite survival in silage and stored forages, with a focus on *Fasciola hepatica* metacercariae, *Grass Forage Sci.* 74(2):211–217. https://doi.org/10.1111/gfs.12429.

John, B.C., Davies, D.R., Williams, D.J. and Hodgkinson, J.E. (2020), Anaerobic fermentation results in loss of viability of *Fasciola hepatica* metacercariae in grass silage, *Vet. Parasitol.* 285:109218. https://doi.org/10.1016/j.vetpar.2020.109218.

Johnsen, J.F., Zipp, K.A., Kälber, T., Passillé, A.Mde, Knierim, U., Barth, K. and Mejdell, C.M. (2016), Is rearing calves with the dam a feasible option for dairy farms?–current and future research, *Appl. Anim. Behav. Sci.* 181:1–11.

Jones, R.A., Brophy, P.M., Davis, C.N., Davies, T.E., Emberson, H., Stevens, P.R. and Williams, H.Y. (2018), Detection of *Galba truncatula*, *Fasciola hepatica* and *Calicophoron daubneyi* environmental DNA within water sources on pasture land, a future tool for fluke control? *Parasit Vectors* 11(1):342. https://doi.org/10.1186/s13071-018-2928-z.

Jones, R.A., Williams, H.Y., Dalesman, S., Ayodeji, S., Thomas, R.K. and Brophy, P.M. (2017), The prevalence and development of digenean parasites within their intermediate snail host, *Galba truncatula*, in a geographic area where the presence of *Calicophoron daubneyi* has recently been confirmed. *Vet. Parasitol.* 240:68–74. https://doi.org/10.1016/j.vetpar.2017.03.021.

Jones, R.A., Williams, H.Y., Dalesman, S. and Brophy, P.M. (2015), Confirmation of *Galba truncatula* as an intermediate host snail for *Calicophoron daubneyi* in Great Britain, with evidence of alternative snail species hosting *Fasciola hepatica*. *Parasit Vectors* 8:656. https://doi.org/10.1186/s13071-015-1271-x.

Jørgensen, R.J. (1980), Epidemiology of bovine dictyocaulosis in Denmark, *Vet. Parasitol.* 7:153–167.

Jørgensen, R.J., Rønne, H., Helsted, C. and Iskander, A.R. (1982), Spread of infective *Dictyocaulus viviparus* larvae in pasture and to grazing cattle: Experimental evidence of the role of *Pilobolus* fungi, *Vet. Parasitol.* 10:331–339. https://doi.org/10.1016/0304-4017(82)90085-1.

Jørgensen, R.J., Satrija, F., Monrad, J. and Nansen, P. (1992), Effect of feeding lucerne pellets on trichostrongyle infections in grazing heifers, *Vet. Rec.* 131:126–127.

Kristensen, T., Thamsborg, S. M., Andersen, H.R., Søegaard, K. and Nielsen, A. L. (2006), Effects of grazing system on production and parasitism of dairy breed heifers and steers grazing wet marginal grasslands, *Anim. Sci.* 82:201–211.

Kuris, A.M., Hechinger, R.F., Shaw, J.C., Whitney, K.L., Aguirre-Macedo, L., Boch, C.A., Dobson, A.P., Dunham, E.J., Fredensborg, B.L., Huspeni, T.C., Lorda, J., Mababa, L., Mancini, F.T., Mora, A.B., Pickering, M., Talhouk, N.L., Torchin, M.E. and Lafferty, K.D. (2008), Ecosystem energetic implications of parasite and free-living biomass in three estuaries, *Nature* 454:515–518. https://doi.org/10.1038/nature06970.

Langkjær, R. B., Vigre, H., Enemark, H. L., & Maddox-Hyttel, C. (2007), Molecular and phylogenetic characterization of *Cryptosporidium* and *Giardia* from pigs and cattle in Denmark. *Parasitology* 134(3):339–350.

Larsson, A., Dimander, S.O., Rydzik, A., Uggla, A., Waller, P.J. and Höglund, J. (2006), A 3-year field evaluation of pasture rotation and supplementary-feeding to control parasite infection in first-season grazing cattle: effects on animal performance, *Vet. Parasitol.* 142:197–206.

Lassen, B., Viltrop, A. and Järvis, T. (2009), Herd factors influencing oocyst production of *Eimeria* and *Cryptosporidium* in Estonian dairy cattle, *Parasitol Res.* 105(5):1211–22. https://doi.org/10.1007/s00436-009-1540-8.

Lund, V. (2006), Natural living: a precondition for animal welfare in organic farming, *Livest. Sci.* 100:71–83.

Maddox-Hyttel, C., Langkjaer, R.B., Enemark, H.L. and Vigre, H. (2006), *Cryptosporidium* and *Giardia* in different age groups of Danish cattle and pigs–occurrence and management associated risk factors, *Vet. Parasitol.* 141(1–2):48–59. https://doi.org/10.1016/j.vetpar.2006.04.032.

Magaya, A., Mukaratirwa, S., Willingham, A.L., Kyvsgaard, N. and Thamsborg, S. (2000), Effects of anthelmintic treatment and feed supplementation on grazing Tuli weaner steers naturally infected with gastrointestinal nematodes, *J. S. Afr. Vet. Assoc.* 71:31–37.

Mangwe, M., Bryant, R. and Gregorini, P. (2020), Rumen fermentation and fatty acid composition of milk of mid lactating dairy cows grazing chicory and ryegrass, *Animals* 10:169. https://doi.org/10.3390/ani10010169.

Marley, C.L., Fychan, R., Davies, J., Scollan, N., Richardson, R., Theobald, V., Genever, E., Forbes, A. and Sanderson, R. (2014), Effects of chicory/perennial ryegrass swards compared with perennial ryegrass swards on the performance and carcass quality of grazing beef steers. *PLoS One* 9:e86259. https://doi.org/10.1371/journal.pone.0086259.

May, K., Brügemann, K., König, S. and Strube, C. (2018), The effect of patent *Dictyocaulus viviparus* (re)infections on individual milk yield and milk quality in pastured dairy cows and correlation with clinical signs, *Parasites Vectors* 11:24.

May, K., Raue, K., Blazejak, K., Jordan, D. and Strube, C. (2022), Pasture rewetting in the context of nature conservation shows no long-term impact on endoparasite infections in sheep and cattle, *Parasites Vectors* 15:33. https://doi.org/10.1186/s13071-022-05155-4.

McLeonard, C. and Dijk, J. (2017), Controlling lungworm disease (husk) in dairy cattle, *In Practice* 39:408–419. https://doi.org/10.1136/inp.j4038.

Merlin, A., Chauvin, A., Madouasse, A., Froger, S., Bareille, N. and Chartier, C. (2016), Explaining variability in first grazing season heifer growth combining individually measured parasitological and clinical indicators with exposure to gastrointestinal nematode infection based on grazing management practice, *Vet. Parasitol.* 225:61–69.

Merlin, A., Chauvin, A., Madouasse, A., Froger, S., Bareille, N. and Chartier, C. (2017), End-season daily weight gains as rationale for targeted selective treatment against gastrointestinal nematodes in highly exposed first-grazing season cattle, *Prev. Vet. Med.* 138:104–112.

Mezo, M., González-Warleta, M., Castro-Hermida, J.A., Ubeira, F.M. (2008), Evaluation of the flukicide treatment policy for dairy cattle in Galicia (NW Spain), *Vet. Parasitol.* 157(3–4):235–243. https://doi.org/10.1016/j.vetpar.2008.07.032.

Michaud, A., Cliozier, A., Bec, H., Chassaing, C., Disenhaus, C., Drulhe, T., Martin, B., Pomi`es, D. and Le Cozler, Y. (2018), D´el´eguer l'allaitement des veaux laitiers aux vaches? R´esultats d'enqu^etes aupr`es des ´eleveurs. *Renc. Rech. Ruminants* 24:66–69.

Michel, J.F., Mackenzie, A., Bracewell, C.D., Cornwell, R.L., Elliot, J., Hebert, C.N., Holman, H.H. and Sinclair, I.J.B. (1965), Duration of the acquired resistance of calves to infection with *Dictyocaulus viviparus*, *Res. Vet. Sci.* 6:344–402. https://doi.org/10.1016/S0034-5288(18)34745-3.

Molina-Hernández, V., Mulcahy, G., Pérez, J., Martínez-Moreno, Á., Donnelly, S., O'Neill, S.M., Dalton, J.P. and Cwiklinski, K. (2015), *Fasciola hepatica* vaccine: we may not be there yet but we're on the right road, *Vet. Parasitol.* 208(1–2):101–11.

Muir, S., Ward, G. and Jacobs, J. (2014), Milk production and composition of mid-lactation cows consuming perennial ryegrass-and chicory-based diets, *J. Dairy Sci.* 97:1005–1015. https://doi.org/10.3168/jds.2013-7183.

Naidu, D.G.T., Roy, S. and Bagchi, S. (2022), Loss of grazing by large mammalian herbivores can destabilize the soil carbon pool, *PNAS*, 119(43):1–7. https://doi.org/10.1073/pnas.2211317119.

Nansen, P., Foldager, J., Hansen, J.W., Henriksen, S.A. and Jørgensen, R.J. (1988), Grazing pressure and acquisition of *O. ostertagi* in calves, *Vet. Parasitol.* 27:325–335.

Nansen, P., Steffan, P., Monrad, J., Grønvold, J. and Henriksen, S.A. (1990), Effects of separate and mixed grazing on trichostrongylosis in first- and second-season grazing calves, *Vet. Parasitol.* 36:265–276.

Nielsen, B., Thamsborg, S.M. and Kristensen, T. (2003), Feed supplements for young dairy breed calves after turn-out to pasture: effect on weight gain and subclinical coccidiosis in organic production systems, *Acta Agric. Scand. Sect. A Anim Sci.* 53:1–10.

Novobilsky. A. and Höglund, J. (2015), First report of closantel treatment failure against *Fasciola hepatica* in cattle, *Int. J. Parasitol Drugs Drug. Resist.* 5(3):172–177. https://doi.org/10.1016/j.ijpddr.2015.07.003.

Novobilsky, A., Sollenberg, S. and Höglund, J. (2015), Distribution of *Fasciola hepatica* in Swedish dairy cattle and associations with pasture management factors. *Geospat. Health.* 9(2):293–300. https://doi.org/10.4081/gh.2015.351.

Novobilsky, A., Rustas B., Grandi, G., Högberg, N. and Höglund, J. (2020), Selective flukicide treatment of non-lactating cows and the corresponding production impact of *Fasciola hepatica* in dairy herds in Sweden. *Vet. Parasitol.* 283:109180. https://doi.org/10.1016/j.vetpar.2020.109180.

Odden, A., Enemark, H.L., Ruiz, A. Robertson, L.J., Ersdal, C., Nes, S.K., Tømmerberg, V. and Stuen, S. (2018), Controlled efficacy trial confirming toltrazuril resistance in a field isolate of ovine *Eimeria* spp., *Parasites Vectors* 11:394. https://doi.org/10.1186/s13071-018-2976-4.

Olsen, A., Frankena, K., Bodker, R., Toft, N., Thamsborg, S.M., Enemark, H.L. and Halasa, T. (2015), Prevalence, risk factors and spatial analysis of liver fluke infections in Danish cattle herds, *Parasit Vectors* 8:160.

Orjales, I., Mezo, M., Miranda, M., González-Warleta, M., Rey-Crespo, F., Vaarst, M., Thamsborg, S., Diéguez, F.J., Castro-Hermida, J.A. and López-Alonso, M. (2017), Helminth infections on organic dairy farms in Spain, *Vet. Parasitol.* 243:115–118.

O'Shaughnessy, J., Garcia-Campos, J., McAloon, C.G., Fagan, S., de Waal, T., McElroy, M., Casey, M., Good, B., Mulcahy, G., Fagan, J., Murphy, D. and Zintl, A. (2018), Epidemiological investigation of a severe rumen fluke outbreak on an Irish dairy farm, *Parasitol* 145(7):948–952. https://doi.org/10.1017/S0031182017002086.

Peña-Espinoza, M., Thamsborg, S.M., Desrues, O., Hansen, T.V.A. and Enemark, H.L. (2016), Anthelmintic effects of forage chicory (*Cichorium intybus*) against gastrointestinal nematode parasites in experimentally infected cattle, *Parasitology* 143: 1279–1293. https://doi.org/10.1017/S0031182016000706.

Peña-Espinoza, M., Valente, A.H., Bornancin, L., Simonsen, H.T., Thamsborg, S.M., Williams, A.R. and López-Muñoz, R. (2020), Anthelmintic and metabolomic analyses of chicory (*Cichorium intybus*) identify an industrial by-product with potent in vitro antinematodal activity. *Vet. Parasitol.* 280:109088.

Peña-Espinoza, M., Valente, A., Thamsborg, S.M., Simonsen, H.T., Boas, U., Enemark, H.L., Lopez-Munoz, R. and Williams, A. (2018), Antiparasitic activity of chicory (*Cichorium intybus*) and the role of its natural bioactive compounds: a review, *Parasites Vectors* 11:475.

Phillips, H.N. and Heins, B.J. (2022), Alternative practices in organic dairy production and effects on animal behavior, health, and welfare, *Animals* 12:1785. https://doi.org/10.3390/ani12141785.

Ploeger, H.W. (2002), *Dictyocaulus viviparus*: re-emerging or never been away? *Trends Parasitol.* 18:329–332. https://doi.org/10.1016/S1471-4922(02)02317-6.

Ploeger, H.W. and Eysker, M. (2000), Simulating *Dictyocaulus viviparus* infection in calves: the parasitic phase, *Parasitol.* 120:S3–S15. https://doi.org/10.1017/S0031182099005831.

Ploeger, H.W., Verbeek, P.C., Dekkers, C.W.H., Strube, C., Van Engelen, E., Uiterwijk, M., Lam, T.J.G.M. and Holzhauer, M. (2012), The value of a bulk-tank milk ELISA and individual serological and faecal examination for diagnosing (sub)clinical *Dictyocaulus viviparus* infection in dairy cows, *Vet. Parasitol.* 184:168–179. https://doi.org/10.1016/j.vetpar.2011.08.032.

Poulin, R. and Morand, S. (2004), *Parasite Biodiversity*. Smithsonian Institution Scholarly Press, Washington.

Pritchard, G.C., Forbes, A,B. Williams, D.J.L., Salimi-Bejestani, M.R. Daniel, R.G. (2005), Emergence of fasciolosis in cattle in East Anglia, *Vet. Rec.* 157(19):578–582. https://doi.org/10.1136/vr.157.19.578.

Rondelaud, D., Vignoles, P. and Dreyfuss G. (2022), Changes in the populations of two lymnaeidae and their infection by *Fasciola hepatica* and/or *Calicophoron daubneyi* over the past 30 years in Central France, *Animals*, 12(24):3566. https://doi.org/10.3390/ani12243566.

Robinson, J. (1962), *Pilobolus* spp. and the translation of the Infective larvæ of *Dictyocaulus viviparus* from fæces to Pastures, *Nature* 193:353–354.

Sargison, N. (2007), The unpredictable epidemiology of *Dictyocaulus viviparus* lungworm infection of cattle, *Vet. Ire. J.* 6:387–391.

Shaw, D.J., Vercruysse, J., Claerebout, E. and Dorny, P. (1998). Gastrointestinal nematode infections of first-grazing season calves in Western Europe: associations between parasitological, physiological, physiological and physical factors, *Vet. Parasitol.* 75:133–151.

Shepley, E., Vasseur, E., Bergeron, R., Villeneuve, A. and Lachance, S. (2015), Short Communication: Birdsfoot trefoil as a preventative treatment for gastrointestinal nematodes in pastured dairy heifers. *Can. J. Anim. Sci.* 95:533–537. https://doi.org/10.4141/cjas-2014-169.

Sievers, G. and Nannig, S. (2006), Effect of the supplementary feeding with *Plantago lanceolata* on the egg output of gastrointestinal nematodes in calves, *Arch. Med. Vet.* 38:233–238. https://doi.org/10.4067/S0301-732X2006000300006.

Silverås, C. and Blanco-Penedo, I. (2013). *Cryptosporidium* spp. in calves and cows from organic and conventional dairy herds, *Epidemiol. Infect.* 141(3):529–539.

Sorge, U.S., Moon, R.D., Stromberg, B.E., Schroth, S.L., Michels, L., Wolff, L.J., Kelton, D.F. and Heins, B.J. (2015), Parasites and parasite management practices of organic and conventional dairy herds in Minnesota, *J. Dairy Sci.* 98:3143–3151.

Springer, A., Jordan, D., Kirse, A., Schneider, B., Campe, A., Knubben-schweizer, G., Müller, K.E., Hoedemaker, M. and Strube, C. (2021), Seroprevalence of major pasture-borne parasitoses (Gastrointestinal Nematodes, Liver Flukes and Lungworms) in German Dairy Cattle Herds, Association with Management Factors and Impact on Production Parameters, *Animals* 11:2078.

Svensson, C., Enemark, H.L., Lauridsen, G.H. and Jensen, A.M. (1997), Græsmarkscoccidiose – årsag til diarre hose førstegangsgræssende kreaturer straks efter udbinding [Coccidiosis on pasture], *Danish Vet. J.* 80(10):447–454.

Svensson. C., Hessle, A. and Höglund, J. (2000), Parasite control methods in organic and conventional dairy herds in Sweden, *Livest. Prod. Sci.* 66:57–69.

Takeuchi-Storm, N., Denwood, M., Hansen, T.V.A., Halasa, T., Rattenborg, E., Boes, J., Enemark, H.L. and Thamsborg, S.M. (2017), Farm-level risk factors for *Fasciola hepatica* infection in Danish dairy cattle as evaluated by two diagnostic methods, *Parasites Vectors*, 10:555.

Takeuchi-Storm, N., Denwood, M., Petersen, H.H., Enemark, H.L., Stensgaard, A.-S., Sengupta, M.E., Beesley, N.J., Hodgkinson, J., Williams, D. and Thamsborg, S.M. (2018), Patterns of Fasciola hepatica infection in Danish dairy cattle: implications for on-farm control of the parasite based on different diagnostic methods, *Parasites Vectors*, 11:1–18.

Takeuchi-Storm, N., Moakes, S., Thüer, S., Grovermann, C., Verwer, C., Verkaik, J., Knubben-Schweizer, G., Höglund, J., Petkevičius, S., Thamsborg, S. and Werne, S. (2019),

Parasite control in organic cattle farming: management and farmers' perspectives from six European countries, *Vet Parasitol. Reg. Stud. Rep.* 18:100329.

Takeuchi-Storm, N., Thamsborg, S.M., Enemark, H.L., Boes, J., Willams, D. and Denwood, M. (2021), Association between milk yield and milk anti-*Fasciola hepatica* antibody levels, and the utility of bulk tank milk samples for assessing within-herd prevalence on organic dairy farms, *Vet. Parasitol.* 291:109374.

Thamsborg, S.M. and Roepstorff, A. (2003), Parasite problems in organic livestock production systems and options for control, *J. Parasitol.* 89:S277–S284.

Thamsborg, S.M., Roepstorff, A. and Larsen, M. (1999), Integrated and biological control of parasites in organic and conventional production systems, *Vet-. Parasitol.* 84:169–186.

Tomazic, M.L., Garro, C. and Schnittger, L. (2018), *Cryptosporidium*. In: *Parasitic Protozoa of Farm Animals and Pets*, eds. Florin-Christensen, M., Schnittger, L. Springer, Cham. https://doi.org/10.1007/978-3-319-70132-5_2.

Valente, A.H., de Roode, M., Ernst, M., Peña-Espinoza, M., Bornancin, L., Bonde, C.S., Martínez-Valladares, M., Ramünke, S., Krücken, J., Simonsen, H.T., Thamsborg, S.M. and Williams, A.R. (2021), Identification of compounds responsible for the anthelmintic effects of chicory (*Cichorium intybus*) by molecular networking and bio-guided fractionation, *Int. J. Parasitol. Drugs Drug Resist.* 15:105–114. https://doi.org/10.1016/j.ijpddr.2021.02.002.

Vanhecke, M., Charlier, J., Strube, C. and Claerebout, E. (2021), Risk factors for lungworm-associated milk yield losses in grazing dairy cattle, *Vet Parasitol.* 292:109414. https://doi.org/10.1016/j.vetpar.2021.109414.

van Leeuwen, C.H.A., van der Velde, G., van Lith, B. and Klaassen, M. (2012), Experimental quantification of long distance dispersal potential of aquatic snails in the gut of migratory birds, *PLoS One* 7(3):e32292. https://doi.org/10.1371/journal.pone.0032292.

Wagenaar, J.P. and Langhout, J. (2007), Suckling systems in calf rearing in organic dairy farming in the Netherlands. In: *Proceeding of QLIF Congress, Hohenheim, Germany*, eds. Niggli, U., Leifert, C., Alföldi, T., Lück, L., Willer, H. Hohenheim, Germany, 356–360.

Wall, R. and Strong, L. (1987), Environmental consequences of treating cattle with the antiparasitic drug ivermectin, *Nature* 327:418–421.

Waller, P.J., Bernes, G., Thamsborg, S.M., Sukura, A., Richter, S.H., Ingebrigtsen, K. and Höglund, J. (2001), Plants as de-worming agents of livestock in the Nordic countries: historical perspective, popular beliefs and prospects for the future. *Acta Vet. Scand.* 42:31–44.

Waller, P.J. and Thamsborg, S.M. (2004), Nematode control in "green" ruminant production systems. *Trends Parasitol.* 20:493–497. https://doi.org/10.1016/j.pt.2004.07.012.

Wynn, S., Fougère, B. (2007), *Veterinary Herbal Medicine*. Elsevier Mosby, St. Louis, MO.

Younie, D., Thamsborg, S.M., Ambrosini, F. and Roderick, S. (2004), Grassland Management and Parasite Control. In: *Animal Health and Welfare in Organic Agriculture*, eds. Vaarst, M., Roderick, S., Lund, V., Lockeretz, W. CABI Pub, Wallingford, Oxon, UK ; Cambridge, MA, USA, 309–328.

Zapa, D.M.B., Couto, L.F.M., Heller, L.M., Cavalcante, A.S. de A., Nicaretta, J.E., Cruvinel, L.B., Maciel, W.G., Teixeira, W.F.P., Felippelli, G., Gomes, L.V.C., Ferreira, L.L., Soares, V.E., da Costa, A.J., Borges, F. de A. and Lopes, W.D.Z. (2021), Association between fecal egg count and weight gain in young beef cattle, *Livest. Sci.* 244:104335.

Zvinorova, P.I., Halimani, T.E., Muchadeyi, F.C., Matika, O., Riggio, V. and Dzama, K. (2016), Breeding for resistance to gastrointestinal nematodes - the potential in low-input/output small ruminant production systems, *Vet. Parasitol.* 225:19–28. https://doi.org/10.1016/j.vetpar.2016.05.015.

Chapter 10

An ethical framework for the responsible use of technology in organic dairy farming

David Christian Rose, Harper Adams University, UK; Juliette Schillings, University College Dublin, Ireland; and Eden Keily-Thurstain, University of Cambridge, UK (former university)

1 Introduction: organic farming values and standards

Before exploring the potential role of precision technology in organic dairying, it is important to trace the origin of the organic movement and the evolution of both the values and standards that play a role in the adoption or non-adoption of technology. Organic farming standards and values vary across time and space and thus defining it can be challenging. The FAO and WHO (1999) define organic agriculture as 'a production management system which promotes and enhances agroecosystem health, including biodiversity, biological cycles, and soil biological activity'. Key features of organic livestock farming are the avoidance of synthetic fertilisers, pesticide, and routine use of antibiotics and wormers, achieving the highest possible animal welfare standards (e.g. space, outdoors where possible), and feeding a natural, organic diet (grass-rich, non-GM animal feed) (Soil Association, 2023).

http://dx.doi.org/10.19103/AS.2024.0138.11

We start with the evolution of organic values which are highly diverse (Duram, 2000). Leduc *et al.* (2023, 1) state 'values act as driving forces for individuals to behave in a certain way or to choose certain actions'. In Europe, Padel *et al.* (2009) found that many organic producers acted on the basis of values, going beyond minimum standards. Organic farming originated as a counter-movement and is still widely considered to be the antithesis of conventional, intensive, capitalist forms of production (Holt-Giménez, 2019; Kledal, 2003; Kuepper, 2010). Kuepper (2010) describes how early pioneers set out to address problems caused by agriculture, including soil erosion, low-quality food and livestock feed, rural poverty, and soil depletion. The values of the organic movement have been associated with resistance to global capitalised agriculture (Schreer and Padmanabhan, 2020), localised food production (Belasco, 2014), and shortened supply chains (Belasco, 2014; Pollan, 2006). The movement has come to represent a more holistic approach compared to its origins as a movement focusing on soil health.

Pollan (2006) identifies a distinctly countercultural vision for organic agriculture, popularised in the 1960s and 1970s in the context of chemical scares and a growing environmental consciousness, which incorporates an alternative production mode, alternative delivery systems, and a 'countercuisine' with minimal processing. Bronson (2005, 20) argues that 'organic production ... could be seen as part of a wider ... social movement involving farmers, consumers and others in a challenge against the assumptions and practices of unbridled economic growth, and the development ideologies at the epicentre of contemporary capitalist society.' In a study in Saskatchewan, Canada, Bronson (2005, ii) describes 'a hope and determination amongst organic farmers who see themselves as resisting the erosion of the rural landscape at the hands of powerful corporations and a dominant industrial model of food production.'

These values ultimately structured organic farmers in particular ways, favouring short supply chains, small farms, employment of family and local labour, and little to no chemical use. Studies of the motivations of organic farmers around the world find that psycho-behavioural and psychosocial factors (Sapbamrer and Thammachai, 2021; Yazdanpanah *et al.*, 2021, Iran), including strong anti-consumption values (Dalmoro *et al.*, 2020, Brazil; Vail, 2022, USA), a dislike of agricultural policy that favours large-scale conventional farming (Duram, 2000, Illinois, USA), a negative attitude towards chemicals (Egri, 1999, Canada), a wish to live a more natural lifestyle (Schösler *et al.*, 2013, The Netherlands), caring for the natural environment and family health (Kociszewski, 2020, Poland; Läpple, 2013, Ireland; Tress, 2001, Denmark), and as an economic option (including as a means of development in the Global South) through the higher prices that organic produce can attract (Seufert *et al.*, 2017). A study in the Czech Republic found organic farming to be a way of life, an occupation, and a way of producing food of perceived higher quality (and eating more 'naturally') than conventional farming (Zagata, 2010).

From this, four principles of organic farming have come to the fore:

- Health – of the natural environment, livestock, and humans.
- Ecology – protecting the natural environment.
- Care – adopting a precautionary principle to new innovations.
- Fairness – ensuring benefits of production are evenly shared and support local communities, families, and workers, rather than benefiting big corporations.

As the organic farming movement has changed over time, incorporating standards as well as basing activities on values, concerns have been raised. Just as the alternative food movement literature is critical of the centrality of market logics, the organic literature has identified a recent trend of the 'conventionalisation' of organic agriculture, where market forces have hollowed out the essence of the movement (Obach, 2007, 2015).

By setting minimum organic standards, it has been argued that organic certification allows large, corporate players to 'enter the game', as well as creating 'ceilings', where farms can do the bare minimum to reach organic status and fall short of agro-ecological ideals (Guthman, 2000, 265), resulting in an 'organic lite' (Guthman, 2004, 301). Padel *et al.* (2009) describe how the growth of the organic market and globalisation have been identified by some organic farmers as problematic. Research on the structural aspects of organic farming in California depicts highly capitalised, large-scale, industrial production that increasingly mirrors conventional agriculture (Buck *et al.*, 1997), a situation which Konstantinidis (2014) argues has also occurred in Europe as organic has become 'capitalism in green disguise'. Further research in California indicates that as agribusiness makes in-roads into organic production, sustainability may be sacrificed (Guthman, 1998). This reinforces the point made at the start of this chapter – organic farming as a movement has become highly diverse and this complicates decision-making over adoption of the technologies discussed below.

Standards for organic farming are based on values and equally vary in time and space. Taking the EU as an example, which aims to have 25% of its agricultural land under certified organic farming by 2030 (EU Commission, 2020), current rules on livestock in organic systems cover:

1. Abiding by organic principles – such as 100% organic feed, suckling animals being fed by natural milk, and natural methods of reproduction.
2. Animal welfare – e.g. personnel on the farm having the right skills to look after animals, low stocking densities, open air, grass grazing where

possible, no routine use of antibiotics, no use of hormones or similar substances.
3. Rules for the food chain – e.g. separation of organic and non-organic products, a minimum organic content of 95% of organic ingredients, clear rules on labelling.
4. Permitted substances in organic production – e.g. substances used in organic production must be pre-approved by the EU Commission etc.

The full rules can be found here.[1]

2 Organic dairy farming and technology: key principles

We recognise that 'technology' is more than just a physical manifestation of science and engineering and that it is important to understand what technology 'does' rather than just what it 'is' (Carolan, 2020). Taking such a view is important when considering how technology intersects with the values of organic farming, which is the basis for adoption decisions by farmers. We also reiterate the diversity of organic farmers so adoption decisions are difficult to predict and generalisations should be treated with caution. A useful report has started to consider the values-based decisions that organic farmers take with regards to technology (Mason *et al.*, 2024).

With this caveat in mind, however, it is fair to say that agricultural technology is sometimes stereotyped as benefitting high-tech, intensive visions of farming over low-input systems such as organic (Lassen, 2015; Castell *et al.*, 2021; Sullivan, 2023). Critical social scientists have argued that agricultural technology has been traditionally designed to support intensive, large-scale production by replacing labour ('fairness'), increasing emissions and outputs, and (by claiming to improve efficiency) masking the environmental impacts of intensive farming that has been made possible by many of the same technologies ('ecology') (Bronson and Sengers, 2022; Carolan, 2020; Ditzler and Driessen, 2022; Fairbairn *et al.*, 2022; Sullivan, 2023).

These concerns intersect with the principles of organic farming. There is scepticism about whether agricultural technology is being used to reduce farmer autonomy by legitimising algorithmic rationality and surveillance capitalism (Brooks, 2021; Gardezi and Stock, 2021; Miles, 2019; Stock and Gardezi, 2021), a process potentially being driven by larger players to benefit the status quo ('fairness') (Birner *et al.*, 2021; Clapp, 2023; Hackfort, 2023; Duncan *et al.*, 2021). The potential for technological lock-in and uneven power relations has been identified (Bronson and Sengers, 2022; Clapp and Ruder,

1 https://modernfarmer.com/2018/06/by-the-numbers-state-of-the-independent-farmer/ (2018).

2020) ("fairness"). In addition, there remain concerns that existing and emergent gene-based technologies (modification or editing) are still being primarily developed, owned, and distributed by powerful, corporate actors who lock producers into a dependent relationship and create a 'biohegemony' (Escobar, 2016) ('fairness'). Related to this are concerns over the lack of transparency and corporate control over emergent technologies and lack of adherence to responsible innovation principles from some industry players, including genomics experts (Clapp, 2023; Hackfort, 2023; Guthman and Biltekoff, 2021; Ruder and Kandlikar, 2023) ('fairness'). With these concerns have come worries over data ownership ('fairness') (Rotz *et al.*, 2019) and the potential facilitation of more intensive farming, including in livestock systems ('health' and 'ecology' (Schillings *et al.*, 2021a). It is also apparent that the full set of consequences of emergent technologies is often difficult to predict ('care') (Rose and Chilvers, 2018). The concerns listed above can set off 'alarm bells' within the organic sector, particularly if we recall the origins of the movement rooted in counterculture and rejection of neoliberal, corporate control over the food system.

However, it is also clear that technologies, including in the dairy sector, can support low-input (including organic) farming. In certain cases, technology may offer more to organic farming, particularly if precision equipment can work more effectively than conventional machinery and other methods in such systems (Daum, 2021). Organic farming is not anti-technology per se, but these farmers may question what technology 'does' (in terms of its wider social and ethical impacts) more than conventional farmers. Some organic farmers may be concerned about abandoning 'naturalness' and take a precautionary approach (Milestad *et al.*, 2020). If technologies are designed with the smaller-scale farmer in mind, are used to support environmentally- and welfare-friendly visions of farming (Daum, 2021), and benefit producers as well as corporations. then alignment with organic principles may be achieved. This would involve a shift in embedded principles of technology development towards values-based, equitable design.

3 Precision and other technologies in organic dairy farming

Existing technology used in organic dairying (set out below) is helping in a number of areas, including health and ecology, and emergent precision technologies have the potential to enhance animal welfare by minimising negative welfare outcomes, although opportunities to encourage positive animal welfare are currently limited (though developing all the time) (Schillings *et al.*, 2021b). This means that precision technologies tend to focus on reducing harm to animals, rather than setting out to explicitly enhance welfare, which is a key organic principle.

We focus first in this section on the potential benefits of technology for organic dairying, noting that many, if not all, of the positive claims below could be a double-edged sword. Long before the digital technology age, various technologies were being widely used (and continue to be used) on organic dairy farmers, which is also true in other farm enterprises. Whilst earlier technologies might have been superseded on some farms by digital equipment, they still continue to perform a useful function on many other farms that are easily forgotten by the rush to consider digitalisation (Rose *et al.*, 2022). Technologies for farmers can include 'lower-tech' products like Wellington boots, farm vehicles, a white/blackboard, pen, and paper to facilitate planning, mobility scoring charts for lameness, and right through to 'higher-tech' equipment such as mobile phone apps and other digital tools (see Mason *et al.*, 2024). Examples of useful technologies for organic dairying include plate meters to measure grass growth, conventional milking parlours, traditional fencing, and so many more. There are increasing calls for organic farming to re-examine its relationship with biotechnology approaches, such as gene editing (Fernandes *et al.*, 2022; Husaini and Sohail, 2018; Wickson *et al.*, 2016). As for breeding, there appears to be no global consensus about the selection strategies that align with organic principles, most likely because the practices and principles of organic agriculture vary across time and space (Nauta *et al.*, 2005).

Whilst developments in precision technologies can be geared towards more intensive production systems, many digital tools can be applied to extensive systems. Precision technologies can be used to optimise grassland management and ensure economic sustainability whilst mitigating the environmental impacts of pasture-based dairy systems (Murphy *et al.*, 2021). For example, some tools can be used to accurately quantify herbage mass, pasture heterogeneity or grass quality, and to advise farmers on grassland management thanks to decision support tools that have been developed in Europe (Murphy *et al.*, 2021). Other decision support tools are being used by organic dairy farmers to access markets and reduce time and waste, for example the FarmSuite initiative in Italy (https://www.farmsuite.it/en/).

A technology that is gaining increased interest for pasture-based livestock is virtual fencing, which makes it possible to contain cattle out on pasture within a given area without the need for physical fences. Using a mobile application, a virtual fence is drawn, and GPS-equipped neck collars are used to track animal location in real-time. When cows approach the virtual fence, they receive an audio warning. If they continue to walk towards the boundaries, the intensity of the signal increases, and if they fail to turn around and cross the boundaries, they receive an electric pulse that will encourage them to move back to the designated area. The use of virtual fencing technology is associated with many benefits, including animal welfare, biodiversity, and landscape conservation (Stampa *et al.*, 2020). Virtual fencing can encourage farmers to keep their cows out on pasture, sometimes on lands that are difficult to access and to physically fence (e.g. upland farming systems).

Higher animal welfare standards and a reduced use of antibiotics are some of the key features of organic farming. There are numerous digital solutions that can theoretically help organic dairy farmers adhere to higher welfare standards. These range from oestrus, calving, and lameness detection to precision feeding, automated body condition scoring, monitoring gastrointestinal health, feeding behaviour, or location. Some tools even aim to move beyond a sole focus on animal health to focus on animal emotional expressivity, a dimension that is integral to animal welfare, with mobile applications developed to conduct qualitative behavioural assessments and promote positive welfare (Schillings *et al.*, 2023a). The use of precision tools could help organic dairy farmers to reach higher welfare goals by allowing the early detection of health issues, thus helping farmers to act in a timely and efficient manner. Indeed, digital tools could allow the detection of health and welfare problems before they are detected by farmers or veterinarians (Taneja *et al.*, 2020; Berckmans *et al.*, 2015). In turn, this could help prevent negative impacts on animal welfare and minimise reliance on antibiotics, thus helping in addressing the core principles of organic farming. Many precision livestock technologies have been developed to monitor dairy cattle welfare and productivity, with devices such as sensors, cameras, boluses, radio-frequency identification (RFID) tags or microphones that are either commercially available or at earlier development stages (Schillings *et al.*, 2021b; Tzanidakis *et al.*, 2023).

By monitoring rumination activity, it is possible to detect feeding deficiencies, which can be an indicator of potential illnesses (Sowell *et al.*, 1998; Weary *et al.*, 2009). In their study, for example, Weary *et al.* (2009) found that cows with decreased feeding times were more likely to have clinical metritis. Similarly, monitoring lying activity with digital technologies can provide helpful indications as to cows' welfare states, since the length and frequency of lying behaviour are useful indicators of cow comfort, which can be influenced by the type of housing or the presence of painful conditions such as sole ulcers (Haley *et al.*, 2000; Vasseur *et al.*, 2012). There are several other mobile applications for planning, market access, communication, animal welfare, and other tasks that are being used on farm to varying degrees.

The use of automated milking systems is also increasing in organic farming for smaller herd sizes and can offer reduced physical workload and flexible working hours (Butler *et al.*, 2012). Automated milking systems are sometimes associated with higher welfare and improved human–animal relationship, since they can allow cows to 'choose' when to be milked, thus potentially reducing stressful handling. Other robots, such as unmanned aerial vehicles, can be used to locate and monitor animals out on pasture.

Many of the potential benefits of technology are, of course, theoretical, especially where newer technologies are at an early stage of development. Technologies that enable farmers to implement practices for enhanced health and welfare require are not enough in themselves. To be effective, the farmer must implement an effective change of practice. It is thus also important to assess whether technologies are enabling these shifts in practice, a topic which is rarely researched (Schillings *et al.*, 2021b).

Adoption factors for farm technology in general are well-known and include issues related to trust and data ownership, attitudes and values, ease and cost of adoption (which is affected by scale and skills), reliability, peer recommendation, regulations, interoperability, and connectivity (Rose *et al.*, 2016). In this context, it is useful to highlight the scale of organic farming, and organic dairying in particular. Organic farms tend to be smaller than conventional farms, although not always. In the USA, for example, the average size of an organic farm is 285 acres compared to an average of 444 acres for all farms (USDA stats reported by Modern Farmer[2]). Across the EU, low-input dairy farms, including organic, tend to be smaller than their higher-input counterparts (Scollan *et al.*, 2017), although there is considerable variation between countries (Wallenbeck *et al.*, 2019), while some authors have questioned whether organic agriculture in the EU is substantially differently structured than conventional farming (Konstantinidis, 2014).

Despite the wide range of technologies available, their adoption in extensive systems is lower than in more conventional systems, often due to connectivity issues related to the large and sometimes remote areas used. In addition, a lack of validation, doubts around reliability, and the high cost associated with precision technology, often represent an important barrier to their adoption (Heins *et al.*, 2023). Other challenges include familiarisation with new technologies, willingness to adopt, difficulties interpreting the data, and lack of technical support. As precision technologies often operate independently, a lack of integrated system can make the use of multiple systems difficult to manage for farmers.

Overall, technology adoption on organic dairy farms is, therefore, influenced by a variety of factors, including:

1. Values and trust
2. Size/scale of operation
3. Standards/rules within a jurisdiction

Turning now to the other side of the 'double-edged sword', many of the claims made above could be challenged depending on how technologies are used.

2 https://modernfarmer.com/2018/06/by-the-numbers-state-of-the-independent-farmer/ (2018)

In terms of values and trust, animal welfare is an important part of the organic 'health' principle. Different organic farmers will hold different perceptions of the role of technology in improving animal welfare, for which there are often conflicting points of view (Schillings *et al.*, 2021a). Whilst technologies may enhance animal welfare, there are concerns over the negative implications of some wearable technology, technologies that may provide a shock to livestock, or that increase the distance between stock persons and their stock (Butler *et al.*, 2012), or that are uncomfortable to wear (Schillings *et al.*, 2021b). Concerns have been raised regarding the potential of automated milking robots to lead to restricted pasturing and increased use of concentrates (to attract cows into the milking system), and changing human-animal relationships (Butler *et al.*, 2012), which may not always be suitable to organic standards (Bühlen *et al.*, 2014). Furthermore, forced traffic systems can actually restrict the freedom of choice for the cow (Munksgaard *et al.*, 2011). In terms of human 'health', technologies like robotic milking may free up time and enhance the lifestyle of farmers, whilst others will experience enhanced stress and reduced enjoyment from 24/7 technology-based monitoring of livestock (Lundström and Lindblom, 2021; Martin *et al.*, 2022). Ultimately, the decision may come down to how each organic farmer defines the notion of "care" that they give to their stock (Lundström and Lindblom, 2021).

As stated above, in terms of other organic principles, technologies may lead to the reduction of environmental footprint through emissions or materials (e.g. fencing) reduction ('ecology'). In terms of 'fairness', technologies could increase transparency in the supply chain for the benefit of farmers, consumers and other actors, but there are concerns over ownership and use of data collected and distribution of benefits across the supply chain, as well as control exerted by private companies and reduced farmer autonomy (Schillings *et al.*, 2021b). Technology could help to facilitate better data collection for organic certification. There may be realistic concerns over labour displacement, the erosion of farmer autonomy and experienced-based knowledge. Precaution is likely to be important to organic dairy farmers with clear evidence needed of how each technology affects the four principles of organic farming before adoption. Important to all of the above is the need for a farmer to take effective action based on data, which is aligned with organic principles/standards. If data is too difficult to interpret, or the suggested actions are too hard or costly to implement, then the promises of technology may not actually be delivered in practice (Schillings *et al.*, 2021b).

Looking at size/scale of operation, not all organic dairy farms are small, but stocking rates tend to be smaller than average. This may reduce the need for some precision livestock technologies which developers argue can help monitor individual animals more easily in larger herds. The size of the farm can also affect ability to invest or skills availability to use equipment if there is a less diverse workforce. Alternatively, however, new technologies could free up time for small-scale farmers to perform other roles. Finally, standards and regulations within some countries and regions may prohibit the use of certain technologies, perhaps most obviously genetic modification or engineering products.

4 Establishing a framework for responsible innovation

Since technology provides both opportunities and risks for organic dairying, a framework is needed to assist in structuring thought processes when deciding on whether to adopt or not. The application of the four organic principles – health, ecology, fairness, and care (precaution) – within a responsible innovation framework can be used for these structured deliberations.

Responsible innovation is defined as 'taking care of the future through collective stewardship of science and innovation in the present' (Stilgoe *et al.*, 2013). Innovation can be made more responsible through the processes of

- inclusion (inviting stakeholders to contribute)
- anticipation (considering the potential impacts of innovation)
- reflexivity (listening and challenging assumptions)
- responsiveness (learning from new knowledge, and changing course)

Innovation that is not responsible runs the risk of causing problems that were not anticipated when they could have been. The framework has been applied to agriculture in several research studies with a number of different methods being proposed to implement its four elements (see e.g. Eastwood *et al.*, 2019; Rose and Chilvers, 2018). We propose that this framework is ideal in considering if and how dairying technologies align with the values and standards of organic farming. The four pillars align with the principles of organic farming:

- firstly, substantive inclusion of all affected stakeholders can lead to a fairer, or more just, development process.
- secondly, anticipating the consequences of dairy technologies, both good and bad, and elucidating uncertainties requiring further research can help organic stakeholders to understand whether impacts underpin the principles of ecology, and health, as well as allowing a precautionary approach to be taken if there is limited evidence.
- thirdly, listening and being prepared to challenge assumptions and consider moral responsibility indicates a fairer development process, rather than one that is dominated, for example, by the interests of private technology developers.

- lastly, having systems in place to respond to new knowledge ensures that the contribution of dairy technology to ecology and health can be consistently monitored and a precautionary stance can be taken in the face of new information.

Below, we demonstrate how a responsible innovation framework might be applied in the consideration of an emergent technology which organic dairy farmers are increasingly able to implement: virtual fencing (VF). We argue that this process could be applied to any technology for dairying to allow organic farmers to judge whether or not it matches their values and standards (see Fig. 1). Ensuring the responsible use of technologies such as VF requires further investigation that can help to inform policy and allow to build recommendations that align with the principles of organic farming.

5 Applying a responsible innovation framework: the example of virtual fencing

As mentioned previously, virtual fencing is an emergent technology that allows livestock farmers to create 'unseen' boundaries to keep animals in designated areas. These areas can be controlled by the farmer through a digital app and constantly changed, with animals learning to stay within boundaries by wearing GPS-equipped neck collars which give an alert or shock when boundaries are nearly (or actually) crossed (Fig. 2).

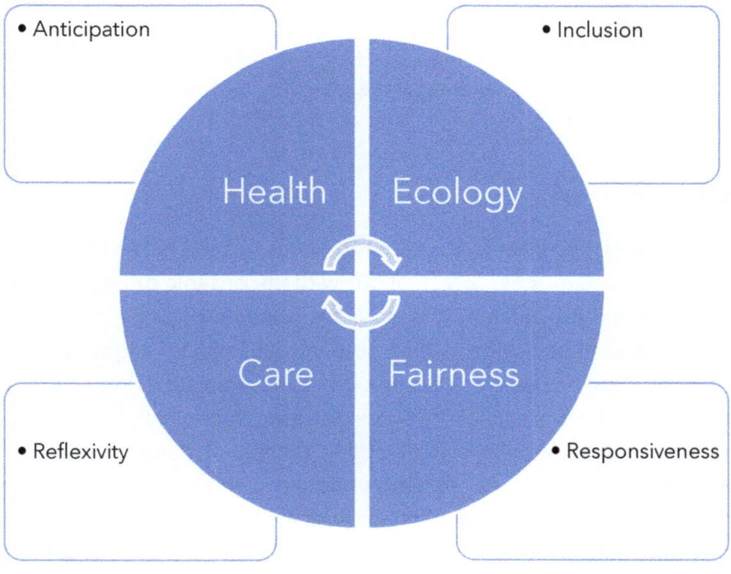

Figure 1 Principles of organic farming combined with elements of the responsible innovation framework (from Stilgoe *et al.* 2013).

Figure 2 Virtual fencing neck collars worn by cows and sheep. These collars generally produce a sound when an animal is approaching a virtual boundary to discourage them from crossing it. A shock may be given in the case of non-compliance (Photos used with kind permission from AFBI, NI).

5.1 Inclusion

The first task in responsible innovation is to include a full range of stakeholders in the development process, which aligns with the organic principle of fairness. Without substantive inclusion, it is hard to gather the range of insights and knowledge needed to judge if and how virtual fencing aligns with organic principles of care, ecology, and health. When considering food system interventions, inclusion can be difficult because, technically, everyone has a stake in the food system (Ayris *et al.,* 2024). Whilst full inclusion of everyone is always likely to be impractical, research has shown that engagement exercises often focus on a narrow group of stakeholders: e.g. a select number of farmers only, missing 'harder to reach' farmers, rural communities, consumers, and other stakeholders (Ayris *et al.,* 2024; Bronson and Sengers, 2022). In the case of virtual fencing, the necessary stakeholders to include (not a comprehensive list) will be livestock farmers (who implement the technology), veterinarians, NGOs and animal welfare scientists (for evidence of impact on animals), other supply chain actors (who play a hand in implementation and standards), rural communities (who need to understand how the technology works e.g. to prevent reports or panic around livestock escape), and policy-makers (who use policy instruments for implementation e.g. incentives, regulation).

The inclusion of citizens in early discussions is an important aspect of responsible innovation. In the context of virtual fencing, uncertainties around the potential impacts on animal welfare could represent an important barrier regarding public acceptance of the technology, since animal welfare is increasingly a subject of concern (European Commission, 2016). Whilst citizens generally lack knowledge of livestock farming practices, their involvement in early discussions on the use of virtual fencing technology should not be neglected, as a lack of acceptance could hinder the widespread use of virtual fencing technology (similar to what has been observed with food technologies such as GMOs or nanotechnology (Siegriest and Hartmann, 2020). In the smart dairying sector, there has been little citizen participation, generally due to difficulties in engaging with community actors and limited perceived benefits (Eastwood *et al.*, 2019). The importance of involving the public to gather their perceptions on the development of these types of technologies was further emphasised by Brier *et al.* (2020) to ensure a responsible innovation process.

5.2 Anticipation

To judge whether the technology aligns with organic principles requires a concerted effort to understand its easily and less easily foreseen benefits and risks. The development and use of virtual fencing technology, as described in the previous section, is often associated with a range of benefits, including financial and environmental advantages. However, there are also concerns relating to their potential socio-ethical consequences, including perceived negative impacts on animal welfare or difficulties for farmers in adjusting to and making effective use of virtual fencing (Brier *et al.*, 2020). Whilst there are many potential benefits associated with virtual fencing (e.g. decreased labour and costs, facilitating strategic grazing management, encouraging conservation grazing…), there are also concerns relating to its use (e.g. impact on animal welfare, negative public perception, impacts on stockmanship). It is important that these potential benefits and risks are assessed at an early stage to minimise uncertainties and negative consequences whilst being able to decide on the direction to take so as to achieve better and more desirable outcomes. A variety of methods can be used to perform inclusive anticipation exercises, including foresighting or scenario building (Eastwood *et al.*, 2019), within which a range of knowledge and views need to be brought together.

Crucial to these processes, however, is openness and transparency. Critical social science literature on emergent agricultural technologies, such as cultured meat and robotics, has argued that powerful food system actors are consolidating their control and developing solutions in secret without offering enough information to stakeholders and without including them (Clapp, 2023; Hackfort, 2023; Guthman and Biltekoff, 2021). If, during the process of discussing potential future effects, stakeholders do not know how and why technologies are being developed, as well as how they work and what they do, it can be nearly impossible to make an informed judgement over alignment with organic standards and principles. However, it could be argued that, in a closed innovation process in which 'big' food system actors dominate the development and distribution of technology, organic dairy farmers would decide not to adopt.

Several studies have focused on virtual fencing's efficiency in containing animals and its impacts on animal welfare. Studies that have compared virtual fencing with conventional electric fences so far have suggested minimal differences in terms of impacts on animal behaviour and welfare (Umstatter *et al.*, 2015; Verdon *et al.*, 2020; Aaser *et al.*, 2022; Sonne *et al.*, 2022). In their study comparing virtual fencing and physical fences, Hamidi *et al.* (2022) found no significant differences in cattle behaviour, live weight gain, herbage consumption, and cortisol levels. This concurs with findings from Ranches *et al.* (2021), who found no differences in cortisol levels or animal behaviour. However, a study conducted by McSweeney *et al.* (2020) on dairy cows revealed a reduction in the time animals spent grazing and ruminating, suggesting that negative impacts on animal welfare are a possible outcome. This lack of clarity underlines the need for more research on the long-term impacts of VF on animal behaviour and welfare.

Whilst several studies have investigated impacts on animal welfare, fewer studies have explored stakeholders' acceptance and perceptions of virtual fencing for livestock grazing management. In a study conducted by Brier *et al.* (2020) foresight exercises were undertaken with farming practitioners in New Zealand to anticipate the potential consequences of using this technology, and key benefits and risks were identified. These included better environmental protection and feed allocation, the possibility of using previously inaccessible grazing locations, decreased labour, and improved individual animal management. Making more efficient use of pasture and being able to reduce labour and maintenance cost by not having to set up physical fences is, indeed, an important benefit often associated with virtual fencing. A study conducted in Ireland also highlighted the potential of virtual fencing to promote the efficiency and sustainability of livestock grazing systems, particularly in upland areas – although important aspects remain to be addressed to enable efficient redesign of these systems (Schillings *et al.*, 2024).

The benefits of virtual fencing may not be possible if proper support and training is not provided to farmers, as the technology may not be suited to certain types of production systems and may require specific skills to navigate the technology. Other key barriers also identified by Brier *et al.* (2020) were a lack of value proposition as perceived by farmers, perceived negative consequences on animal welfare, farmers' lack of feed budgeting skills and training time, further highlighting the need to adopt an inclusive approach to better understand end-users' and other stakeholders' expectations and ensure the technology is mutually beneficial to farmers, animals, and the farming industry.

5.3 Reflexivity

Reflexivity means to listen, learn, and challenge one's assumptions in light of the inclusive anticipation exercises mentioned previously. If closed innovation processes are dominated by private technology companies who are not willing to change trajectories or re-consider assumptions in response to feedback from stakeholders, then innovation cannot be responsible. Where possible, a values-based design process should be followed. For virtual fencing technology, the views of stakeholders in the organic dairy sector should be taken into account when developing and implementing virtual fencing. However, the outcome for organic dairy farmers could be as simple as a particular community deciding that virtual fencing does not currently align with its values or standards, leading to the choice not to adopt it. This may not affect development and implementation in other farm enterprises.

5.4 Responsiveness

To be seen as suitable for implementation in organic systems, mechanisms must be in place to ensure that virtual fencing aligns with organic principles and standards. If organic farmers are satisfied in principle that the likely consequences of virtual fencing are acceptable, various rules and regulations could be put in place to ensure this. For example, minimum standards around virtual fencing (e.g. how the technology works in regards to electric shocks) could be put in place to influence how it is developed for and used in an organic dairying context. Organic produce from farms using virtual fencing should be required to follow these basic minimum standards decided by the organic community. Monitoring systems must also be put in place so that unforeseen problems can be addressed quickly.

It will also be necessary to understand how the potential use of virtual fencing aligns with existing regulations on organic produce and wider legislation. In the UK for example, the use of shock collars for cats or dogs is banned in Wales and applying a sustained shock to an animal is illegal under the Animal Welfare Regulations in the Republic of Ireland. A lack of clarity regarding legal and regulatory aspects was also highlighted by DEFRA in England. It is crucial to clarify these aspects before the widespread adoption of virtual fencing technology, as a lack of regulatory framework could have serious implications for the welfare of farm animals. Setting up codes of conduct and standards based on scientific knowledge to encourage the responsible use of virtual fencing technology could help facilitate reflexivity and build trust between different stakeholders and make sure it is right for organic dairying.

6 Using a responsible innovation framework to accept or reject new technology

The responsible innovation framework above could be applied to individual or suites of technology in organic dairying systems and can be used as a mechanism for farmers to accept, conditionally accept, or reject them. In working through the framework, conclusions can be reached about whether technology aligns with organic principles (or standards) held by individuals or groups of producers. One overarching concern of organic producers about emergent precision technologies could be that they currently tend to focus on reducing harm to animals, rather than explicitly enhancing welfare, which is a key organic principle (Schillings *et al.*, 2021a).

Using the framework in the virtual fencing example, some organic producers might decide not to adopt until further evidence is available regarding the stress and welfare impacts caused to livestock during the training phase ('health' and 'care' principle). Or, organic farmers could decide to adopt if the shock mechanism for training was replaced with an alternative or conclude that the impact on the animal is not sufficiently negative as to endanger the principle of 'health'.

Other examples can be worked through. For example, artificial intelligence technologies are increasingly available to enable automated body condition scoring or earlier detection of diseases such as lameness (Schillings *et al.*, 2023b). When anticipating possible consequences of this technology, farmers have raised concerns over data ownership and privacy, specifically who would have access to data and benefit from it. Other queries related to a change in the human–animal relationship and concerns over lack of reliability (Schillings *et al.*, 2023b). In the anticipation, responsiveness, and reflexivity phases, therefore, organic farmers may decide that the technology does not align with principles of fairness or, for example, may/may not be satisfied that developers have listened to their concerns and acted to implement transparent and acceptable governance of data. Alternatively, organic farmers may not think they have been satisfactorily involved in the process of developing and implementing the technology such that they cannot be sure over alignment with organic principles.

Assessing automated milking (Fig. 3) using a responsible innovation framework could also lead to more informed decisions by organic farmers about whether and how to adopt. In anticipating consequences, concerns may be raised over the human–animal relationship (e.g. potential impact on stress of handling stock if contact is limited), 24/7 data stress affecting farmer/worker mental health, balanced with other evidence about the possible improvement to farmer lifestyles and working conditions, as well as being able to more closely monitor animal health through data. Again, if organic farmers do not feel that they can adapt the technology to align with their values or do not see enough evidence that negative impacts on animals and workers are being avoided (or indeed opportunities for enhanced health and welfare), a non-adopt decision could be reached.

We also believe that the responsible innovation framework should be used by developers of technology who intend to market products to organic dairy producers. It would firstly encourage them to include organic stakeholders in the design and implementation of their technology, helping to anticipate the effects of intended and unintended consequences. In our experience, developers do not always consider the unintended or less easily foreseen consequences of their technology beyond their 'sales pitch', and may not fully appreciate that technology is not always used exactly as intended in practice or that technology itself shapes farms, farmers, and animals in ways not foreseen

Figure 3 A cow being milked (photograph from two angles) by an automated milking robot (author owned copyright [David Rose]).

(Rose *et al.*, 2018). Thus, they may legitimately think that their technology aligns with organic standards and values, but this view may not be based on a full appreciation of on-farm human and non-human impacts. Using a reflexive and responsive process, the design and delivery of the technology could be undertaken in a way that addresses stakeholder concerns and ensures better alignment with not only organic standards, but values. This should ideally involve a re-orientation of the purpose of technology, not only to reduce harm to animals and people, but explicitly to enhance health and welfare for the animal, farmer, and worker.

7 Conclusion

Whether digital or non-digital, old or new, technology will continue to play an important role in organic dairying. This chapter has indicated, however, that decisions over the implementation of existing and emergent technologies on organic dairy farms are affected by two components:

- firstly, whether the use of a technology matches organic standards in a given place
- secondly, whether it aligns with the values and principles held by organic farmers

Making a judgement on the latter, however, may not be easy. The impacts of emergent technologies are not always certain and, therefore, there is a need for a structured decision-making process to consider suitability for use in organic systems. We propose that the responsible innovation process can be worked through by the organic community, in collaboration with technology developers, supply chain actors, policy-makers, and the research community, to anticipate whether or not individual technologies match organic values and standards.

While it may be impractical to do this collectively for every technology, a set of standards and valued-based positions could be drawn up as a guide for the organic community to judge new or existing technologies against. Where 'red flags' are raised, for example the technology may cause animal welfare problems, is controlled by large corporate actors without including stakeholders, has uncertain impacts, or is not considered natural, a non-adoption decision could be reached. A clear message for those interested in increasing the adoption of technologies in organic dairying, including technology companies and policy-makers, is to ensure that development is responsible. This will require interventions from private companies and policy-makers, including opening up development processes, deliberately designing for the small-scale organic farmer, and using standards and regulations appropriately.

8 Where to look for further information

- Guthman, J. (2004). The trouble with 'organic lite' in California: a rejoinder to the 'conventionalisation' debate. *Sociologia Ruralis*, 44(3), 301–316.
- Rose, D.C. and Chilvers, J. (2018). Agriculture 4.0: broadening responsible innovation in an era of smart farming. *Frontiers in Sustainable Food Systems*, 2, 1–7.
- Schillings, J., Bennett, R. and Rose, D. C. (2021). Exploring the potential of Precision Livestock Farming technologies to help address farm animal welfare. *Frontiers in Animal Science*. https://doi.org/10.3389/fanim.2021.639678.
- Stilgoe, J., Owen, R. and Macnaghten, P. (2013). Developing a framework for responsible innovation. *Research Policy*, *42*(9), 1568–1580.

9 References

Aaser, M.F., Staahltoft, S.K., Korsgaard, A.H., Trige-Esbensen, A., Alstrup, A.K.O., Sonne, C., Pertoldi, C., Bruhn, D., Frikke, J. and Linder, A.C. (2022). Is virtual fencing an effective way of enclosing cattle? Personality, herd behaviour and welfare. *Animals*, 12(7), 842.

Ayris, K., Jackman, A., Mauchline, A. and Rose, D.C. (2024). Exploring inclusion in UK agricultural robotics development: who, how, and why? *Agriculture and Human Values*, 41, 1257–1275.

Belasco, W.J. (2014). *Appetite for change: how the counterculture took on the food industry / Warren J. Belasco* (Second updated edition). Cornell University Press.

Berckmans, D., Hemeryck, M. and Berckmans, D. (2015). Animal sound … talks ! In Real-time sound analysis for health monitoring in livestock. *International Symposium on Animal Environment and Welfare*, 1–8.

Birner, R., Daum, T. and Pray, C. (2021). Who drives the digital revolution in agriculture? A review of supply-side trends, players and challenges. *Applied Economics Perspectives and Policy*, 43(4), 1260–1285.

Brier, D., Eastwood, C.R., Dela Rue, B.T. and Viehland, D.W. (2020). Foresighting for responsible innovation using a Delphi approach: a case study of virtual fencing innovation in cattle farming. *Journal of Agricultural and Environmental Ethics*, 33(3–6), 549–569.

Bronson, K. (2005). *Of global cause and consequence: the organic farmers of saskatchewan versus monsanto and bayer* (Thesis). Saskatoon: University of Saskatchewan.

Bronson, K. and Sengers, P. (2022). Big tech meets big ag: diversifying epistemologies of data and power. *Science as Culture*, 31(1), 15–28.

Brooks, S. (2021). Configuring the digital farmer: a nudge world in the making? *Economy and Society*, 50(3), 374–396.

Buck, D., Getz, C. and Guthman, J. (1997). From farm to table: the organic vegetable commodity chain of Northern California. *Sociologia Ruralis*, 37(1), 3–20.

Bühlen, F., Ivemeyer, S., Krutzinna, C. and Knierim, U. (2014). Potential effects of automatic milking systems on grazing in organic dairy farming. *Organic Agriculture*, 4, 301–308.

Butler, D., Holloway, L. and Bear, C. (2012). Milking systems and the changing role of the stockperson. *Journal of the Royal Agricultural Society of England*, 173(January), 1–6.

Carolan, M. (2020). Automated agrifood futures: robotics, labor and the distributive politics of digital agriculture. *The Journal of Peasant Studies*, 47(1), 184–207.

Castell, S., Clemence, M., Kamvar, R. and Reynolds, M. (2021). *Living landscapes. Public dialogue on the future of land use*. Conducted on behalf of the Royal Society. https://royalsociety.org/-/media/policy/Publications/2021/23-03-21-living-landscapes-public-dialogue.pdf

Clapp, J. (2023). The rise of big food and agriculture: corporate influence in the food system. In C. L. Sage (Ed.), *A research agenda for food systems* (pp. 45–66). Edward Elgar Publishing.

Clapp, J. and Ruder, S.-L. (2020). Precision technologies for agriculture: digital farming, gene-edited crops, and the politics of sustainability. Global Environmental Politics, 20(3), 49–69. https://doi.org/10.1162/glep_a_00566

Dalmoro, M., de Matos, C. A. and de Barcellos, M.D. (2020). Anticonsumption beyond consumers: the role of small organic producers in environmentally oriented anticonsumption. *Psychology and Marketing*, 37(2), 291–307.

Daum, T. (2021). Farm robots: ecological utopia or dystopia?, *Trends in Ecology & Evolution*, 36(9), 774–777.

Ditzler, L. and Driessen, C. (2022). Automating agroecology: how to design a farming Robot without a monocultural mindset? *Journal of Agricultural and Environmental Ethics*, 35(1), 2.

Duncan, E., Glaros, A., Ross, D. Z. and Nost, E. (2021). New but for whom? Discourses of innovation in precision agriculture. *Agriculture and Human Values*, 38, 1181–1199.

Duram, L.A. (2000). Agents' perceptions of structure: how Illinois organic farmers view political, economic, social, and ecological factors. *Agriculture and Human Values*, 17, 35–48.

Eastwood, C., Ayre, M., Nettle, R. and Dela Rue, B. (2019) Making sense in the cloud: farm advisory services in a smart farming future. *NJAS - Wageningen Journal of Life Sciences*, 90–91, 100298.

Egri, C.P. (1999). Attitudes, backgrounds and information preferences of Canadian organic and conventional farmers: implications for organic farming advocacy and extension. *Journal of Sustainable Agriculture*, 13(3), 45–72.

Escobar, L.G. (2016). The Red de Semillas Libres: contesting biohegemony in Colombia. *Journal of Agrarian Change*, 16(4), 711–719.

EU Commission. (2020). Farm to fork strategy. For a fair, healthy and environmentally-friendly food system. https://food.ec.europa.eu/system/files/2020-05/f2f_action-plan_2020_strategy-info_en.pdf

European Commission (2016). Special Eurobarometer 442: Attitudes of Europeans Towards Animal Welfare. Animal Welfare. https://doi.org/10.2875/884639

Fairbairn, M., Kish, Z. and Guthman, J. (2022). Pitching agri-food tech: performativity and non-disruptive disruption in Silicon Valley. *Journal of Cultural Economy*, 15(5), 652–670. https://doi.org/10.1080/17530350.2022.2085142

FAO, WHO. (1999). *Guidelines for the production, processing, labelling and marketing of organically produced foods*. FAO. www.fao.org/input/download/standards/360/cxg_032e.pdf

Fernandes, P.M.B., Favaratto, L., Fernandes, A.A.R., Vicien, C., Capalbo, D.M.F. and Zerbini, F.M. (2022). To become more sustainable organic agriculture needs genome editing technology. *Frontiers in Bioengineering and Biotechnology*, 10. https://doi.org/10.3389/fbioe.2022.912793

Gardezi, M. and Stock, R. (2021). Growing algorithmic governmentality: interrogating the social construction of trust in precision agriculture. *Journal of Rural Studies*, 84, 1–11.

Guthman, J. (1998). Regulating meaning, appropriating nature: the codification of California organic agriculture. *Antipode*, 30(2), 135–154.

Guthman, J. (2000). Raising organic: an agro-ecological assessment of grower practices in California. *Agriculture and Human Values*, 17(3), 257–266. https://doi.org/10.1023/A:1007688216321

Guthman, J. (2004). The trouble with 'organic lite' in California: a rejoinder to the 'conventionalisation' debate. *Sociologia Ruralis*, 44(3), 301–316.

Guthman, J. and Biltekoff, C. (2021). Magical disruption? Alternative protein and the promise of de-materialization. *Environment and Planning E: Nature and Space*, 4(4), 1583–1600.

Hackfort, S. (2023). Unlocking sustainability? The power of corporate lock-ins and how they shape digital agriculture in Germany. *Journal of Rural Studies*, 101, 103065.

Haley, D.B., Rushen, J. and de Passillé, A.M. (2000). Behavioural indicators of cow comfort: activity and resting behaviour of dairy cows in two types of housing. *Canadian Journal of Animal Science*, 80(2), 257–263.

Hamidi, D., Grinnell, N.A., Komainda, M., Riesch, F., Horn, J., Ammer, S., Traulsen, I., Palme, R., Hamidi, M. and Isselstein, J. (2022). Heifers don't care: no evidence of negative impact on animal welfare of growing heifers when using virtual fences compared to physical fences for grazing. *Animal*, 16(9), 100614.

Heins, B.J., Pereira, G.M. and Sharpe, K.T. (2023). Precision technologies to improve dairy grazing systems. *Presented as part of the Joint ADSA Midwest Branch/Forages and Pastures Symposium: Grazing to Improve Profitability of Midwest Dairy Farms held at the ADSA Annual Meeting. JDS Communications*, 4(4), 318–323.

Holt-Giménez, E. (2019). Capitalism, food, and social movements: the political economy of food system transformation. *Journal of Agriculture, Food Systems, and Community Development*, 9(Suppl. 1), 23–35.

Husaini, A.M. and Sohail, M. (2018). Time to redefine organic agriculture: can't GM crops be certified as organics? *Frontiers in Plant Science*, 9. https://doi.org/10.3389/fpls.2018.00423

Kledal, P.R. (2003). *The political economy of organic foods*. Paper at: SAFO: Sustaining Animal Health and Food Safety in Organing Farming. 'Socio Economic Aspects of Health and Food Safety in Organic Livestock", Florence Italy, 5–7 September 2003.

Kociszewski, K., Graczyk, A. and Mazurek-Lopacinska, S.M. (2020). Social values in stimulating organic production involvement in farming – The case of Poland. *Sustainability*, 12(15), 5945.

Konstantinidis, C. (2014). Capitalism in green disguise: the political economy of organic farming in the European Union. Working Paper 2014-01. http://repec.umb.edu/RePEc/files/2014_01.pdf

Kuepper, G. (2010). *A brief overview of the history and philosophy of organic agriculture*. Kerr Center for Sustainable Agriculture.

Läpple, D. (2013). Comparing attitudes and characteristics of organic, former organic and conventional farmers: evidence from Ireland. *Renewable Agricultural and Food Systems*, 28(4), 329–337.

Lassen, J. (2015). *Technologies in organic farming: consumers' values and acceptance.* EurSafe2015 Conference: Know your food! – Food Ethics and Innovation. Cluj-Napoca, Romania, May 27–30, 2015.

Leduc, G., Billaudet, L., Engström, E., Hansson, H. and Ryan, M. (2023). Farmers' perceived values in conventional and organic farming: a comparison between French, Irish and Swedish farmers using the Means-end chain approach. *Ecological Economics*, 207, 107767.

Lundström, C. and Lindblom, J. (2021). Care in dairy farming with automatic milking systems, identified using an activity theory lens. *Journal of Rural Studies*, 87, 386–403.

Martin, T., Gasselin, P., Hostiou, N., Feron, G., Laurens, L., Purseigle, F. and Ollivier, G. (2022). Robots and transformations of work in farm: a systematic review of the literature and a research agenda. *Agronomy for Sustainable Development*, 42, 66. https://doi.org/10.1007/s13593-022-00796-2

Mason, A., Thomas, P. and Woodward, L. (2024). Agroecological intelligence establishing criteria for agroecological appropriate technology (Interim Report). https://abiggerconversation.org/wp-content/uploads/2023/06/Agroecological-Intelligence_Interim-Report_ABC_Final-280623.pdf

McSweeney, D., O'Brien, B., Coughlan, N.E., Férard, A., Ivanov, S., Halton, P. and Umstatter, C. (2020). Virtual fencing without visual cues: design, difficulties of implementation, and associated dairy cow behaviour. *Computers and Electronics in Agriculture*, 176, 105613.

Miles, C. (2019). The combine will tell the truth: on precision agriculture and algorithmic rationality. *Big Data & Society*, 6(1), 1–12.

Milestad, R., Röös, E., Stenius, T. and Wivstad, M. (2020). Tensions in future development of organic production–views of stakeholders on organic 3.0. *Organic Agriculture*, 10(4), 509–519.

Munksgaard, L., Rushen, J., de Passillé, A.M. and Krohn, C.C. (2011). Forced versus free traffic in an automated milking system. *Livestock Science*, 138(1–3), 244–250.

Murphy D.J., Murphy M.D., O'Brien B. and O'Donovan, M. (2021). A review of precision technologies for optimising pasture measurement on Irish Grassland. *Agriculture*, 11(7), 600.

Nauta, W.J., Groen, A.F., Veerkamp, R.F., Roep, D. and Baars, T. (2005). Animal breeding in organic dairy farming: an inventory of farmers' views and difficulties to overcome, *NJAS - Wageningen Journal of Life Sciences,* 53(1), 19–34.

Obach, B.K. (2007). Theoretical interpretations of the growth in organic agriculture: agricultural modernization or an organic treadmill? *Society & Natural Resources*, 20(3), 229–244.

Obach, B.K. (2015). *Organic struggle: the movement for sustainable agriculture in the United States*. MIT Press.

Padel, S., Röcklinsberg, H. and Schmid, O. (2009). The implementation of organic principles and values in the European Regulation for organic food. *Food Policy,* 34, 245–251.

Pollan, M. (2006). *The Omnivore's Dilemma: the search for a perfect meal in a fast-food world*. Bloomsbury.

Ranches, J., Connor, R., Johnson, D., Davies, K., Bates, J., Boyd, C., Bohnert, D.W. and Parker, T. (2021) Effects of virtual fence monitored by global positioning system on beef cattle behavior. *Translational Animal Science*, 5(Supplement_S1), S144–S148.

Rose, D.C., Barkemeyer, A., de Boon, A., Price, C. and Roche, D. (2022). The old, the new, or the old made new? Everyday counter-narratives of the so-called fourth agricultural revolution. *Agriculture and Human Values*, 1–17. https://doi.org/10.1007/s10460-022-10374-7

Rose, D.C. and Chilvers, J. (2018). Agriculture 4.0: broadening responsible innovation in an era of smart farming. *Frontiers in Sustainable Food Systems*, 2, 1–7.

Rose, D.C., Morris, C., Lobley, M., Winter, M., Sutherland, W.J. and Dicks, L.V. (2018). Exploring the spatialities of technological and user re-scripting: the case of decision support tools in UK agriculture. *Geoforum*, 89, 11–18.

Rose, D.C., Sutherland, W.J., Parker, C., Lobley, M., Winter, M., Morris, C., Twining, S., Ffoulkes, C., Amano, T. and Dicks, L.V. (2016). Decision support tools for agriculture: towards effective design and delivery. *Agricultural Systems*, 149, 165–174.

Rotz, S. Gravely, E., Mosby, I., Duncan, E., Finnis, E., Horgan, M., LeBlanc, J., Martin, R., Neufeld, H. T., Nixon, A., Pant, L., Shalla, V. and Fraser, E. (2019). Automated pastures and the digital divide: how agricultural technologies are shaping labour and rural communities. *Journal of Rural Studies*, 68, 112–122.

Ruder, S. L. and Kandlikar, M. (2023). Governing gene-edited crops: risks, regulations, and responsibilities as perceived by agricultural genomics experts in Canada. *Journal of Responsible Innovation*, 10(1), 2167572. https://doi.org/10.1080/23299460.2023.2167572

Sapbamrer, R. and Thammachai, A. (2021). A systematic review of factors influencing farmers' adoption of organic farming. *Sustainability*, 13(7), 3842.

Schillings, J., Bennett, R. and Rose, D.C. (2021a). Animal welfare and other ethical implications of Precision Livestock Farming technologies. *CABI Agriculture and Bioscience*, 2, 17.

Schillings, J., Bennett, R. and Rose, D.C. (2021b). Exploring the potential of Precision Livestock Farming technologies to help address farm animal welfare. *Frontiers in Animal Science*. https://doi.org/10.3389/fanim.2021.639678

Schillings, J., Bennett, R., Wemelsfelder, F. and Rose, D.C. (2023a). Digital Livestock Technologies as boundary objects: investigating impacts on farm management and animal welfare. *Animal Welfare*. https://doi.org/10.1017/awf.2023.16

Schillings, J., Bennett, R. and Rose, D.C. (2023b). Perceptions of farming stakeholders towards automating dairy cattle mobility and body condition scoring in farm assurance schemes. *Animal*. https://doi.org/10.1016/j.animal.2023.100786

Schillings, J., Holohan, C., Lively, F., Arnott, G. and Russell, T. (2024) The potential of virtual fencing technology to facilitate sustainable livestock grazing management. *Animal*, 101231. https://doi.org/10.1016/j.animal.2024.101231

Schösler, H., De Boer, J. and Boersema, J.J. (2013). The organic food philosophy: a qualitative exploration of the practices, values, and beliefs of Dutch organic consumers within a cultural–historical frame. *Journal of Agricultural and Environmental Ethics*, 26, 439–460.

Schreer, V. and Padmanabhan, M. (2020). The many meanings of organic farming: framing food security and food sovereignty in Indonesia. *Organic Agriculture*, 10(3), 327–338.

Scollan, N. Padel, S., Halberg, N., Hermansen, J., Nicholas, P., Rinne, M., Zanoli, R., Zollitsch, W. and Lauwers, L. (2017). Organic and low-input dairy farming: avenues to enhance sustainability and competitiveness in the EU. *Eurochoices*. https://doi.org/10.1111/1746-692X.12162

Seufert, V., Ramankutty, N. and Mayerhofer, T. (2017). What is this thing called organic? – How organic farming is codified in regulations. *Food Policy*, 68, 10–20.

Siegrist, M. and Hartmann, C. (2020). Consumer acceptance of novel food technologies. *Nature Food*, 1(6), 43–350.

Soil Association. (2023). Organic principles. https://www.soilassociation.org/who-we-are/organic-principles/

Sonne, C., Alstrup, A.K.O., Pertoldi, C., Frikke, J., Linder, A.C. and Styrishave, B. (2022). Cortisol in manure from cattle enclosed with nofence virtual fencing, *Animals*, 12(21), 3017.

Sowell, B.F., Bowman, J.G.P., Branine, M.E. and Hubbert, M.E. (1998). Radio frequency technology to measure feeding behavior and health of feedlot steers. *Applied Animal Behaviour Science*, 59(4), 277–284.

Stampa, E., Zander, K. and Hamm, U. (2020). Insights into German Consumers' Perceptions of virtual fencing in grassland-based beef and dairy systems: recommendations for communication. *Animals*, 10(12), 2267.

Stilgoe, J., Owen, R. and Macnaghten, P. (2013). Developing a framework for responsible innovation. *Research Policy*, 42(9), 1568–1580.

Stock, R. and Gardezi, M. (2021). Make bloom and let wither: biopolitics of precision agriculture at the dawn of surveillance capitalism. *Geoforum*, 122, 193–120.

Sullivan, S. (2023). Ag-tech, agroecology, and the politics of alternative farming futures: the challenges of bringing together diverse agricultural epistemologies. *Agriculture & Human Values*. https://doi.org/10.1007/s10460-023-10454-2

Taneja, M., Byabazaire, J., Jalodia, N., Davy, A., Olariu, C. and Malone, P. (2020). Machine learning based fog computing assisted data-driven approach for early lameness detection in dairy cattle. *Computers and Electronics in Agriculture*, 171, 105286.

Tress, B. (2001). Converting to organic agriculture – Danish farmers' views and motivations. *Danish Journal of Geography*, 101(1), 131–143.

Tzanidakis, C., Tzamaloukas, O., Simitzis, P. and Panagakis, P. (2023). Precision Livestock Farming Applications (PLF) for grazing animals. *Agriculture*, 13(2), 288.

Umstatter, C., Morgan-Davies, J. and Waterhouse, T. (2015). Cattle responses to a type of virtual fence. *Rangeland Ecology & Management*, 68(1), 100–107.

Vail, D.D. (2022). A counterculture agriculture: organic farming in a commercial food age. In R. D. Hurt (Ed.), *A companion to American agricultural history*. John Wiley & Sons Ltd.

Vasseur, E., Rushen, J., Haley, D.B. and de Passillé, A.M. (2012). Sampling cows to assess lying time for on-farm animal welfare assessment, *Journal of Dairy Science*, 95(9), 4968–4977.

Verdon, M., ALee, C., Marini, D. and Rawnsley, R. (2020). Pre-Exposure to an electrical stimulus primes associative pairing of audio and electrical stimuli for dairy heifers in a virtual fencing feed attractant trial. *Animals*, 10(2), 217.

Wallenbeck, A. Rousing, T., Sørensen, J.T., Bieber, A., Spengler Neff, A., Fuerst-Waltl, B., Winckler, C., Peiffer, C., Steininger, F., Simantke, C., March, S., Brinkmann, J., Walczak, J., Wójcik, P., Ribikauskas, V., Wilhelmsson, S., Skjerve, T. and Ivemeyer, S. (2019).

Characteristics of organic dairy major farm types in seven European countries. *Organic Agriculture*, 9, 275–291.

Weary, D.M., Huzzey, J.M. and von Keyserlingk, M.A.G. (2009). Board-invited review: using behavior to predict and identify ill health in animals. *Journal of Animal Science*, 87(2), 770–777.

Wickson, F., Binimelis, R. and Herrero, A. (2016). Should organic agriculture maintain its opposition to GM? New techniques writing the same old story. *Sustainability*, 8(11), 1105.

Yazdanpanah, M., Moghadam, M.T., Zobeidi, T., Turetta, A.P.D., Eufemia, L. and Sieber, S. (2022). What factors contribute to converstion to organic farming? Consideration of the health belief model in relation to the uptake of organic farming by Iranian farmers. *Journal of Environmental Planning and Management*, 65(5), 907–929.

Zagata, L. (2010). How organic farmers view their own practice: results from the Czech Republic. *Agriculture and Human Values*, 27(3), 277–290.

Part 3

Improving sustainability

Chapter 11

Integration of organic dairy cattle into a whole farm context

Stephen Roderick, Duchy College, UK; Lindsay Whistance, Organic Research Centre, UK

1 Introduction

The organic principle of ecology proposes that organic production should be based on ecological processes, fitting with natural cycles and being adapted to local conditions, ecology, culture and scale. Inputs should be reduced to reused, recycled and conserved natural resources that maintain and improve environmental quality (IFOAM, 2005). For organic dairy farms, this means that the nourishment and nutritional well-being of the animals should rely on feed produced from living soil occurring within the farm ecosystem, with nutrients recycled within closed cycles. On close examination, many of our organic dairy farms appear removed from that particular aspirational pathway in that they are often based on adaptations of conventional and intensive models of production, albeit with legislative limits imposed on the percentage of concentrate feeds in the diet. For example, in England during 2020–21, more than half of the variable costs associated with the production of organic milk were attributed to purchased concentrate feeds (Scott, 2023).

The design of agroecological systems is partly based on the principle of enhancing beneficial biological interactions and synergy between diverse agricultural activities and components (Altieri, 2000). Greater integration within natural cycles and between farmed enterprises is a fundamental route

http://dx.doi.org/10.19103/AS.2024.0138.12

through which we can re-think and re-purpose organic dairy farming into a more ecologically principled and sustainable form of food production that also enables the achievement of the other broader key organic principles of health, care and fairness.

2 Defining integration

2.1 Synergy and complementarity

Integration within a farm requires relationships of varying degrees of synergy, cooperation and complementarity between two or more of the farm elements or enterprises. These relationships can range from simple systems with no more than two enterprises focused on prescribed inputs aimed at optimising a single output to more dynamic and diverse systems that integrate enterprises in time and space, where there is synergistic sharing of resources between enterprises (Peel, 2013). Levels of management complexity tend to increase with the complexity of the relationship. To avoid confusion with farms that have diverse enterprises operating in isolation, Peel (2013) defines truly mixed crop and dairy farms as systems operating with complementarities.

Complementarity can be defined as occurring when an increase in one 'output contributes to increasing the marginal product of the other' (Villano et al., 2010). This occurs when changes in the characteristics or output of one enterprise result in positive changes to the output of other activities and the overall output. This can be expressed as the system effects that arise when the whole is greater than the sum of its synergy parts (Parmigiani and Mitchell, 2009). Functional complementarity occurs, for example, when cover crops are grown to improve soil fertility and these also provide high-quality forage for animals or improve the quality of straw from a cereal crop that in turn improves feed quality (Peel, 2013).

Synergies between enterprises can help reduce competition between them, and in many farm situations, there are opportunities to exploit the diversity between synergistic enterprises (Villano et al., 2010). The types of synergy that exist within mixed dairy and crop systems have been classified by Peel (2013) as functional complementarity, augmentation, combining of farm resources, information-sharing and joint decision-making and cost and risk sharing.

2.2 Diversity and complexity

The evolution of crop–animal systems is in response to factors such as population pressure, market opportunities and changes in environmental circumstances such as pollution, soil quality and nutrient loading (Entz et al., 2005). Considering the complexities of animal–crop integration, Sumberg (2003) conceptualises integration as potentially occurring across the four dimensions of space, time, management and ownership, and notes that the benefits of integration can be achieved along one, some or all of these. Five categories of farming systems with increasing levels of diversity, complexity and management intensity have been identified by Peel (2013), from a basic system with no more than two enterprises to dynamic integrated systems with multiple species that interact in space and/or time to create interactions resulting in synergetic resource transfer between the enterprises. As diversity increases, so do 'opportunities for coexistence and beneficial interactions between species that can enhance agroecosystem sustainability' (Altieri, 2000).

2.3 Integration and circularity

The aims of integrated organic dairy farming can be regarded as being aligned with the key principles of a circular economy: the elimination of waste and pollution; the circulation of resources and materials and the regeneration of natural systems (Ellen MacArthur Foundation, 2015). To varying degrees, the categories of integration outlined in the following section can be viewed as contributing to this circularity concept. For example, the use of food waste and crop by-products as a source of nutrition, the integration of cattle within crop rotations enabling nutrient cycling and the symbiotic ecological and behavioural relationship between trees and dairy cows all represent varying degrees of integration with elements of circularity.

3 Within farm integration

Within farm integration can be categorised according to spatial land use, or the extent to which enterprises physically meet in relation to the land that they occupy. Hilimire (2011) categorises the extent of this integration as:

1. Spatially separated, where animals are always kept on a separate uncropped part of the farm but are receiving inputs, especially feeds, that are grown on the farm. This type of integration is perhaps most typical of European organic dairy farms and an example is the use of so-called waste crop material as a feed source;
2. Rotational, whereby animals and crops may occupy the same area but at different times, for example when animals utilise crop residues or short-term soil fertility building fodder crops or grass leys; and
3. Fully combined, when animals are grazing underneath or between crops, for example, silvopastoral systems.

In this section, we consider these various forms of within-farm integration and the contribution they make to achieving organic farming objectives, summarised in Table 1.

Table 1 Summary of the key types of integrating of dairy farming into the whole farm or region and their contribution to achieving key organic objectives

Type of integration	Adherence to organic aims	Biodiversity conservation	Animal welfare	Socio-economic issues	Resource efficiency
Rotational crop–animal integration – same area but at a different time	+ Good opportunity for a forage-based system + Based on sound crop rotation to control weeds, fertility, pests – Reliance on cultivation	+ Opportunities for multi-species pastures promoting biodiversity	+ Opportunity for high levels of nutrition e.g. protein-rich legume leys	+ Self sufficiency + Multi-enterprise, spreading economic risk and benefits – Complex management – May require mobile infrastructure	+ Opportunity to utilise fertility-building crops + Paddock grazing opportunity to improve manure use efficiency + Carbon capture opportunities in plants and soil – Nutrient cycling is not necessarily synchronised to crop requirements
Spatially separated crop–animal integration using crop residues and wastes	– Moves away from forage-based systems – Issues concerned with non-certified feeds	– Little opportunity for ecological interactions	– Animals consuming feeds that they may not be physiologically adapted to – Increased need for confinement – Risk of toxicity – Risk of irregular feed supply	+ Focus on food, not feed – Issue of consumer acceptability – Seasonality of feed supply	+ Resource efficient, circular system + Reduced wastage

Multi-species co-grazing (more than one animal species utilising the same area at the same time)	+ Increased diversity + Pasture-based	+ Biodiversity benefits from diverse manuring	+ Can be complementary but also independent of each other and not integrated + Opportunity for parasite control	+ Aesthetically pleasing – opportunity to mimic natural, wild herbivore grasslands + Shared inputs and spreading of risk + Increased overall performance compared with single species	+ Improved forage utilisation + Positive impact on soil quality
Multi-species sequential grazing (more than one animal species utilising the same area at different times)	+ Increased diversity + Pasture-based	– Reduced biodiversity benefits compared with co-grazing	+ Opportunity for parasite control – No inter-species behavioural interaction	+ Potentially easier management compared with co-grazing	+ Improved forage utilisation
Between farm cooperation	– Limits opportunity for closed farm nutrient cycling	– Reduced opportunity for biodiversity	– Biosecurity risk	+ Increases opportunity for integration + Can help build social networks – Requires infrastructure, cooperation – Reliance on transport	+ Reduced wastage through sharing + Regional/local utilisation of resources – Increased fuel requirement
Regional integration	– Limits opportunity for closed farm nutrient cycling	+ Opportunity for wildlife corridors, easier migration, etc.	– Biosecurity risk	+ Opportunity for aesthetically pleasing, landscape-scale integration	+ Opportunities for concerted efforts to reduce diffuse water pollution + Regional/local utilisation of resources

+ Positive contribution; – negative contribution.

3.1 Rotational integration: crops and animals

3.1.1 Integrated crop rotations

In organic farming, the use of crop rotations is a key practice concerned with the manipulation and control of nutrient supply for crops, as well as other important functions, especially weed and disease control. Crop–animal systems that incorporate biological nitrogen fixation and manure recycling are especially important (Wilkins, 2008). Farm animals integrated within a crop rotation is a classic example of complementarity, particularly through the utilisation of fertility-building crops and crop residues as animal feeds and the strategic use of manure as a crop fertiliser.

Nutrient-building crops, especially legumes, are interspersed with nutrient-depleting crops, which are generally the income-generating element of the rotation. Treating a fertility-building crop as forage for animals provides an opportunity for economic return from a crop that otherwise has no harvestable value and whose presence can add considerable cost to production and removal. A neat circular system occurs with legumes increasing soil fertility through nitrogen fixation and also providing high-quality grazed or conserved forage for animals, with the animal manure produced being used as a primary source of crop nutrients.

The design of a crop-dairy rotation can be of critical importance, with numerous options and potential trade-offs. The balance between the number and type of crop species grown, the number and production level of dairy cows, and their potential interaction are key factors influencing the achievement of the multiple objectives of organic farming, including economic efficiency, increasing soil organic matter, efficient nutrient recycling and optimising water usage (Liang et al., 2018), as well as ensuring of high standards of animal welfare.

In terms of the greenhouse gas emissions from integrated organic crop-dairy farms, the bulk of these come from milking cows and are related to methane emissions from enteric fermentation. The emissions from the crop elements can vary considerably between crop types and are largely concerned with nitrous oxide emissions from the application of manures as a crop fertiliser, as well as the use of fuel for cultivations. In terms of endeavours to reduce emissions, a number of options have been proposed, including intercropping to improve resource efficiency and the use of anaerobic digestion of manures and crop residues (Jeswani et al., 2018).

3.1.2 Manures and nutrient recycling

Animals on organic farms can be a key route through which nutrients are provided to crops. One of the key features of integrated crop–animal systems is the opportunity created for locally controlled nutrient recycling, either through fixing of nitrogen from legumes grown in a crop rotation, as discussed above, or from manures excreted from animals grazing fertility building crops within a crop rotation or from manures produced from other parts of the farm but from animals fed the products or by-products from crops produced on the same farm. It has been established that the use of farmyard manures, along with soil fertility-building leys and legume cash crops grown in rotations, should provide organic farms with a sufficient supply of nitrogen to meet crop requirements (Berry et al., 2002; Foissy et al., 2013).

The degree to which nutrients from grazing animals are returned to subsequent crops is dependent on a number of factors, including pasture or ley composition, grazing system and grazing intensity. Animals grazing short-term leys within a crop rotation contribute positively to the production from following crops, whereas the removal of fodder as conserved forage also removes nutrients with a subsequent negative impact on the following crop (Entz et al., 2005). There are additional factors such as the potential imbalance in the deposition of nutrients through uneven field use e.g. excessive use by animals of specific areas offering shade or shelter for thermoregulation.

A further consideration is the balance between the positive contribution that incorporating crop residues can have on soil quality, particularly soil carbon, and the benefits of feeding crop residues to animals. Although feeding residues can in effect be a re-cycling of nutrients and biomass to the soil via manures, the efficiency with which this is achieved is reduced as a consequence of nutrient losses that occur during the processes (Duncan et al., 2016). Infrastructural and labour considerations associated with the transport and spreading of manures are also important. Conversely, manure has a lower carbon-to-nitrogen ratio than crop residue and the nutrients tend to be more readily available for plant production than those in crop residues.

Although the nutrient content of manures can vary considerably depending on a range of factors, including rainfall, straw content and storage, it has been estimated in England that farm animals excrete 50–95% of the nitrogen (N), phosphorus (P) and potassium (K) they consume. Of that, approximately 90% of K and 40–60% of P is made available to subsequent crops when applied as farmyard manure or slurry. For N, this ranges from 10% to 50% in the year following application, with more becoming available in subsequent years (Wilkins, 2008).

A practical example of an economically successful integration of dairy cattle into a crop rotation is provided by Macaroni Farm in England, described below. This example clearly demonstrates nutrient cycling whilst also achieving environmental and animal welfare benefits. The wider environmental impacts are discussed in Chapter 13, where the same case-study farm is used to demonstrate how the combination of various factors positively impacts on farm biodiversity.

3.1.3 The Cotswold Mobile Dairy Ltd: an example of an integrated dairy and arable system

An innovative and integrated dairy system, the Cotswold Mobile Dairy Ltd, has been developed through a partnership between two organic farms (Macaroni Farm and Bradwell Grove Estate) located in the Cotswolds region of England, with the aim of utilising the fertility-building phase of a crop rotation to produce milk. The crop rotation on the farms lasts for eight years, with four years of cereal cropping followed by four years of fertility building and dairy production.

The rotation on the farm has always had an animal component, but over time it was felt that more was being taken from the soil than being replaced. Previously, a herd of beef suckler cows were set-stocked and housed in winter and were not felt to be fully utilising the potential offered by the cropping rotation. The drivers for moving to dairy farming with beef offspring were to ensure greater resilience through achieving better financial returns, improved nutrient cycling and soil quality.

The fertility-building phase of the rotation involves growing a multi-species mixture containing legumes that fix nitrogen, grasses and other deep-rooting plants that improve soil structure in preparation for the subsequent cereal crop. Other plants included within the fertility-building species mixture are selected for their farm-specific benefits, such as their drought tolerance characteristics. Manure produced by the grazing dairy herd provides important additional soil nutrients.

The 350-cow herd is a three-way cross of Jersey, Flekvieh and New Zealand Friesian, suitable for a pasture-based system. It is a forage-based system with no concentrate feeding. Having multi-species (herbal) leys and cover crops is viewed as an essential element, adding the necessary nutrients and giving enhanced resilience. As well as being critical to the success of the subsequent cereal crops in the rotation, the multi-species composition encourages biodiverse invertebrate populations that assist in decomposition and the release of nutrients from the manure. The leys provide a four-year break, which allows control of arable weeds such as black grass by burying the weed plants and preventing the production of seed.

In summer, the cows are rotationally grazed on the herbal leys following the 'thirds' principle of one-third grazed, one-third trampled and one-third left (to encourage fertility and regrowth). The herd has daily access to paddocks of approximately 6 hectares and is moved twice daily to fresh pasture, once after morning milking and once in the afternoon between 2:00 pm and 4:00 pm. The mobile milking parlour follows the herd around the farm, towed on a tri-axled trailer. All cattle are kept outdoors, year-round, and during the winter (100–120 days), the herd is grazed on the arable cover crops, including forage rye, vetch and radish, alongside silage bales that have already been spaced out for daily access in each of the paddocks. In summer, there is also access to woodland grazing. Being mobile allows the spreading of fertility across 1800 hectares (4500 acres) of arable land, which would not have been possible with a static system.

The main challenges in this system are not specifically related to being organic and include weather, water supply and forage production. With both increasingly longer dry and wet periods, it is a challenge to develop a forage surplus to buffer the system. In this regard, the inclusion of deeper-rooted plants within the herbal leys and paying specific attention to stocking rates have been important factors.

Water access and infrastructure are critical. Water is provided via a borehole and pumped around the farm, and there are two mobile water troughs that are moved to follow the cows. All of the pipes are above ground and can be drained in sections if necessary. The most recent winter only had four days of frost, so freezing pipes are not currently a problem.

The partnership between the two adjacent farms is important to provide sufficient land, and over time the capacity is building, enabling the dairy enterprise, alongside beef and sheep enterprises, to be sustained by adding to nutrient recycling and profitability. Ensuring that the management of the system suits both partners equally is a key factor in ensuring success. The milking herd has access to 75% of available land across the two farms, with access to the remaining land being limited by geographical features as the mobile dairy is not movable through a valley. The dairy cows have priority over other animal groups (beef, sheep and dairy replacements) in terms of receiving the best grazing to help maintain milk yield.

A number of key benefits have emerged as this system has developed. It has been noticeable that the animals are particularly healthy, requiring minimal veterinary inputs. Mineral and trace element balances are much improved, better covering animal and crop needs, although magnesium and sulphur are still in deficit. The use of herbal leys grazed in the rotation has improved biodiversity, soil organic matter and water infiltration. It has been noted that the swards are becoming increasingly dense.

3.2 Fully combined integration: animals and trees

Silvopasture is a system of integrated farming that involves the combination of grazing animals, pasture and trees in a mutually beneficial way and has been proposed as a potential solution to some of the environmental challenges associated with intensive dairy farming (Irish Agroforestry Forum, 2020). As a crop in their own right, trees can yield additional economic benefits from the harvest of timber, fruits, nuts and other tree products such as biofuel and medicines (Stobart et al., 2023). The type and value of benefits to the animals can be challenged by these other roles that trees may play on-farm, where the value of their contribution to the farm economy is high and the management techniques employed to optimise harvests involve restricting access.

3.2.1 Environmental benefits

The environmental benefits of having trees situated within pasture include making a significant contribution to the farm carbon footprint through the sequestration of carbon (Feliciano et al., 2018) as well as improvements in soil quality that are likely to accrue (Torralba et al., 2016). Silvopasture has been shown to be more productive than open pasture when other crops and benefits are considered (Orefice et al., 2019; Pent, 2020) and, in general, provides better pollination coverage of crops compared to monocultural systems, further promoting yields (Varah et al., 2020).

3.2.2 Nutritional benefits

Browse and tree fodder are readily eaten by ruminant animals and can provide good sources of nutrients including protein, vitamins and minerals. Browse becomes a particularly important source of feed in dry or drought conditions when grass is scarce. Trees originally planted to help reduce soil erosion in New Zealand have become a valuable fodder crop in periods of minimal grass growth (Charlton et al., 2003). Browse up to a height of 2 m can be accessible to adult cattle, though if browse is to be considered as a meaningful part of their diet, it should be offered at a more easily accessible and comfortable feeding height (Whistance, 2018a).

There are a range of factors that affect the feasibility of using tree leaves to supplement ruminant diets, including availability, slow regrowth and the species of tree influencing palatability (Woodland Grazing Toolbox, n.d.), along with seasonal variation in available nutrients. In general, leaves are a good source of metabolizable energy and crude protein, though both season and tree species influence the levels available to browsing animals (Kendall et al., 2021). The presence of condensed tannins in browse supports the delivery of good quality rumen bypass protein (Patra and Saxena, 2010) alongside helping to control worm burdens in grazing animals (Novobilský et al., 2011).

Variations between common tree species in the concentration of certain minerals, such as zinc, cobalt and magnesium, have been found, whereas other minerals, such as selenium, may be influenced by the site on which they are growing and yet others, such as iodine, by season (Kendall et al., 2021). Calcium levels across tree species have been found to be significantly higher than those found in grassland (Kendall et al., 2021). Some species are accumulators of certain minerals, e.g. willow and zinc (a property which is already utilised for bioremediation purposes), and so these are likely to be high regardless of site and soil type, though levels in leaves can still change with season. Developing an understanding of the mineral availability of different trees grown on farms could help to support cow health and further promote closed-cycle farming practices.

3.2.3 Animal welfare benefits

Beyond nutritional benefits, there are further benefits to providing cattle access to trees within dairy farm environments. Trees, in well-designed silvopastoral systems, act as an important buffer against temperature fluctuations, cooling the air in summer and warming the ground in winter, along with slowing wind speeds. The provision of shelter in cold, wet and windy weather reduces the need to provide compensatory feed in cold weather (which may maintain production levels but, by itself, does not offer good welfare conditions). As a source of shade, tree species can significantly reduce the solar radiation received by grazing cattle and, depending on the spacing of trees within a pasture, they can also reduce the negative impact of cattle crowding together to seek shade (Whistance, 2018b).

Negative impacts of heat stress include disrupted behaviour patterns, such as lying and feeding behaviour, leading to reduced yield. In open pasture, or where any shade offered is far from a water source, cattle tend to stay close to the water source, leading to an uneven use – and deposition – of resources. With increased heat stress, both digestive and reproductive tract health are compromised, with the negative consequences to health, growth and yield being measurable in three generations (Laporta et al., 2020). In well-designed silvopastoral systems, cattle can better maintain their daily behaviour and utilise their landscape more evenly. Trees and shrubs also provide a place for newly born calves to hide, thus enabling the expression of an important behaviour pattern for both cow and calf (Rørvang et al., 2018).

3.3 Spatially separated integration: utilisation of residues

A key principle of circular food systems is that resources should only be used for animal feed as a secondary priority after human food, and that these feeds should be grown with minimal reliance on external inputs and in closed nutrient cycles. Adopting this approach requires animals to be fed on processed by-products and wastes, as well as forages grown in areas unsuitable for crop production or as the fertility-building elements of crop rotations. If such an approach were to be adopted, animals would then complete the closed loop by contributing nutrients in the form of manures back to the production of human food (van Zanten et al., 2019). Diets that rely on consuming only animal-sourced food produced solely from these so-called low-opportunity-cost feedstuffs could have potentially significant positive impacts on greenhouse gas emissions, as well as nitrogen and phosphorus losses (van Zanten et al., 2018).

The use of low-opportunity-cost feedstuffs, including the by-products of crop production and processing, as well as grass resources, is a very land-efficient method of production that can provide nutrient-dense foods that do not compete with land required for the production of plant-based foods for humans (van Zanten et al., 2018). Such opportunities also enable the extension of integration beyond the farm gate and into the whole food chain. In some regions of the world, such as sub-Saharan Africa, systems that rely on crop–animal interactions are commonplace, and crop residues are often a key component of animal diets, although there are potential trade-offs with their use as soil enhancers (Baudron et al., 2015; Duncan et al., 2016). The use of crop residues and waste as low-cost feed elements completing circular economies of animals integrated with crops is less commonly found in European dairy farming systems, where systems based on circularity are largely dependent on forage-based feeds.

There are two significant issues that we face when considering the use of by-products in organic dairying specifically. The first relates to the fact that the aspiration in organic farming is to produce food from ruminants that rely on forages as the primary source of nutrition. This does not necessarily eliminate the use of by-products, but it can limit the extent to which they are used. Secondly, many of the by-products that are available to farmers may be from non-organic sources, either because of a lack of certification or because an aspect of the processing is unacceptable from an organic perspective. Escribano (2018) recommends the use in organic farming of local non-certified by-products, provided they have demonstrably not been treated with inorganic fertilisers and pesticides, as a further means by which organic farming contributes to circular economies through recycling and reducing wastage.

Other issues that do not specifically relate just to organic farms but are of relevance to all dairy farms are as follows:

- Products require feeding at specific levels or under certain management conditions to avoid causing undesirable animal health or wider welfare responses;
- Products that are in short or irregular supply, such as brewer's grain, that can cause metabolic disturbances and yield fluctuations resulting from frequent or dramatic changes in dietary constituents; and
- Lack of seasonal synchronisation between by-product availability and nutritional requirements.

The use of regionally available by-products can be limited by poor understanding of their nutritional value, including palatability and toxicity, as well as their economic value (Schroeder, 1999). Supplementation of forage-based diets with products from the food processing industry offers the potential to improve the efficiency of organic dairy herds without competing with humans for food. For example, Ertl et al. (2017) demonstrated that adding wheat bran and dried sugar beet pulp at a rate of 25% of dietary dry matter to a forage-only diet increased intake and milk yield, as well as helping to maintain a positive energy balance in mid-lactating cows. Consistency of supply and quality are also important considerations, as is ease of collection and proximity to animals (Klopfenstein and Owen, 1981).

Should feeding by-products be restricted to those that are produced during the growing, harvesting and on-farm processing of farm products, or should the integration between animals and crops also include the use of by-products that have been produced after the primary product has left the farm for processing elsewhere, e.g. brewer's grain? This adds to the complexity of considerations, as the circular economy to which it relates starts to include other industries, potentially greater distances and less opportunity for the development of ecologically balanced systems.

The nutritional value of crop by-products can vary significantly, e.g. soybean meal has a higher nutritional value than rapeseed meal, and as a potential component of a circular system, it may be argued that the value of a crop should be based on the value of its by-products as well as its human nutritional quality (van Zanten et al., 2018), although in the case of soybean meal, competition with food crops for land is also an important consideration (Mottet et al., 2017). It should be noted that, in order for animal products solely produced using this circular approach to meet the national nutritional needs of most European countries, a significant reduction in animal products, including milk, would be required (Frehner et al., 2022).

3.3.1 Examples of low-cost-feed opportunities

The production and harvest of vegetables can result in very high levels of waste with significant nutritional feed value, and for some vegetable species, the majority of the plant material is not harvested and therefore available as a feed source (Monllor et al., 2020). The existence of such potentially valuable feeds can occur in both spatially segregated systems and within rotational integration, as described above. The economic and environmental benefits of feeding fresh vegetable wastes and crop residues to animals are recognised, although their preservation as a feed is limited by their high-moisture content (de Evan et al., 2020). Some wastes have beneficial bioactive compounds. The inclusion of tomato and cucumber waste in the diet of dairy goats has been shown to have positive impacts on milk quality and methane production (Romero-Huelva et al., 2022). The dry matter residue of brassicas such as cauliflower contains high concentrations of digestible protein and sugars (de Evan et al., 2020), although there are potential palatability, animal health and product quality risks associated with over-feeding (Bischoff, 2019).

The wastes produced following the processing of crops can also offer viable alternatives to concentrated feeds. For example, berry seed residues such as blackcurrant have been shown to be a potentially economically and nutritionally profitable feed for dairy cattle production (Bryszak et al., 2019). The quality of meat and milk from sheep and goats fed on alternative feeds commonly grown in the Mediterranean region, including crop by-products, to replace concentrate feeds is reviewed by Vasta et al. (2008).

3.4 Multi-species integration

Farming systems that integrate different animal species also offer opportunities for functional diversity that can enhance environmental, economic and social sustainability. These benefits arise largely as a result of species differences in factors such as morphology, physiology and feeding preference (Martin et al., 2020). Differences between species in relation to, for example, feeding behaviours enable greater resource efficiency as a consequence of the complementary consumption of different parts of plants, and in some cases preferences for different plant species. Grazing dairy cattle alongside other species such as goats and/or sheep in situations where forage is not excessively limited can result in a reduced dietary overlap compared with grazing cattle only at similar stocking rates. This impact is narrowed as forage becomes limited and species compete for a more restricted plant population.

There is strong evidence to show the advantages of co-grazing of different species on total productivity (Prache et al., 2023, D'Alexis et al., 2014; Fraser et al., 2007), although the benefits are often accrued in only one of the species. In the case of mixed cattle and sheep systems, it is most often sheep performance that appears to benefit (Benoit et al., 2023; Wright et al., 2006) and this has been related to the higher competitive ability of sheep, higher rates of prolificacy and a reduced helminth challenge under co-grazing (Marley et al., 2006), with the effect being influenced by species ratio, stocking rate and grazing management practice (Martin et al., 2020). Mixed grazing also has potential economic benefits through complementary opportunities for shared inputs and spreading of risk, although there are trade-offs concerning the increased management requirements and likely differences in economic returns between species systems and products e.g. sheep farming may be less profitable than dairy farming.

Grazing dairy heifers alongside pigs has also been shown to enhance the efficiency of utilisation of pasture, with both species growing faster when grazed together compared to single species systems (Sehested et al.,

2004). Combined dairy and pig systems also offer an opportunity for closed loop systems whereby young pigs are fed whey obtained from the on-farm processing of cheese (Martin et al., 2020). It has also been observed that other species grazing alongside cattle are more likely to consume the higher quality forage that grows close to areas where cattle have previously defecated, again increasing feed quality and utilisation efficiency (Martin et al., 2020).

Pasture composition can enhance the benefits of mixed-grazing of cattle and sheep, with more diverse populations of plant species providing greater opportunity to benefit from the complementarity variations in feeding preferences (Jerrentrup et al., 2020), as well as potential yield benefits associated with diversity (Connolly et al., 2009; Jing et al., 2017). In multi-species pastures, when cattle and sheep are mixed, it is cattle that are responsible for maintaining plant diversity as they have less selective and less destructive grazing behaviours (Su et al., 2023). Mixed species grazing can also result in significant biodiversity impacts from the combined effects of different feeding behaviours on vegetation structure and the wider range of manure types and plant litter on populations of insects and soil organisms (Wang et al., 2019).

Integrating different species of ruminant farm animals sequentially i.e. using the same area but at different times, is also referred to as cross-grazing or a leader-follower system and is an established method of controlling internal parasites in cattle and sheep (Tapia-Escarate et al., 2021). The normal sequence is for sheep to follow cattle. Although effective in controlling roundworms in sheep compared with single species systems, the best results have been shown when cattle and sheep are grazed together (Marley et al., 2006).

4 Between farm integration

4.1 Linking farms and cooperation

Integration of crops and animals within a single farm scenario can often be difficult to achieve, and cooperation between farms may be a more practical route to enabling greater utilisation of resources, including by-products. Improving efficiency over large areas could be increased through greater linkages between farms with different, but potentially complementary enterprises, both in terms of the supply of feed and of manures (Wilkins, 2008). However, the extent to which this is achieved may depend on the nature of the arrangement, as there is a distinction between cooperation as an economic arrangement and integration between organic farms where nutrient flows and ecosystem benefits are prioritised (Regan et al., 2019).

This type of cooperation can be organised structurally on a larger scale and without some of the complexity of crop rotations, additional skills and infrastructure on individual farms (Entz et al., 2005; Asai et al., 2018), although without diversity within a system, opportunities for ecological integration are diminished and the same enhanced biodiversity benefits of within-farm integration cannot be achieved. Sharing of resources between two or more specialised farms may result in productivity improvements, but the same nutrient-sharing benefits and ecological complementarities are less likely than a more intimate and symbiotic co-sharing of resources between animals, plants and soils within a single organic farm environment.

Regan et al. (2019) examined a number of farm cooperation scenarios in Europe aimed at resource sharing and showed that under some arrangements, intensification increased as a result of cooperation and this type of integration did not guarantee the achievement of environmental benefits. Case studies of dairy farmers cooperating locally and regionally with crop producers in the Netherlands, the United States, France and Japan, presented by Asai et al. (2018), identify critical factors and strategies that can determine the success of this form of integration. These relate to the requirements for policy and support structures to strengthen social networks as a key to lowering the costs of information gathering and collective decision-making, as well as legal frameworks and the role of third parties. Operational costs of arrangements were found to be largely dependent on the spatial proximity of farms, although this depends on the scale of the area and the number of farmers participating.

Asai et al. (2018) identify three broad categories of between-farm integration: simple-farm-to-farm exchange of raw materials (e.g. feed, manures and bedding); local groups of animal and arable farmers agreeing on land-use patterns; and thirdly a situation whereby regional contracts between groups of farmers are agreed with the involvement of third-party entities, such as cooperatives or land agents, with minimal communication between individual farmers. The key drivers are often concerned with regulations on manure application and input use in organic farming (Asai et al., 2018).

The first and simplest form is a partnership between specialised crop and animal farms (farm-to-farm), where they exchange raw materials (manure, grain, fodder and straw). The sharing of manures between farms enables a mutually beneficial enhancement of soil quality and reduction in purchased fertilisers on the receiving side (crops) and a reduced risk of pollution and nutrient run-off on the supply side (animals). Such manure-sharing initiatives are only really feasible when the movement of manures is limited to relatively short distances, as it is a bulky product and a potentially undesirable pollutant. Cattle manures are much less energy efficient (the energy value per unit of energy expended) than the much drier and more energy-dense pig and poultry manures (Wilkins, 2008).

A second type of direct exchange can be organised by local groups of crop and animal farmers negotiating land-use allocation patterns. Collectives, whereby nearby farmers work together to share land and resources such as manures, feedstuffs and bedding, allow opportunities for integration beyond the traditional farm boundary (Martin et al., 2016). Entz et al. (2005) describe a system in India where nomadic herders seasonally graze their sheep and goats on arable land to make use of crop residues and stubble whilst also contributing nutrients via manure to cropland. A similar system of 'flying flocks' is common in the United Kingdom, where flocks of sheep are brought to arable farms for fattening off crops. Such integration can have significant impacts and some of the more synergistic arrangements result in significant positive impacts on ecosystem services, including soil quality, as well as farmer empowerment and other shared benefits. This type of collaboration can lead to significant landscape-level enhancements, which may be impossible when carried out within the artificial boundaries of a single economic entity of a farm (see below).

A third type involves upscaling to, for instance, a regional scale where spatially separated groups of specialised animal and crop farmers integrate through coordination by a third party (e.g. an agricultural cooperative or firm). Here, the farmers involved are not necessarily communicating directly. For certified organic farms, the extent of collaboration between farms will be limited by legislative restrictions pertaining to the certification of materials as well as their origin and compliance with organic standards.

4.2 Integration at the landscape level

The wider integration of activities between farms also offers opportunities for greater environmental and ecosystem benefits, especially at the landscape level. Whereas on-farm integration offers greater social impact to individual farmers who maintain control of production, area-wide integration can, in some circumstances, provide greater ecological benefit as well as potentially better labour and economic efficiency (Entz et al., 2005). Across vast areas of the world's rangelands, pastoralism, which is a form of integrated management, enables mixed populations of farmed and wild animal species to co-exist, and where a number of core organic farming objectives are satisfied (Roderick, 2019). Without the constraints of artificial boundaries, integration at the landscape level can address regional ecological issues, and therefore potentially more attractive for policymakers and the provision of greater public good. Area-wide collaborations provide opportunities for environmental protection at, for example, the catchment level, where land-use activities that integrate animals and crops can influence the extent of diffuse water pollution through nitrogen leaching and phosphorus loss in particular (Wilkins, 2008).

Operating agriculture at this regional scale requires cooperation between farmers, other community members and stakeholders. There is often also a need for legislation and forms of governance, and facilitation and technical support by external agents is often a key driver (Asai et al., 2018). From a more formal organic farming perspective, whilst landscape-level integration offers multiple potential benefits, there are barriers that include issues around certification. In Alpine settings, where mixed farming may be limited by topographic and climatic constraints, dairy farming is frequently integrated within forest environments to create physical and cultural landscapes that also benefit from eco-tourism (Wanner et al., 2021). A system of closed nutrient cycling operating at a regional level within a population of Alpine organic dairy farms was shown to have lower environmental impacts that relate to site-specific management adaptations, including reduced stocking rates and reduced purchased feeds (Grassauer et al., 2022).

5 Discussion

5.1 Multiple benefits of integrated dairy farming

It is now increasingly recognised that a return to mixed farming, whereby animals and crops are integrated, provides a better opportunity to achieve key positive ecological objectives (Weisner et al., 2020) as well as greater agronomic resilience and options for resource circularity. The overall efficiency of mixed crop–animal systems is increased by harnessing the functional complementarities between the components (Puech and Stark, 2023). One of the main advantages is the possibility of soil carbon and nitrogen accumulation through the actions of grass-clover leys and animal manures (Reinsch et al., 2021). Wilkins (2008) highlights this potential for improved nutrient re-cycling, reduced diffuse pollution and emissions and improved soil quality, all contributing to greater functional biodiversity but also recognising that success requires a degree of output guarantee as well as resilience regarding the ability to respond to changes in economic, social and physical conditions. This efficient and sustainable use of resources in farm production and land management is referred by Wilkins (2008) as eco-efficiency.

Compared with both confined and pasture-based systems, fully grazed integrated crop-dairy farming has been shown, in terms of nitrogen balance and greenhouse gas emissions, through carbon sequestration and plant biomass accumulation, to have a lower environmental impact per unit area and per unit of product. They also have the added advantage of lowering the environmental footprint of subsequent cash crops (Reinsch et al., 2021).

It has also been shown that integrated farms have greater agro-ecological potential than specialised dairy farms (Funes-Monzote et al., 2009). More diverse, integrated crop–animal farming enables greater ecological interaction between the different elements of the farm as well as enabling greater resource efficiency associated with crop fertilisation from animal manures and animal feeds from crops. Under European conditions, mixed crop–animal systems tend to be found in more unfavourable areas, and in this context, they have also been shown to be more economically resilient and less polluting (Ryschawy et al., 2012). Carbon, nitrogen and phosphorus cycling are enhanced as a consequence of more effective fertilising, including manure application (Wiesner et al., 2020).

5.2 Constraints to the integration of dairy farming

A key question relates to the degree to which dairy farming can be truly integrated compared with the potentially more flexible meat-producing systems. Integrating milk production can be more complex than for other species, partly as meat animals in particular tend to be more transient, which in turn enables greater flexibility, whereas dairy farming is more permanent in nature, with associated technical infrastructure and equipment.

The prevalence and benefits of truly integrated, mixed farming systems with dairy are more evident in tropical regions and particularly in small-holder situations. For example, in India, the ecological benefits of integration of dairy, fish and poultry with diversified cropping in coastal lowland ecosystems in India have been shown to offset the negative impacts of continuous rice cropping (Paramesh et al., 2019). Tropical smallholder systems in particular frequently have complementary diverse crop and dairy production that enable the circularity of nutrients, as is the case in countries such as Kenya (Odhong et al., 2014).

5.2.1 Farm infrastructure and access

The design and infrastructure of many dairy farms can limit the opportunities for greater integration and utilisation of the farm resources, including within crop rotations. Mobile milking facilities offer an alternative whereby the milking system follows cows to the pasture rather than requiring cows to travel distances to be milked, either as a semi-permanent unit on a trailer towed by a tractor (Oudshoorn, 2008; De Koning, 2010; Dufrasne et al., 2010, 2012) or as a stand-alone mobile, robotic unit (de Haan et al., 2010). The case study from Macaroni Farm above provides a practical example of how mobile milking equipment, along with mobile and flexible water infrastructure, has been used to enable a successful, fully integrated circular dairy system.

Pasture-based automatic milking systems (AMS) have been shown to be positively associated with pasture utilisation efficiency, as well as having some animal welfare and biodiversity benefits (Shortall et al., 2016). Although mobile AMS adds further flexibility, the technology requires more research and development (Bosona and Gebresenbet, 2021) and there are potential production and management issues (Lessire et al., 2015; Gaworski and Kic, 2017) that could limit their current widespread usage.

5.3 Social and economic factors

There are important social considerations related to farm integration and diversity, including those associated with labour and management requirements, social capital and skills and knowledge acquisition. Crop–animal integration tends to require greater management attention (Bell and Moore, 2012), and the desire of the farmer, as well as their management capabilities and capacity, may be a determining factor influencing the extent to which integration occurs within a farm (Low et al., 2023).

Diversified farms tend to require more farming infrastructure, machinery and labour (Low et al., 2023), along with a greater consideration of farm function and flow within and between the different enterprises. Greater diversity within a farm requires a potentially greater number of tasks that need to be completed, as well as a wider spread of skills. Although on solely organic dairy farms, the labour requirement is similar to that on non-organic dairy farms, generally, labour use per unit area tends to be higher on organic farms. In part, this is because there are more labour-intensive activities associated with complex crop rotations and the mixture of crops and animals on some organic farms (Morison et al., 2005), and the organisation of labour tends to be more complex (Orsini et al., 2018). For similar reasons, organic farms are also considered more 'knowledge-intensive' (Jansen, 2000). There are also increased prospects of clustering and clashing of tasks, both seasonally and routinely, on farms that closely integrate more than one enterprise. A more integrated approach to management and work practice, based on flexibility, has been proposed as an effective strategy for dealing with workload peaks (Martin et al., 2016). Ulukan et al. (2022) identified the need for forward planning and the effective distribution and task sharing to deal with the issue of complex and diverse workloads that can occur on mixed organic and non-organic farms.

It has been suggested that higher crop yields on specialist farms, compared with crop–animal farms, may be associated with differences in farmers' knowledge, with farmers with mixed enterprises tending to have more generalist knowledge about cropping and animals compared with the more focused knowledge of solely arable farmers (Marton et al., 2016) and solely dairy farmers (Peel et al., 2014). Generally, the decline in the popularity of integrated farming may have resulted in the loss, particularly amongst younger farmers, of the broad knowledge required for this type of farming (Hilimire, 2011).

At the farm level, greater integration between crops and animals can enhance economic resilience, through resource sharing, greater efficiency, reduced wastage and the spreading of risk. The extent and success of integration, in terms of production and marketing benefits, has also been related to increased coordination between farmers, partly through greater sharing of resources and risk. Social interactions between farmers and with other stakeholders can be a significant factor influencing the nature and outcomes of between-farm integration at the territorial level (Moraine et al., 2014). At the community level, the presence of diverse farms has been shown to positively contribute to social capital, with greater local and consumer appreciation of farming practices (Boody et al., 2005).

Greater integration between farms can also result in clashes in management style and attitudes, as well as a sense of reduced ownership of processes and loss of control (Low et al., 2023). In some situations, the decline in social capital in many rural areas may be a barrier to achieving the enhanced labour, knowledge and skills requirement of crop–animal farming systems (Low et al., 2023).

Successful crop–animal integration beyond the farm level, as discussed above, can be a complex process relying on social learning, collective thinking and negotiation (Martin et al., 2016). However, risk aversion and other factors, such as trust, differing motivations and legal concerns, may limit the extent to which some are prepared to participate (Martin et al., 2016). In Denmark, successful partnerships between organic dairy and organic arable farmers, involving manure exchange, were shown to be based on strong social connections resulting from high-quality communications and trusted and reciprocal relationships, although well-functioning institutional arrangements and the existence of environmental regulations were also key factors (Asai and Langer, 2014).

5.4 Integration across the food chain

Whilst recognising the resource efficiency of mixed cropping at the farm level, Wilkins (2008) viewed the concept of integration as requiring consideration across the whole food chain, including the life cycle of inputs to the farm and the marketing and processing of outputs, and highlights the contradiction of the airfreighting of organic farm products produced under resource-efficient, mixed farming conditions.

6 Conclusion

In this chapter, we have sought to identify the key attributes of different approaches to integrating organic dairy herds into the wider farm, and under some conditions, beyond the farm. These opportunities extend across different dimensions and intensity. They range from those systems that utilise limited inputs obtained from other farming or industrial processes to those that endeavour to whole-heartedly link the production of milk and other dairy products, including meat, to other agricultural activities through the complementary sharing of resources, resulting in improved efficiencies, enhanced environmental attributes and positive contributions to the welfare of dairy cattle. The relative contribution that each of these pays to the achievement of societal objectives can be judged by the extent to which they contribute to the re-balancing of the food versus feed situation and by the extent to which they contribute to a circular economy that relies on either the re-using, re-purposing or recycling of resources and products. Integration can also be measured by the enhancement of the social and economic conditions of farmers, farming families, employees and local communities, as well as improvements in the environment and hence the lives of cows and calves, and other associated animal species. Biodiversity and soil and water quality benefits are also imperative and evidential of success.

Whilst the organic principles from which we can measure success are largely focussed on integration at the farm level, we can see from the descriptions and analyses above that between-farm and wider regional integration should also form part of our ambitions for more sustainable organic dairy farming into the future, as should the notion of integration across the food chain. The marketing of organic milk needs to reflect the positive benefits of integrated forms of production and thus lift one of the biggest constraints to the development of organic dairy farming.

7 Where to look for further information

The Ellen MacArthur Foundation is a charity committed to creating a circular economy designed to eliminate waste and pollution, circulate products and materials, and regenerate nature. https://ellenmacarthurfoundation.org/ covers a wide range of topics and provides many relevant resources, including case studies and publications.

The Mix Enable project, funded via the ERA-net CORE Organic Cofund, was a project exploring the conditions of sustainability and robustness of European organic mixed livestock farms, integrating two or more animal species. The project webpage https://projects.au.dk/coreorganiccofund/core-organic-cofund-projects/mix-enable/ provides links to resources and publications, including an educational toolkit.

The Woodland Grazing Toolbox https://forestry.gov.scot/woodland-grazing-toolbox is a guide for the development of plans for livestock as a woodland management tool to achieve biodiversity and/or cultural heritage objectives.

Agricology.co.uk is a website dedicated to sharing practical information about sustainable farming based on agroecological principles, including farmer profiles highlighting integrated farming practices, an extensive resource library and details of research projects.

8 References

Altieri, M. A. (2000). *Agroecology: principles and strategies for designing sustainable farming systems. Agroecology in action*. New York: Hayworth Press.

Asai, M. and Langer, V. (2014). Collaborative partnerships between organic farmers in livestock-intensive areas of Denmark. *Organic Agriculture - Official Journal of The International Society of Organic Agriculture Research*, *4*(1), 63–77. https://doi.org/10.1007/s13165-014-0065-3

Asai, M., et al. (2018). Critical factors for crop-livestock integration beyond the farm level: a cross-analysis of worldwide case studies. *Land use policy*, *73*, 184–194.

Baudron, F., et al. (2015). Multi-scale trade-off analysis of cereal residue use for livestock feeding vs. soil mulching in the Mid-Zambezi Valley, Zimbabwe. *Agricultural Systems*, *134*, 97–106.

Bell, L. W. and Moore, A. D. (2012). Integrated crop–livestock systems in Australian agriculture: trends, drivers and implications. *Agricultural Systems*, *111*, 1–12.

Benoit, M., et al. (2023). Combining beef cattle and sheep in an organic system. II. Benefits for economic and environmental performance. *Animal*, *17*(4), 100759.

Berry, P. M., et al. (2002). Is the productivity of organic farms restricted by the supply of available nitrogen?. *Soil Use and Management*, *18*, 248–255.

Bischoff, K. (2019). Glucosinolates and organosulfur compounds. In R. C. Gupta, et al. (Eds.), *Nutraceuticals in veterinary medicine* (pp. 113–119). Cham: Springer.

Bosona, T. and Gebresenbet, G. (2021). Multipurpose simulation model for pasture-based mobile automated milking and marketing system, part-I: pasture, milk yield, and milk marketing characteristics. *Computers and Electronics in Agriculture*, *190*, 106212.

Boody, G., et al. (2005). Multifunctional agriculture in the United States. *BioScience*, *55*(1), 27–38.

Bryszak, M., et al. (2019). Effects of berry seed residues on ruminal fermentation, methane concentration, milk production, and fatty acid proportions in the rumen and milk of dairy cows. *Journal of Dairy Science*, *102*, 1257–1273.

Charlton, J. F. L., et al. (2003). Farmer experience with tree fodder. *NZGA: Research and Practice Series*, *10*, 7–15.

Connolly, J., et al. (2009). Effects of multi-species swards on dry matter production and the incidence of unsown species at three Irish sites. *Irish Journal of Agricultural and Food Research*, *48*, 243–260.

D'Alexis, S., et al. (2014). Mixed grazing systems of sheep and cattle to improve liveweight gain: a quantitative review. *The Journal of Agricultural Science*, *152*(4), 655–666.

de Evan, T., et al. (2020). Potential of recycling cauliflower and romanesco wastes in ruminant feeding: in vitro studies. *Animals*, *10*(8), 1247.

de Haan, M. H. A., et al. (2010). Better grazing opportunities with a mobile milking robot. In *23th General Meeting of the European Grassland Federation, Kiel, Germany, 29 August-02 September, 2010* (pp. 178–180).

De Koning, C. J. A. M. (2010). *Automatic milking–common practice on dairy farms*. The First North American Conference on Precision Dairy Management 2010.

Dufrasne, I., et al. (2010). Grazing with a mobile milking robot. In *23th General Meeting of the European Grassland Federation*.

Dufrasne, I., et al. (2012). Effects of environmental factors on yield and milking number in dairy cows milked by an automatic system located in pasture. In *Proceedings of the 24th General Meeting of the European Grassland Federation* (Vol. 3, pp. 231–233).

Duncan, A. J., et al. (2016). Crop residue allocation to livestock feed, soil improvement and other uses along a productivity gradient in Eastern Africa. *Agriculture, Ecosystems & Environment*, *228*, 101–110.

Ellen MacArthur Foundation. (2015). Schools of Thought – Performance Economy. EMF (Accessed on 01/06/2023). https://www.ellenmacarthurfoundation.org/circular-economy/concept/schools-of-thought.

Entz, M. H., et al. (2005). Evolution of integrated crop-livestock production systems. In D. A. McGilloway (Ed.), *Grassland: A global resource* (pp. 137–148). Wageningen, the Netherlands: Wageningen Academic Publications.

Ertl, P., et al. (2017). Effects of supplementation of a forage-only diet with wheat bran and sugar beet pulp in organic dairy cows. *Renewable Agriculture and Food Systems*, *32*(5), 446–453.

Escribano, A. J. (2018). Organic feed: a bottleneck for the development of the livestock sector and its transition to sustainability? *Sustainability*, *10*(7), 2393.

Feliciano, D., et al. (2018). Which agroforestry options give the greatest soil and above ground carbon benefits in different world regions?. *Agriculture, Ecosystems & Environment*, *254*, 117–129.

Foissy, D., et al. (2013). Managing nutrient in organic farming system: reliance on livestock production for nutrient management of arable farmland. *Organic Agriculture*, *3*, 183–199.

Fraser, M. D., et al. (2007). Effects on animal performance and sward composition of mixed and sequential grazing of permanent pasture by cattle and sheep. *Livestock Science*, *110*(3), 251–266.

Frehner, A., et al. (2022). The compatibility of circularity and national dietary recommendations for animal products in five European countries: a modelling analysis on nutritional feasibility, climate impact, and land use. *The Lancet Planetary Health*, *6*(6), e475–e483.

Funes-Monzote, F. R., et al. (2009). Agro-ecological indicators (AEIs) for dairy and mixed farming systems classification: identifying alternatives for the Cuban livestock sector. *Journal of Sustainable Agriculture*, *33*(4), 435–460.

Gaworski, M. and Kic, P. (2017). Improvement of mobile milking parlours in small dairy farms including technical and functional aspects. In *Proceedings of the 16th International Scientific Conference Engineering for Rural Development, Jelgava, Latvia, 24th–26th May* (Vol. 2017, pp. 402–407).

Grassauer, F., et al. (2022). Environmental assessment of austrian organic dairy farms with closed regional production cycles in a less favorable production area. *Frontiers in Sustainable Food Systems*, *6*, 817671.

Hilimire, K. (2011). Integrated crop/livestock agriculture in the United States: a review. *Journal of Sustainable Agriculture*, *35*(4), 376–393.

IFOAM. (2005). *Principles of organic agriculture international federation of organic agriculture movements*. Bonn: IFOAM.

Irish Agroforestry Forum. (2020). Why dairy farming and silvopastoral agroforestry could be the perfect match. *Irish Farm Business Dairying*, *7*(2), 38–41.

Jansen, K. (2000). Labour, livelihoods and the quality of life in organic agriculture in Europe. *Biological Agriculture & Horticulture*, *17*(3), 247–278.

Jerrentrup, J. S., et al. (2020). Diverse swards and mixed-grazing of cattle and sheep for improved productivity. *Frontiers in Sustainable Food Systems*, *3*, 125.

Jeswani, H. K., et al. (2018). Life cycle greenhouse gas emissions from integrated organic farming: a systems approach considering rotation cycles. *Sustainable Production and Consumption*, *13*, 60–79.

Jing, J., et al. (2017). Species diversity effects on productivity, persistence and quality of multispecies swards in a four-year experiment. *PLoS One*, *12*(1), e0169208.

Kendall, N., et al. (2021). Tree leaves as supplementary feed for ruminant livestock. http://dx.doi.org/10.13140/RG.2.2.25731.27685

Klopfenstein, T. and Owen, F. G. (1981). Value and potential use of crop residues and by-products in dairy rations. *Journal of Dairy Science*, *64*(6), 1250–1268.

Laporta, J., et al. (2020). Late-gestation heat stress impairs daughter and granddaughter lifetime performance. *Journal of Dairy Science*, *103*(8), 7555–7568.

Lessire, F., et al. (2015). Rumination time, milk yield, milking frequency of grazing dairy cows milked by a mobile automatic system during mild heat stress. *Advances in Animal Biosciences*, *6*(1), 12–14.

Liang, Y., et al. (2018). Multi-objective economic-resource-production optimization of sustainable organic mixed farming systems with nutrient recycling. *Journal of Cleaner Production*, *196*, 304–330.

Low, G., et al. (2023). Mixed farming and agroforestry systems: a systematic review on value chain implications. *Agricultural Systems*, *206*, 103606.

Marley, C. L., et al. (2006). The effect of mixed or sequential grazing of cattle and sheep on the faecal egg counts and growth rates of weaned lambs when treated with anthelmintics. *Veterinary Parasitology*, *142*, 134–141.

Martin, G., et al. (2020). Potential of multi-species livestock farming to improve the sustainability of livestock farms: a review. *Agricultural Systems*, *181*, 102821.

Martin, G., et al. (2016). Crop–livestock integration beyond the farm level: a review. *Agronomy for Sustainable Development*, *36*, 53. https://doi.org/10.1007/s13593-016-0390-x

Marton, S. M., et al. (2016). Comparing the environmental performance of mixed and specialised dairy farms: the role of the system level analysed. *Journal of Cleaner Production*, *124*, 73–83.

Monllor, P., et al. (2020). Ensiling process in commercial bales of horticultural by-products from artichoke and broccoli. *Animals*, *10*(5), 831.

Moraine, M., et al. (2014). Farming system design for innovative crop-livestock integration in Europe. *Animal*, *8*, 1204–1217.

Morison, J., et al. (2005). Survey and analysis of labour on organic farms in the UK and Republic of Ireland. *International Journal of Agricultural Sustainability*, *3*(1), 24–43.

Mottet, A., et al. (2017). Livestock: on our plates or eating at our table? A new analysis of the feed/food debate. *Global Food Security*, *14*, 1–8.

Novobilský, A., et al. (2011). Condensed tannins act against cattle nematodes. *Veterinary Parasitology*, *182*(2–4), 213–220.

Odhong, C., et al. (2014). Challenges of conversion to organic dairy production and prospects of future development in integrated smallholder farms in Kenya. *Livestock Research for Rural Development*, *26*(7), 14–21.

Orefice, J., et al. (2019). Forage productivity and profitability in newly-established open pasture, silvopasture, and thinned forest production systems. *Agroforestry Systems*, *93*, 51–65.

Orsini, S., et al. (2018). Labour use on organic farms: a review of research since 2000. *Organic farming*, *4*(1), 7–15.

Oudshoorn, F. (2008). Mobile milking robot offers new grazing concept. *Grassland Sci Europe*, *13*, 721–723.

Parmigiani, A. and Mitchell, W. (2009). Complementarity, capabilities, and the boundaries of the firm: the impact of within-firm and interfirm expertise on concurrent sourcing of complementary components. *Strategic Management Journal*, *30*(10), 1065–1091.

Paramesh, V., et al. (2019). Sustainability, energy budgeting, and life cycle assessment of crop-dairy-fish-poultry mixed farming system for coastal lowlands under humid tropic condition of India. *Energy*, *188*, 116101.

Patra, A. K. and Saxena, J. (2010). Exploitation of dietary tannins to improve rumen metabolism and ruminant nutrition. *Journal of the Science of Food and Agriculture*, *91*, 24–37.

Peel, S. (2013). *Investigating crop and dairy complementarities within a Canterbury farming system case studies from the Mid-Canterbury region, New Zealand*. Masters thesis, University of Lincoln.

Peel, S. (2013). *Investigating crop and dairy complementarities within a Canterbury farming system case studies from the Mid-Canterbury region, New Zealand*. Masters thesis, University of Lincoln. Available at: https://researcharchive.lincoln.ac.nz/server/api/core/bitstreams/52048b5a-78e7-428e-bb21-29f75642866d/content

Pent, G. J. (2020). Over-yielding in temperate silvopastures: a meta-analysis. *Agroforestry Systems*, *94*, 1741–1758.

Prache, S., et al. (2023). Combining beef cattle and sheep in an organic system. I. Co-benefits for promoting the production of grass-fed meat and strengthening self-sufficiency. *Animal*, *17*(4), 100758.

Puech, T. and Stark, F. (2023). Diversification of an integrated crop-livestock system: agroecological and food production assessment at farm scale. *Agriculture, Ecosystems & Environment*, *344*, 108300.

Regan, J. T., et al. (2019). Does the recoupling of dairy and crop production via cooperation between farms generate environmental benefits? A case-study approach in Europe. *European Journal of Agronomy*, *82*, 342–356.

Reinsch, T., et al. (2021). Toward specialized or integrated systems in Northwest Europe: on-farm eco-efficiency of dairy farming in Germany. *Frontiers in Sustainable Food Systems*, *5*, 614348.

Roderick, S. (2019). Pastoralism and organic animal farming: are they complementary? In M. Vaarst and S. Roderick (Eds.), *Improving organic animal farming* (pp. 175–201). Cambridge UK: Burleigh Dodds Science Publishing Limited.

Romero-Huelva, M., et al. (2012). Nutrient utilization, ruminal fermentation, microbial abundances, and milk yield and composition in dairy goats fed diets including tomato and cucumber waste fruits. *Journal of Dairy Science*, *95*(10), 6015–6026.

Rørvang, M. V., et al. (2018). Prepartum maternal behavior of domesticated cattle: a comparison with managed, feral, and wild ungulates. *Frontiers in Veterinary Science*, *5*, 45. https://doi.org/10.3389/fvets.2018.00045

Ryschawy, J., et al. (2012). Mixed crop-livestock systems: an economic and environmental-friendly way of farming?. *Animal*, *6*(10), 1722–1730.

Scott, C. (2023). *Farm Business Survey 2021/2022 organic farming in England*. Rural Business Research, Newcastle University.

Schroeder, J. W. (1999). By-products and regionally available alternative feedstuffs for dairy cattle. http://www.ag.ndsu.edu/pubs/ansci/dairy/as1180w.htm

Sehested, J., et al. (2004). Grazing with heifers and sows alone or mixed: herbage quality, sward structure and animal weight gain. *Livestock Production Science*, *88*(3), 223–238.

Shortall, J., et al. (2016) Investment appraisal of automatic milking and conventional milking technologies in a pasture-based dairy system. *Journal of Dairy Science*, *99*, 7700–7713.

Stobart, A., et al. (2023). Medicinal tree products offer therapeutic benefits and potential diversification for small scale agroforestry and farm woodlands: An overview with indications for further studies. *Medicinal Agroecology*, 105–121.

Su, J., Xu, F. and Zhang, Y. (2023). Grassland biodiversity and ecosystem functions benefit more from cattle than sheep in mixed grazing: a meta-analysis. *Journal of Environmental Management*, *337*, 117769.

Sumberg, J. (2003). Towards a dis-aggregated view of crop-livestock integration in Western Africa. *Land Use Policy*, *20*, 253–264.

Tapia-Escarate, D., et al. (2021). Evaluation of cross-grazing deer with sheep or cattle, as means to reduces anthelmintic usage to control gastrointestinal and pulmonary nematodes in farmed red deer (Cervus elaphus) in New Zealand. *Veterinary Parasitology*, *298*, 109534.

Torralba, M., et al. (2016). Do European agroforestry systems enhance biodiversity and ecosystem services? A meta-analysis. *Agriculture, Ecosystems and Environment*, *230*, 150–161.

Ulukan, D., et al. (2022). Positive deviant strategies implemented by organic multi-species livestock farms in Europe. *Agricultural Systems*, *201*, 103453.

Van Zanten, H. H., et al. (2018). Defining a land boundary for sustainable livestock consumption. *Global Change Biology*, *24*(9), 4185–4194.

Van Zanten, H. H., et al. (2019). The role of farm animals in a circular food system. *Global Food Security*, *21*, 18–22.

Varah, A., et al. (2020). Temperate agroforestry systems provide greater pollination service than monoculture. *Agriculture, Ecosystems and Environment*, *301*, 107031.

Vasta, V., et al. (2008). Alternative feed resources and their effects on the quality of meat and milk from small ruminants. *Animal Feed Science and Technology*, *146*, 223–246.

Villano, R., et al. (2010). Evidence of farm-level synergies in mixed-farming systems in the Australian Wheat-Sheep Zone. *Agricultural Systems*, *103*(3), 146–152.

Wang, L., et al. (2019). Diversifying livestock promotes multidiversity and multifunctionality in managed grasslands. *Proceedings of the National Academy of Sciences*, *116*(13), 6187–6192.

Wanner, A., et al. (2021). The future of Alpine pastures. Agricultural or tourism development? Experiences from the German Alps. *Journal of Outdoor Recreation and Tourism*, *35*, 100405.

Whistance, L. (2018a). Browse, preserved tree fodder and nutrition – How offering access to browse and feeding tree fodder can supplement the diet of domestic animals. AFINET Factsheet 12. zenodo.org/record/2454797

Whistance, L. (2018b). Trees for shade, shelter, survival and body maintenance – How offering access to trees can improve the welfare of domestic animals. AFINET Factsheet 13. zenodo.org/record/2454775

Wiesner, S., et al. (2020). Increasing dairy sustainability with integrated crop–livestock farming. *Sustainability*, *12*, 765.

Wilkins, R. J. (2008). Eco-efficient approaches to land management: a case for increased integration of crop and animal production systems. *Philosophical transactions of the Royal Society of London. Series B, Biological sciences*, *363*(1491), 517–525.

Woodland Grazing Toolbox. (n.d.). https://www.forestry.gov.scot/woodland-grazing-toolbox

Wright, I., et al. (2006). The effect of sward surface height on the response to mixed grazing by cattle and sheep. *Animal Science*, *82*(2), 271–276.

Chapter 12

How can organic dairy farming address and improve biodiversity and healthy ecosystems?

Will Simonson, Organic Research Centre, UK

1 Introduction

Biodiversity, short for 'biological diversity', refers to the variability to be found within living organisms and encompasses different scales: the diversity within species (their genetic diversity), between species (often measured as species richness) and ecosystems. The variability of life in any one geographical area is critical for the healthy state and functioning of its ecosystems. In this chapter, we explore the impact of farming, especially dairy production, on biodiversity and how organic approaches can improve biodiversity and ecosystem health.

Farming has transformed landscapes across the world. Some 38% of global land surface area is farmed (FAO, 2019) and about 75% of Britain's land area (Newton, 2017). The mass of all livestock has been estimated to account for 67% of the global total mass of vertebrate life, compared to 1% wild vertebrate fauna which, 12 millennia earlier, represented 99% (Smil, 2011). This transformation of the land area by humans and their livestock has had an expectedly dramatic effect not just on vertebrate wild animals but the state of biodiversity in its entirety. Farming is considered one of the most significant drivers of species

http://dx.doi.org/10.19103/AS.2024.0138.13

decline and extinction throughout the world, mediated by habitat destruction and pollution but also, indirectly, the contribution of farming to invasions of non-native species and climate change (Sala et al., 2000; Newbold et al., 2015).

At the same time, farmland can harbour important communities of wild fauna and flora, and indeed some traditional farming systems are among the most biodiverse of any habitat. An example is extensively managed wood pasture systems with their complex mosaic of wooded and semi-open areas as well as richness of micro-habitats (Garbarino and Bergmeier, 2014; Plieninger et al., 2015). But modern, increasingly industrialised farming systems are, in general, notoriously barren for wild species. The intensification of farming has led, over time, to a serious decline in species groups such as farmland birds, tracked as a composite indicator of the fortunes of biodiversity in Britain and elsewhere (Gregory et al., 2003). Food production is therefore a key battleground for conservation movements today but with divergent opinion on whether land sparing (prioritising the protection of remaining natural habitat by maximising the productivity of farmland and therefore avoiding further conversion) or land sharing (improving the biodiversity value of farmed land through nature-friendly – usually less productive – farming methods) represents the strategy with best overall benefit to biodiversity.

In this chapter, I will explore the degree to which organic farming, and specifically organic dairy systems, has a part to play in biodiversity conservation in this land sparing–land sharing continuum of approaches. It has been widely stated and assumed that organic farming should increase biodiversity because it is less intensive than conventional farming (Bengtsson et al., 2005). Does this hold out in practice? I first provide a brief review of evidence for organic farming being nature-friendly and the proposed mechanisms behind the relationships documented in literature. I then survey studies focussed on organic dairy production and how some of these investigations have sought to account for the question of scale and demand for land in our 'teleconnected' world. Next, I consider what many of these studies are often failing to capture: the question of farmer attitudes and how these may co-vary with organic practice with important implications for biodiversity. These strands are then brought together into commentary on the way forward, including future research priorities. Throughout, different indicators and proxies of 'biodiversity' are referred to; needless to say, biodiversity has many different facets and the works cited in this chapter necessarily focus on a small part of the whole.

2 Organic farming and biodiversity

The contribution of organic farming to biodiversity is well studied and a number of meta-analyses comparing organic and conventional temperate systems have been published (Bengtsson et al., 2005; Fuller et al., 2005; Gabriel et al.,

2010; Hole et al., 2005; Rahmann, 2011; Stein-Bachinger et al., 2021). These consistently provide strong empirical evidence for organic practices supporting more biodiversity. Rahmann (2011) report 327 of 396 studies finding a higher degree of biodiversity in organic compared to conventional farming, although the relationship is heterogeneous depending on the taxonomic group and scale of focus. Clearly, the response is affected by a number of contextual factors. In the study of Fuller et al. (2005) on English cereal farms, the effect of organic production on plant species richness and abundance was stronger than for invertebrates, birds and bats. Similarly, Stein-Bachinger et al. (2021) found the greatest effect for plants, with average (median) species number of flora on arable land being 95% higher under organic management. Gabriel et al. (2010) investigated in particular the importance of scale, pointing out that whilst many comparisons are at farm level, lots of macro-organisms respond to the environment at substantially larger spatial scales. Hence, in their study based on 10 × 10 km landscapes containing low-to-high amounts of organic farming, they conclude that on-farm biodiversity components depend on farming practices at both farm and landscape scales. The small size and often isolated nature of organic farms can be a strong determinant of the observed biodiversity benefits of organic on those farms (Fuller et al. 2005). Bengtsson et al. (2005) suggest that positive effects can be expected in intensively managed agricultural landscapes but not in small-scale landscapes comprising a mixture of biotopes among the fields.

The characteristics of organic practice that encourage on-farm biodiversity are discussed by many of the studies and reviews on this topic. Hole et al. (2005) summarise them as prohibition/reduced use of chemical pesticides and inorganic fertilisers; sympathetic management of non-cropped habitats; and preservation of mixed farming. Habitat extent, composition and management were the important factors noted by Fuller et al. (2005), whilst versatile crop rotations, reduced applications of nutrients and the ban on pesticides were considered the main drivers of increased soil biological activity (bacteria, Monera; fungi, Mycota; springtails, Collembola; mites, Arachnida; and earthworms, *Lumbricus terrestris*) in organic compared to conventional systems (Hansen et al., 2001). The mechanisms by which individual species groups are benefited will vary. Potts et al. (2009), in their study of pollinator (bumblebee and butterfly) diversity in intensive grasslands, manipulated management practices (application of inorganic fertiliser, cutting frequency, cutting height, and aftermath grazing) in four livestock farms in SW England to produce a gradient of management intensity. They also tested the introduction of sown components into the sward as a specific intervention. They found that bumblebees responded to the sown treatments but not the grass-based treatments, whilst butterflies were seen to increase with more extensive management of the grassland. The assemblage structure of both groups was driven by floral abundance, floral richness, availability of nectar sources and sward structure. These results accord with the conclusion of Vickery et al. (2001) that intensive grasslands – in receiving high fertiliser application rates and frequent intensive defoliation to optimise harvested forage quality – results in degraded species pools and structurally homogenous swards.

The latter study focuses on 'functional' or agriculturally important biodiversity and therefore not only on the drivers of biodiversity change but also its consequences. In this respect, the evenness inherent in species assemblages can be as important as species richness or diversity, given that domination by a few common species can contribute to pest outbreaks. In a study based on potato production in field enclosures, Crowder et al. (2010) demonstrated how organic farming promotes evenness among natural enemies, exerting enhanced pest control and yielding higher plant biomass compared to conventional systems.

Livestock farming, generally, is important for many British birds, and under organic practice, there are opportunities to enhance habitat and prey for bird species. This is further explored in Box 1.

Box 1 Cattle, pastures, dung and birds

The following information is largely taken from Ian Newton's 2017 book Farming and Birds, *published in 2017 as part of The New Naturalist Library, as well as the additional sources cited.*

The cattle population of Britain in around 2015 was estimated to be nearly 10 million animals (Defra, 2016) and more than half of these animals were used for milk and other dairy products. The main impact that cattle have on birds is through their effects on habitat, as large areas of grassland, as well as arable crops, are given over to produce their food. For example, in the past, fields of root crops were grown as winter fodder. These supported an abundance of weeds which in turn fed many finches and other seed eaters. Although this is still practised, such crops have largely been replaced by silage, which under intensive management is perhaps the least bird-friendly crop grown in Britain. Intensive silage grass is cut too frequently to allow ground-nesting birds to breed successfully, and on intensive pastures, cattle are grazed at such high densities that the nests of ground-nesting birds are trampled before they can produce young.

Nevertheless, pastureland provides an important feeding habitat for birds such as corvids, thrushes, lapwing and starlings that eat soil-dwelling invertebrates (Fig. 1). The short grass of grazed pastures enables birds to forage efficiently on soil-dwelling prey through the breeding season. Cattle-grazed pastures are favoured foraging areas for swallows. It has been found that intermediate levels of grazing support more diverse and abundant arthropod communities than

Figure 1 Pastureland provides important feeding habitat for Lapwing, *Vanellus vanellus*, including the insects associated with the animal dung. Credit: Nick Goodrum CC BY 2.0 DEED.

either low- or high-intensity grazing, by creating more heterogeneity in vegetation structure and abiotic condition (van Klink et al., 2015).

Cattle dung pats can hold dozens of invertebrate species including dung beetles in the genus *Aphodius*. These animals play an essential role in the breakdown of organic matter and recycling the nutrients that it contains. Without such recycling, pastures would soon become covered with rank patches of grass of low indigestibility and nutrient value, eventually becoming unsuitable for grazing. Several bird species exploit the insects associated with animal dung: wagtails pick off the surface insects, and corvids, starlings, lapwings, black-headed gulls and others break open the dung pats and turn over the pieces to expose the insect larvae and beetles within. House sparrows and yellowhammers pick out undigested cereal and grass seeds. The warmth of manure makes muck heaps attractive feeding sites when the ground is frozen. Rotted farmyard manure spread over the fields increases the abundance of soil-dwelling invertebrates, including earthworms, in the surface layer within the reach of soil-probing birds. These prey are usually less abundant under inorganic fertilisers.

However, the use of anthelmintics (e.g. ivermectin) to control parasitic worms in the guts of livestock can have harmful effects on bird life. These chemicals persist in dung and populations of dung flies, dung beetles and other invertebrates, including those in the soil, can be killed off as a result. This reduces prey for birds and results in the interruption of the natural processes of break-down and recycling of the nutrients back into the system. The reduction of anthelmintic use in organic production systems can therefore have a direct benefit to the invertebrate fauna and wider food chain.

3 Organic dairy production and biodiversity

As a subset of organic farming, organic dairy systems and their relationship with biodiversity are understandably less studied in the scientific literature. Evidence of a positive relationship in general holds. For example, in an investigation mirroring the aforementioned studies of functional biodiversity, Power and Stout (2011) found that insect pollination networks were larger and more asymmetric (hence potentially more stable) in organic paired with conventional dairy farms in SW England, with abundance and diversity of pollinators being higher. However, the response of biodiversity to this sub-sector of organic farming is heterogeneous and other factors can be more important, as demonstrated in a study of plant biodiversity in mountain grasslands under dairy management in the eastern Alps (Pornaro et al., 2021). Comparing 49 small-scale farms and carrying out plot-level botanical inventories to generate a farm-level characterisation of the flora, these authors observed a pattern of decreasing plant species richness and a number of phytosociological classes with increasing milk yield and livestock density. However, from data on farm management and the use of different management practices, they failed to uncover evidence of a difference between organic and conventional farms in terms of plant species richness.

A number of studies of biodiversity impact of organic dairy system adopt the so-called life cycle assessment (LCA) approach focussed on the product (i.e. milk) and relating the environmental impacts to a functional unit (e.g. a litre of milk) (Schader et al., 2014). A typical LCA analytical approach is exemplified in a comparison of organic and conventional milk production in Sweden (Mueller et al., 2014). This study used characterisation factors for *occupancy* (based on area, time and difference in plant species richness between areas under pasture/foodstuff production and comparative areas of natural habitat) and *transformation* (based on area and time taken to regenerate). Hence, unlike other approaches described above, the LCA methodology takes into account the production of concentrated feed in third countries. In this Swedish study, taking plants as the indicator group for biodiversity more widely, organic was found to have 60% less biodiversity damage potential. A large part of this observed difference in impact was due to organic feedstocks consisting of >60% roughage and with >50% of feed being sourced on farms.

A similar LCA-based comparison of organic and conventional dairy in western Europe included soil carbon changes and freshwater ecotoxicity as well as biodiversity – again measured as plant species richness (Knudsen et al., 2019). Characterisation factors were defined, being based on a data set derived from field recording of plant species diversity in farmland in six European countries within the Temperate Deciduous Forest biome. The impacts of organic milk production on freshwater ecotoxicity, biodiversity and resource depletion were found to be 2%, 33% and 20% of the impacts of conventional management, across three systems of production: lowland mixed crop-livestock, lowland grassland and mountain.

In a further example of a LCA-based approach, the biodiversity and wider environmental performance (e.g. energy consumption, climate impact, nitrate leaching and conservation of soil fertility), animal welfare and milk quality of 27 organic dairy farms in Germany were studied, classified by grassland in total farm area and feeding intensity. Biodiversity was not measured on farms but instead with an expert-based scoring system incorporating a range of factors: intensity of grassland cultivation, biotope quality of crop rotation, and diversity of cultivated crops. In this way, the study focussed on how the potential biodiversity of the specific site is influenced by farm management. The difference in biodiversity potential between the farm types (classified by grassland area and feeding intensity) was found to be small. In none of the farms was productive area found to be an important influence on biodiversity; instead, it was concluded that nature conservation measures, which can be carried out on all farm types (not just organic) generally have a positive effect on biodiversity. This conclusion raises the question: is it organic production, *per se*, or specific management practices that often co-exist with organic, that are important for biodiversity on organic farms?

4 Beyond organic? Farmer attitudes and whole-farm approaches to biodiversity conservation in organic dairy systems

In their 2021 opinion article, Tscharntke et al. (2021) argue that organic practice alone has limited benefits for biodiversity. Whilst on average it raises the local richness of widespread species by about one-third and species abundance by about 50%, it also reduces yield considerably and therefore requires more land to produce the same amount of food. In other words, biodiversity benefits can disappear when measured per unit of land necessary for a defined agricultural output or yield. Instead, their argument goes, diversifying cropland, reducing field size, using green manure, achieving a landscape-level mosaic of natural habitat patches and fine-grained cropland diversification – in both conventional and organic agriculture – is key for promoting biodiversity. At the same time, the authors recognise that such measures are often recommended by organic food organisations and can be more prevalent on organic than conventional farms, even though they are not a formal part of certification regulations.

On this latter observation, certain studies are supportive. For example, in an investigation of ten pairings of organic and conventional farms in SW England, organic farms were found to have greater areas of semi-natural habitat and boundary features (Gibson et al., 2007). The difference was particularly pronounced in the case of woodlands and, importantly, this was attributed to a higher level of tree-planting effort rather than the chance occurrence of larger woodlands on these farmers. The biodiversity of the woodlands, measured in terms of floristic richness, was not higher in the organic farms compared to the conventional farms, but the availability of this habitat was higher. In the case of the arable fields, these were found to be more biodiverse on the organic farms. Taking these results together, one can deduce two important drivers of increased biodiversity under organic production methods: the benign or positive nature of prescribed organic practices with respect to biodiversity (e.g. cereals cultivation without chemicals) and a wider set of nature-friendly, agro-ecological approaches which are often practised on organic farms.

Research into the importance of farmer attitudes helps to explain this association. Power et al. (2013) carried out a sociological survey comparing organic and conventional farms in Ireland and whether farmer attitudes and knowledge on the environment and biodiversity had an influence on the biodiversity (again measured as plant species richness) on their farms. They found that plant species richness was higher on organic farms and higher for farmers with greater knowledge or environmental concern. They commented that 'organic and conventional farmers had similar attitudes to farming achievement and the environment but organic farmers were better informed about environmental issues and carried out more environmentally orientated behaviours'. This wider sustainable farming philosphy is arguably evidence of the four organic principles of health, ecology, fairness and care (Luttikholt 2007) being of more importance than certified organic practice in encouraging biodiversity in dairy and other production systems. Such a 'organic plus' philosophy and practice, driving biodiversity gain, is evident in the Cotswold organic dairy case study of Bradwell Grove Estate.

5 Case study: Bradwell Grove Estate

A joint venture between two Cotswold organic farmers and a dairy farmer from Cornwall has given rise to an organic dairy enterprise looking to achieve triple wins in terms of biodiversity, soil health and profitability. Charles Hunter-Smart, managing the 3000-acre Bradwell Grove Estate near Burford for the Heyworth family, and Sam Phillips of the neighbouring 1950-acre Macaroni Farm joined forces to transition from a rotation of three cereal crops and two fertility-building leys, with a suckler cow and sheep business, to a dairy-based system with a greater margin and stretching the rotation of herbal ley to four years. Cotswold Mobile Diary Ltd was formed with Dave Sanders, a dairy farmer from Cornwall, providing the milk contract and cows. The mobile milking system allows fertility to be spread over as much of the joint estate as possible. The mobile milking parlour moves on after two daily cycles of milking so that the cows, passing through the flatbed trailer-based system, are enticed by the fresh pasture on the other side. The ley grass and wildflower mix provided by Cotswold Seeds is fundamental to the success of this operation and the achievement of benefits for biodiversity and soils, as well as animal health. The mix of species includes N-fixing legumes (e.g. sainfroin, clovers and trefoils) and deep-rooted plants, such as cocksfoot and chicory, which improve soil fertility and structure. Manuring by the animals also drives an essential part of the nutrient cycle, supporting a community of worms, bugs and beetles helping to release the nutrients for plant growth.

Managing the herd of 350 cows (as well as 100 heifers and 100 heifer calves) is by herdsman Rob Richmond. The mob grazing system (grazing a third, trampling a third and leaving a third) and rotation of blocks into the dairy are helping to regenerate the land across the estate, and as a result, Rob has noticed that the biodiversity has been increasing, e.g. flocks of birds observed on the fields in the winter of 2022–2023, including mixed finches, starlings, meadow pipits (one flock of 20) and wagtails, using the soft ground under the cattle. According to Rob, the engine for this richness of birds and indeed the health of the whole ecosystem is the grazing management and use of herbal leys, together creating an abundance and diversity of associated arthropods, as well as plentiful seeds. The more tightly grazed pastures under the previous beef suckler cow and sheep system failed to nurture that essential part of the food web. Nick Adams further comments how the out-wintering of the cattle (compared to the previous suckler herd that was housed in winter) are key here. Insect eaters like meadow pipit and pied/white wagtails are drawn to the insects associated with the herbal leys as well as the disturbed ground and droppings, whilst there is still enough plant cover to provide food for the seed-eaters. Thrush species like fieldfare and redwing forage for invertebrates on the recently grazed areas and starling have returned in their hundreds in the winter, having not been seen for a number of years.

Whilst the managed fields have become a more important habitat for wildlife, the field margins also contribute to farm-level floristic diversity and associated fauna. By standardising the shapes of the fields for easier grazing, large areas of habitat have been created on the field edges. These margins are planted with a variety of Stewardship options like flower-rich grassland (AB8) and wildbird covers (OP2). This suits species like grey partridge for nesting and yellowhammer and linnet for foraging near their nest sites. Previously these might have been cut/grazed to the edge. Most fields in the arable rotation have 12 m wildflower and grass margins, and central field strips (similar to beetle banks) give an infield nesting option for species like grey partridge, skylark and corn bunting, whilst also providing foraging throughout the year.

Nick identifies potential biodiversity downsides of a farming system that optimises the efficiency of the dairy production, whether organic or not. The rotation is based on block farming which reduces the variety of habitats available in any one part of the farmland. This means that species using particular areas are exposed to 'feast-and-famine' habitat availability. An example case is corn bunting, which might find itself nesting in a system dominated by spring oats (excellent) or silage herbal leys (potential disaster). Furthermore, some fields are broken up with permanent fencing so the grazing can be carried out more efficiently. This brings features into fields where there were none, giving predators perches to observe species like corn bunting and skylark and target their nest sites.

Ultimately, for Nick perhaps the most important aspect of wildlife protection of such a large, dynamic farming system – where grazing or cutting of fields is adaptively managed depending on how the growing year is going – is communication. If there is a system, as simple as a map in the farmyard, for people to log sightings of nest sites for important species such as corn bunting and grey partridge, field operations such as silaging can be informed to avoid nest destruction and ensure some habitat is left. The more notice there is, the more accuracy is possible and the less impact there will be on the activity.

6 Impacts of scaling up organic practice

Evidence has been presented of organic dairy farming having more benefits to biodiversity at the local scale than conventional practice. These benefits may indeed be higher than those reported because comparisons of organic and conventional farms cannot control for the small size and isolated context of organic farms in a wider landscape of conventional production (Fuller et al., 2005). For example, differences in vegetation composition between organic and conventional sown pastures tend to be less marked than for arable fields (Hole et al., 2005; Lampkin and Pearce, 2021). Whilst this may be attributed to non-organic pasture management being relatively less intensive (hence there being a higher baseline value for biodiversity), it is also hypothesised that the natural colonisation of grassland to form a diverse sward is a slow and unreliable process, especially where rarer species are largely absent from the seedbank. Isolatedness can matter.

Organic production is a small minority of dairy farming in Britain, and so its national contribution to biodiversity conservation will, overall, be equally small. What contribution could it make to biodiversity if applied at larger scales? And does this have to be a hypothetical question? As we have seen, this scale dimension has attracted a body of research that considers the trade-off between the higher biodiversity of organic dairy farms and the additional land required to achieve the same yield output (Fig. 2). The trade-off is very context specific, and one meta-analysis of studies in which both yield and biodiversity have been measured concluded that organic farming (land sharing) is the

Figure 2 Comparison of conventional (top) and organic (bottom) farming strategies that reach the same total food production. More land may be required for organic farming, but which strategy offers the highest biodiversity-abundance will depend on the relative levels of biodiversity to be found in farmed and unfarmed areas. Adapted from Gong et al. (2022). Credit: Farmland scenes, stockarch.com/stockmedia.cc (top) and Walter Baxter CC BY-SA 2.0 DEED (bottom); woodland scene, IA Photography CC BY-ND 2.0 DEED.

more effective conservation strategy compared to conventional (land sparing) if the unfarmed lands are less than 2.4 times as biodiverse as conventional farms (Gong et al., 2022). Global projections by Barbieri et al. (2021) are also instructive in this context. Their Global Organic Agriculture Nutrient Model predicted that the global 'option space' for organic agriculture is delimited by nitrogen availability, with the production gap between organic and conventional increasing as organic farming covers more area globally. This can be explained by the key role of crop growth nitrogen requirements. However, targeting food supply (via redesign of the livestock sector) and demand (reducing average per capita caloric intake) could enable a transition to organic of 40–60% of global agricultural area even under current nitrogen limitations.

In Europe, a lower but still ambitious target has been set by the European Union for the expansion of organic farming to 25% of agricultural land area by 2030 as part of its Farm to Fork and Biodiversity Strategies intended to deliver on the Green Deal. A preliminary assessment of the environmental impacts of achieving this (Lampkin and Padel, 2022) concluded that it was not unreasonable to expect an increase in farmland biodiversity by 5–10% in total, 20–30% of this increase on the organic area. Furthermore, the authors suggested this gain could be enhanced if the integration of natural habitats and landscape elements could be increased as part of the 10% nature restoration on farmland target set in the EU's Biodiversity Strategy.

7 Conclusion

We have seen how there is strong empirical evidence for organic practices supporting biodiversity when compared to conventional farming. However, many of these studies focus on comparisons of floristic diversity and richness and this points to the need for more comprehensive assessments capturing arthropods, birds and other fauna (Knudsen et al., 2019). The case study of Bradwell Grove Estate illustrates the significance of organic dairy farming practice on bird communities. Some of the comparisons of organic dairy and conventional farming we have surveyed in this chapter have adopted a so-called LCA approach which recognises potential impacts off-farm as well as on-farm and demonstrates that organic production has less biodiversity damage potential. Whilst organic farming is too narrowly practised to change the fortunes of biodiversity at a national or regional level, scenarios of widespread uptake of organic farming show the dramatic impact that this wider adoption could make.

Ultimately, the benefits of organic approaches may derive not only from the certified organic practice itself but also from the wider nature-friendly measures that co-occur on systems under the management of environmentally well-informed and motivated farmers. Rather than meaning that one has to look beyond organic farming for the answer to the nature crisis (Tscharntke et al., 2021), one can instead argue that the four organic principles of health, ecology, fairness and care, to the extent they capture this wider sustainable farming philosphy, are highly effective guiding principles for encouraging biodiversity in dairy and other production systems.

8 Where to look for further information

For a more in-depth understanding of the topic, the reader is pointed to the bibliography below, particularly to the articles of Bengtsson et al. (2005), Fuller et al. (2005), Gabriel et al. (2010), Gong et al. (2022) and Power et al. (2013).

The following initiatives and organisations are also useful sources of information and inspiration for farmers and scientists:

- Conservation Evidence (conservationevidence.com) synthesises the state of the science around types of land management action that conserves and restores biodiversity. Much of this is relevant to farming. It summarises the evidence for the effectiveness of over 3500 different actions and has created synopses on topics including farmland conservation, bird conservation and natural pest control.
- The British Ecological Society (britishecologicalsociety.org), a membership organisation, has a special interest group in Agricultural Ecology.

- Pasture for Life (pastureforlife.org) publish a range of biodiversity case studies featuring the approaches of farmers to improve nature and biodiversity on their farms. The featured farms include organic dairy and other organic production.
- The Nature Friendly Farming Network (nffn.co.uk) is a UK network bringing together farmers with a passion for sustainable farming and nature. They seek to exchange knowledge and secure positive changes in policy. The Farming and Wildlife Group (fwag.co.uk) and Linking Environment and Farming (LEAF; leaf.eco) are two other organisations with a wealth of knowledge and advisory support.
- There are national biodiversity recording schemes and monitoring protocols available in Britain for different taxonomic groups including birds, butterflies and plants. Further information can be obtained from the wildlife charities responsible, namely the British Trust for Ornithology (bto.org), Butterfly Conservation (butterfly-conservation.org) and Plantlife (plantlife.org.uk). The Organic Research Centre (organicresearchcentre.com) has developed and applied biodiversity monitoring protocols for farmland as part of its ongoing research programme into the benefits of organic farming.

9 References

Barbieri, Pietro, et al. 2021. "Global Option Space for Organic Agriculture Is Delimited by Nitrogen Availability." *Nature Food* 2 (5): 363–372. https://doi.org/10.1038/s43016-021-00276-y.

Bengtsson, Janne, et al. 2005. "The Effects of Organic Agriculture on Biodiversity and Abundance: A Meta-Analysis." *Journal of Applied Ecology* 42 (2): 261–269. https://doi.org/10.1111/j.1365-2664.2005.01005.x.

Crowder, David W., et al. 2010. "Organic Agriculture Promotes Evenness and Natural Pest Control." *Nature* 466 (7302): 109–112. https://doi.org/10.1038/nature09183.

Defra. 2016. *Agriculture in the UK 2015*. London: Department for Environment, Food and Rural Affairs.

FAO. 2019. "FAOSTAT." 2019. https://www.fao.org/faostat/en/#home.

Fuller, R. J., et al. 2005. "Benefits of Organic Farming to Biodiversity Vary among Taxa." *Biology Letters* 1 (4): 431–434. https://doi.org/10.1098/rsbl.2005.0357.

Gabriel, Doreen, et al. 2010. "Scale Matters: The Impact of Organic Farming on Biodiversity at Different Spatial Scales." *Ecology Letters* 13 (7): 858–869. https://doi.org/10.1111/j.1461-0248.2010.01481.x.

Garbarino, M. and Bergmeier, E. 2014. "Plant and Vegetation Diversity in European Wood-Pastures." In *European Wood-Pastures in Transition: A Sociological Approach*, edited by T. Hartel and T. Plieninger, 113–131. New York: Earthscan.

Gibson, R. H., et al. 2007. "Plant Diversity and Land Use under Organic and Conventional Agriculture: A Whole-Farm Approach." *Journal of Applied Ecology* 44 (4): 792–803. https://doi.org/10.1111/j.1365-2664.2007.01292.x.

Gong, S., et al. 2022. "Biodiversity and Yield Trade-Offs for Organic Farming." *Ecology Letters* 25 (7): 1699–1710.

Gregory, R. D., et al. 2003. "Using Birds as Indicators of Biodiversity." *Ornis Hung* 12–13: 11–24.

Hansen, Birgitte, et al. 2001. "Approaches to Assess the Environmental Impact of Organic Farming with Particular Regard to Denmark." *Agriculture, Ecosystems and Environment* 83 (1–2): 11–26. https://doi.org/10.1016/S0167-8809(00)00257-7.

Hole, D. G., et al. 2005. "Does Organic Farming Benefit Biodiversity?" *Biological Conservation* 122 (1): 113–130. https://doi.org/10.1016/j.biocon.2004.07.018.

Klink, R. van, et al. 2015. "Effects of Large Herbivores on Grassland Arthropod Diversity." *Biological Reviews* 90 (2): 347–366. https://doi.org/10.1111/brv.12113.

Knudsen, Marie Trydeman, et al. 2019. "The Importance of Including Soil Carbon Changes, Ecotoxicity and Biodiversity Impacts in Environmental Life Cycle Assessments of Organic and Conventional Milk in Western Europe." *Journal of Cleaner Production* 215 (2019): 433–443. https://doi.org/10.1016/j.jclepro.2018.12.273.

Lampkin, N. and Padel, Katrin. 2022. *Environmental Impacts of Achieving the EU's 25% Organic Land by 2030 Target: A Preliminary Assessment*. Brussels: IFOAM.

Lampkin, Nic and Pearce, Bruce. 2021. *Organic Farming and Biodiversity*. Brussels: IFOAM.

Luttikholt, L. W. M. 2007. "Principles of Organic Agroclture as Formulated by the International Federation of Organic Agroculture Movements." *NJAS-Wageningen Journal of Life Sciences* 54 (4): 347–360. https://doi.org/10.1016/S1573-5214(07)80008-X.

Mueller, Carina, et al. 2014. "Comparing Direct Land Use Impacts on Biodiversity of Conventional and Organic Milk - Based on a Swedish Case Study." *International Journal of Life Cycle Assessment* 19 (1): 52–68. https://doi.org/10.1007/s11367-013-0638-5.

Newbold, Tim, et al. 2015. "Global Effects of Land Use on Local Terrestrial Biodiversity." *Nature* 520 (7545): 45–50. https://doi.org/10.1038/nature14324.

Newton, Ian. 2017. *Farming and Birds*. London: William Collins.

Plieninger, Tobias, et al. 2015. "Wood-Pastures of Europe: Geographic Coverage, Social-Ecological Values, Conservation Management, and Policy Implications." *Biological Conservation* 190: 70–79. https://doi.org/10.1016/j.biocon.2015.05.014.

Pornaro, Cristina, et al. 2021. "Plant Biodiversity of Mountain Grasslands as Influenced by Dairy Farm Management in the Eastern Alps." *Agriculture, Ecosystems and Environment* 320: 107583. https://doi.org/10.1016/j.agee.2021.107583.

Potts, S. G., et al. 2009. "Enhancing Pollinator Biodiversity in Intensive Grasslands." *Journal of Applied Ecology* 46 (2): 369–379. https://doi.org/10.1111/j.1365-2664.2009.01609.x.

Power, Eileen F., et al. 2013. "Impacts of Organic and Conventional Dairy Farmer Attitude, Behaviour and Knowledge on Farm Biodiversity in Ireland." *Journal for Nature Conservation* 21 (5): 272–278. https://doi.org/10.1016/j.jnc.2013.02.002.

Power, Eileen F. and Stout, Jane C. 2011. "Organic Dairy Farming: Impacts on Insect-Flower Interaction Networks and Pollination." *Journal of Applied Ecology* 48 (3): 561–569. https://doi.org/10.1111/j.1365-2664.2010.01949.x.

Rahmann, Gerold. 2011. "Biodiversity and Organic Farming: What Do We Know?" *Agricukture and Forestry Research* 61: 189–208.

Sala, O. E., et al. 2000. "Global Biodiversity Scenarios for the Year 2100." *Science* 287 (5459): 1770–1774. https://doi.org/10.1126/science.287.5459.1770.

Schader, C., et al. 2014. "Farm- and Product-Level Biodiversity Assessment of Conventional and Organic Dairy Production in Austria." *International Journal of Biodiversity Science, Ecosystem Services and Management* 10 (1): 20–39. https://doi.org/10.1080/21513732.2013.878752.

Smil, Vaclav. 2011. "Harvesting the Biosphere: The Human Impact." *Population and Development Review* 37 (4): 613–636.

Stein-Bachinger, Karin, et al. 2021. "To What Extent Does Organic Farming Promote Species Richness and Abundance in Temperate Climates? A Review." *Organic Agriculture* 11 (1): 1–12. https://doi.org/10.1007/s13165-020-00279-2.

Tscharntke, Teja, et al. 2021. "Beyond Organic Farming – Harnessing Biodiversity-Friendly Landscapes." *Trends in Ecology and Evolution* 36 (10): 919–930. https://doi.org/10.1016/j.tree.2021.06.010.

Vickery, J. A., et al. 2001. "The Management of Lowland Neutral Grasslands in Britain: Effects of Agricultural Practices on Birds and Their Food Resources." *Journal of Applied Ecology* 38 (3): 647–664. https://doi.org/10.1046/j.1365-2664.2001.00626.x.

Chapter 13

Market structures of organic dairy farming and consumer attitudes towards organic milk processing

Otto Schmid, Research Institute of Organic Agriculture, Switzerland; Hanna Stolz, Research Institute of Organic Agriculture, Switzerland; Karlotta Koch, University of Hohenheim, Germany; and Claudia Meier, Research Institute of Organic Agriculture, Switzerland

1 Introduction

For the past decade, the organic market has grown substantially for both plant and animal products. Also, the organic milk supply has grown substantially in many European countries since 2007. Furthermore, organic milk production and products are part of increasingly complex food system structures. Large retail chains play an important role in most countries where the demand for organic dairy products has increased. However, since 2021, market structures and demand for milk and dairy products have been changing. While the established retail chains still dominate, discount retailers are now gaining market share. Traditional organic and health food stores are facing falling revenues and business closures. In some countries, the demand for organic milk and dairy products is also declining or stagnating. One reason for this is the growing market share of a variety of cheaper labels for dairy products that compete with organic (e.g. labels for regional production, animal welfare, grass-fed-milk, etc.).

http://dx.doi.org/10.19103/AS.2024.0138.14

To find appropriate solutions to these challenges for the organic market, the authors analysed in more detail the role and the potentials of different market channels for organic dairy products in selected European countries with a particular focus on Germany and Switzerland. The regulatory framework for production and processing set by the EU organic food and farming regulation as well as private organic standards were also taken into account in this analysis.

A better understanding of consumer perceptions and their purchasing decisions is particularly important. Since little is known about consumer acceptance of processing technologies, which have different impacts on food quality, the results of recent surveys conducted within the ERA-NET-funded project ProOrg (Developing a Code of Practice for organic food processing) are of particular interest for market development and selection of organic dairy processing technologies. The results may help market actors to adapt their strategies and communications targeting both regular and irregular organic consumers.

2 Regulatory framework for organic milk production and processing

In the European Union (EU), organic food production and processing is regulated by the European Council Regulation (EC) No. 2018/848 from 2022 on. This organic EU regulation sets higher animal welfare standards than are found on conventional farms. According to a study on the measures of the Common Agriculture Policy (CAP) and instruments promoting animal welfare and reduction of antimicrobials use (EC, 2022), organic farming can be considered a relevant systemic approach to improving animal welfare and reducing antimicrobial use. This is because the practices used in organic livestock production (as defined by the EU Organic Regulation) differ significantly from conventional practices. These organic practices are characterised by a greater amount of space, constant access to the outdoors, daily exercise, emphasis on roughage, no preventive use of antimicrobials, stricter treatment management when antimicrobials are used, and fewer mutilations (for safety reasons and under anaesthesia or analgesia). Specific practices are also established for each species. Organic cattle are granted longer weaning times, more roughage, and specific bedding requirements instead of full slatted floors.

Some private organic standards did set stricter rules than the EU for roughage use by cattle, outdoor access, antibiotic use, or dehorning (e.g. Demeter biodynamic standard: no dehorning).

With regard to processing, the 'EU organic regulation' prohibits the use of many preservatives and additives. The regulation remains at a more general level regarding the use of specific processing technologies or methods. It states that organic foods be processed with care and use biological, mechanical, and physical methods that do not alter the true nature of the product. In addition, the EU rules do not allow the restoration of properties lost during processing and storage. However, the EU regulation No. 2018/848 does not clearly regulate what processing with care means. This gives room for interpretation for labelling organisations, processors, and researchers (Kahl et al., 2014).

Consequently, a wide range of (more or less carefully) processed organic quality products, including convenience and ready-to-eat foods are currently available on the marketplace. The question is whether all these products comply with the principle of 'care' and, whether they are perceived as carefully processed from the consumers' point of view. However, studies on consumer understanding and expectations of more or less 'carefully' or 'minimally' processed foods were scarce or even non-existent for the organic consumer segment. Therefore, it was previously not clear whether and what role 'careful' processing played in the food choices of organic consumers. Some recent consumer studies conducted within the ERA-NET-funded project ProOrg (2018–2021) helped to close this knowledge gap (see Section 4).

3 Market development and structures of organic dairy products in Europe

A recent EU report on 'Organic Agriculture in the EU – A Decade of Organic Growth' shows significant growth in the organic sector between 2010 and 2020 (EC, 2023). With 14.8 million ha in 2020, the organically farmed area represented 9.1% of the total agricultural area in the EU. Permanent grassland accounted for the largest share (42%), followed by green fodder (17%), cereals (16%), and permanent crops such as fruits, olives, and vines (11%). Despite significant growth, organic livestock production in the EU accounts for only a small share of total EU livestock production, ranging from 1% to 7%, depending on the livestock sector. There are significant differences between countries, animal species, and product groups (EC, 2023).

Although statistics on the number of animals, supply of products, and market shares are not always complete and up-to-date in all EU countries, the analysis of available studies nevertheless demonstrates interesting developments and changes in structures and trends in the organic dairy sector compared to other animal husbandry sectors such as sheep/goats, pigs, and poultry.

It is estimated that in 2020, 6.0% of cattle and 7.2% (2019) of sheep and goat farms in EU countries were organic. The organic share was estimated to be 3.6% for poultry and 1.0% for pigs. Nevertheless, organic pig and poultry production, albeit from a lower base, show higher annual growth rates (9% and 11%, respectively) than the rest of livestock production (EC, 2023), as shown in *Fig.* 1.

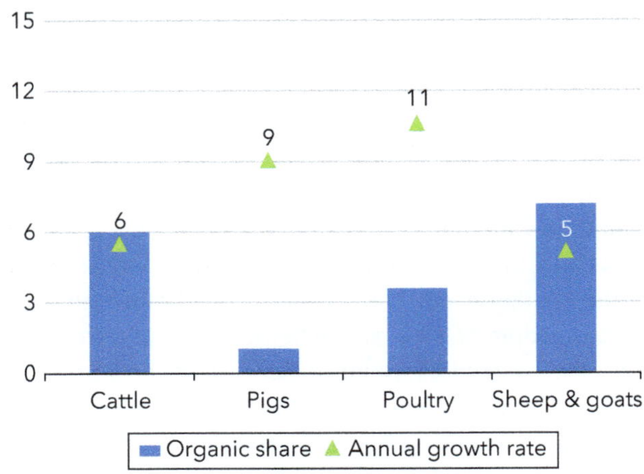

Figure 1 Organic animal production, share in 2020 and annual growth rate 2014–2020, by animal production category (%). Source: Adapted from: DG AGRI calculation based on Eurostat data 2020. DG AGRI estimate for poultry and goats. The share of sheep and goats refers to 2019 (EC, 2023).

3.1 Development of supply of organic milk and milk products in European countries

Unlike other production-related indicators, there is well-established data for the development of organic cow's milk production in Europe. The yearbook *The World of Organic Agriculture* (Willer et al., 2021), published by FiBL & IFOAM – Organics International, reports that the production of organic cow's milk almost doubled between 2007 and 2019 (see *Fig.* 2). In 2019, 6.04 million tons of milk were produced organically in all EU countries, representing 3.4% of the European Union's dairy cow production.

The production of organic milk varies from country to country, with western European countries showing significant growth between 2014 and 2020. Countries with high volumes are Germany and France; their market share in volume (tons) was 4–5% in 2020 (see *Fig.* 3).

One reason for the growth in organic cow milk production is a higher producer price that helps cover lower milk yields. Recent farm accountancy data network (FADN) data on dairy farms in selected EU countries show a premium on the producer price of more than 20% for organic milk, after falling in 2017 following an increase in conventional milk prices. The exceptions are Poland and Latvia, where there is no premium on organic milk (see *Fig.* 4). Based on FADN data for those farms specialising in milk production, milk yields on organic farms are 8–33% lower than on conventional farms (EC, 2023).

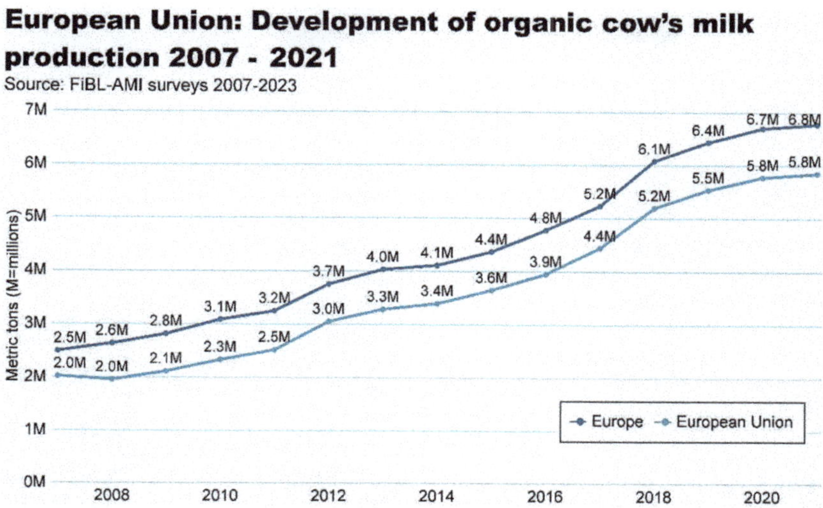

Figure 2 Europe and European Union: development of organic cow's milk production 2007–2021. Source: Adapted from: FiBL Survey 2009–2021; Willer et al. (2021).

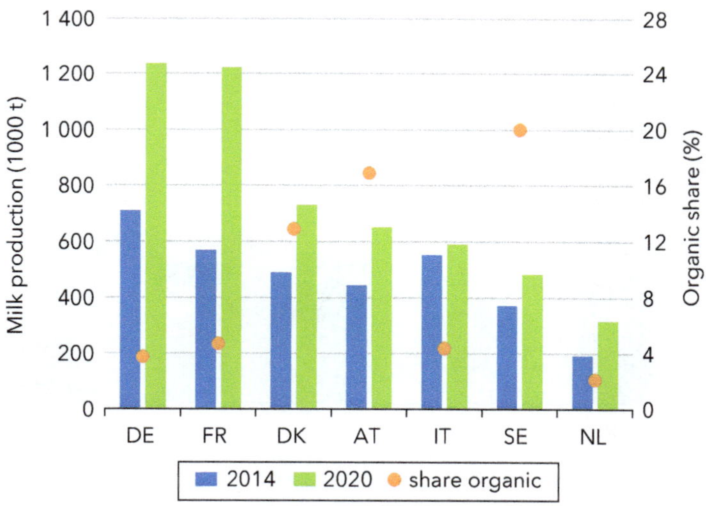

Figure 3 Production of organic milk, volume 2014 and 2020 (thousand tonnes) and share 2020 (%), of main producing countries. Source: Adapted from: DG AGRI calculation based on Eurostat data, provisional data for Italy (EC, 2023).

Drinking milk accounted for more than half of the organic milk production volume in several countries in 2020, followed by cheese and butter (see *Fig.* 5). Recently, the volume of drinking milk has been in decline due to the growing market of plant-based milk substitutes e.g. in Germany (BÖLW report, 2023).

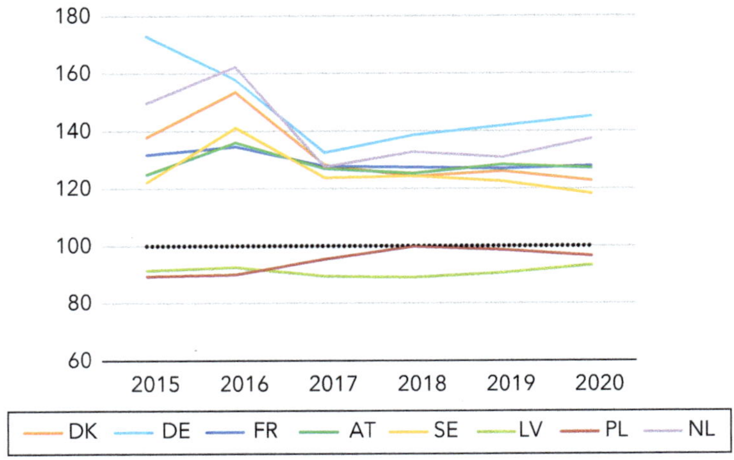

Figure 4 Premium on producer price for organic milk (conventional = 100) 2015–2020 in selected countries. Source: Adapted from: DG AGRI calculation based on EU FADN data; 2020 data are preliminary (EC, 2023). The farm accountancy data network (FADN) of the European Union monitors farms' income and business activities.

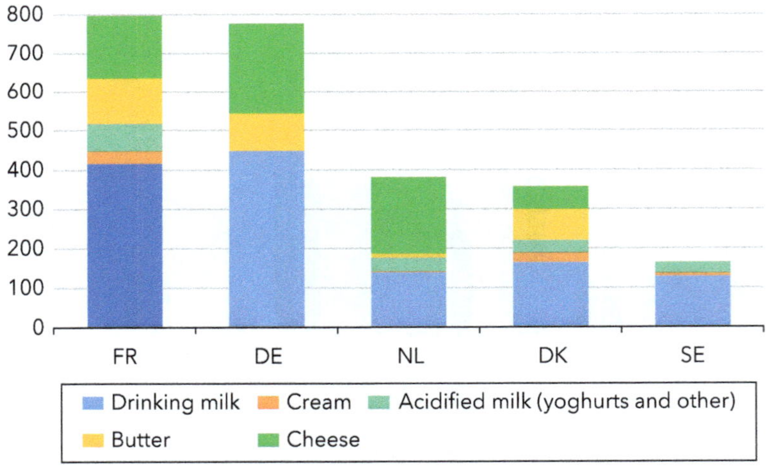

Figure 5 Production of organic dairy products, volume (thousand tonnes milk equivalent) in 2020, of main producing countries. Source: Adapted from: DG AGRI calculation based on Eurostat data. Data are not available for all countries (EC, 2023).

3.2 Development of market shares of organic milk products in selected countries

Over the past several years, FiBL (research institute of organic agriculture), Switzerland, and AMI (Agrarmarkt Informations-Gesellschaft mbH), Germany, have conducted surveys and published them in the 'World of organic agriculture' reports (Willer et al., 2020, 2021, 2022a, 2023), which analyse market shares based on sales values in Euros for selected products. The figures indicate significant differences between the various European countries, with some experiencing low growth and others with no or negative growth (see Table 1).

The organic market share of all milk and dairy products was high in 2020: Austria with 13.1% (relatively stagnant), Germany with 12.3%, and Switzerland with 11.4% (stagnant). The other countries had an organic market share of around 4% of retail sales.

The organic drinking milk market share by value for 2020 was 34.3% in Denmark, 26.0% in Switzerland, 21.2% in Austria, and 16.4% in Germany. Other countries had a market share of about 4–5% (Belgium, Finland, and the UK).

The market share of organic cheese was lower than that of milk in 2020. Countries with high shares were Austria with 11.0% (numbers are stagnant), Denmark with 8.5%, Switzerland with 7.9%, and Germany with 6.0%. Other countries with significant local non-organic cheese production have relatively low shares of organic cheese, such as France with 2.9% and Italy with 1.0% in 2019.

Organic yoghurt had a high retail market share in 2020 in Denmark (34.3%) and Austria (25.3%; 26.2% in 2021). Other countries showed more stagnation, e.g. Germany (10.7%) and the UK (8.2%) in 2020, or no data is available.

Furthermore, the market share of organic butter in 2020 was high in Denmark at 17.8% and in Austria at 11.8%, but much lower in many other countries.

An interesting comparison can be made between the organic milk sector and the leading livestock sector, which is that of organic eggs. In some countries such as Austria, Denmark, Finland, Germany, Sweden, and Switzerland, organic eggs reached a market share of 20–30% of retail sales in 2020/2021. One reason for this relatively larger market share for organic eggs could be the high value placed on animal welfare in these countries (Stolz, 2009; BLE, 2020; EC, 2022). Another reason might be that in some smaller retail outlets, consumers only have the choice between conventional and organic eggs, versus an often larger array of choices competing for consumer attention in the dairy section. This indicates that other factors, such as the retailer's product range and communication policy may play a role.

Another interesting example is the organic meat sector, which, when compared to milk, commands a relatively small share of retail sales (Willer

Table 1 Organic shares in % for retail sales values (Euros) for selected products 2018–2022

	Austria	Denmark (2020)	Finland	France (2019)	Germany (2020)	Italy (2019)	Netherlands	Sweden	Switzerland	UK (2020)
- Milk dairy	14.1			5.8	12.3	3.6	4.0	(10.4)	11.4	3.5
products	13.1			4.4	8.5	-	4.0		11.3	3.5
2021	12.4				-		-		11.0	-
2020	-						5.6		-	
2019										
2018 (2017)										
- Butter	12.1	17.8		7.4	7.5	2.8				
2021	11.8	16.8		5.6	6.9	-				
2020	11.2	(16.6)			4.5					
2019	10.8				4.5					
2018 (2017)										
- Cheese	11.1	8.5	2.0	2.6	6.1	1.0		1.9	8.4	1.1
2021	11.0	6.8	2.0	1.6	6.0	-		-	7.9	-
2020	10.3	(5.0)	-		4.7			-	7.6	-
2019	10.2		-		4.7			-	-	
2018 (2017)										
- Milk	30.2	34.3	4.0	15.7	17.1	8.1		14.1	26.8	5.1
2021	21.9	32.3	4.5	12.7	16.4	-		-	26.0	1.1
2020	21.8	-	4.5		-			-	24.9	1.1
2019	23.2		4.0		-			-	-	
2018										
- Yoghurt	26.3		-	9.1	10.4	6.1				8.2
2021	25.3		-	6.9	10.7	-				7.8
2020	23.7		-		12.4					8.2
2019	21.9		2.0		12.1					
2018										

Note: Due to classifications and nomenclatures differing from country to country, it is not possible to supply data for all product groups, even if data for individual products may be available. Not all countries have data on the market shares of organic products or not every year and category, thus the empty spaces in the table.

et al., 2020, 2021, 2022a, 2023). This is despite the fact that, when producing milk, only a small subset of male calves are used for breeding, while the rest are slaughtered or raised for beef production (Haskell, 2020). Therefore, the supply of meat animals from organic dairy farms is clearly available. Although detailed data on organic meat was difficult to obtain from national authorities and market actors, one reason for the fact that the availability of meat animals from organic farms does not translate directly into sales of organic meat are the results of several studies which show that organic dairy farmers often have difficulties marketing their animals or beef as organic. Therefore, although it is being produced organically, the meat is often sold as conventional (Reiber et al., 2020). Another major factor affecting the organic meat sector could be the relatively high price difference between organic and non-organic meat products, and competition with other labels could also play a role (Stolz, 2019).

3.3 Comparison of market sales channels relevant for organic dairy products in selected countries

With the growth of the organic market, the marketing channels are also changing. *Figure* 6 shows that in many countries, the major retail chains are the most important distribution channel with a share of 60–80% (except Denmark with almost 100%) in 2020 (Willer et al., 2023). In some countries, specialty retailers still play an important role for organic products (Germany, Italy, and the UK). However, they are under economic pressure as their organic products are generally priced higher than supermarket chains or discount retailers (see *Fig.* 6).

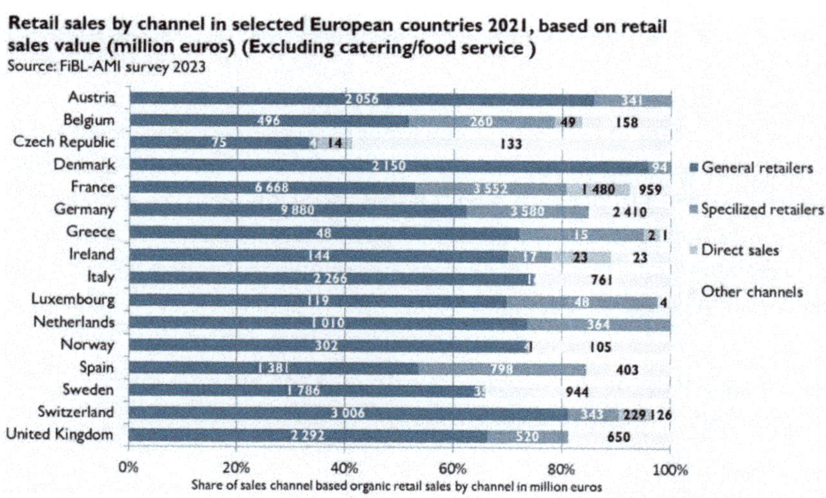

Figure 6 Retail sales by channel for organic products in selected European countries in 2021. Source: Adapted from: FiBL-AMI survey 2023 based on national data sources and Willer et al. (2023).

The growth rate of the market for organic products changed between 2020 and 2022. During the pandemic lockdown, many consumers spent more money on organic products. However, after the end of the lockdown, consumers spent significantly less on consumption at home (Willer et al., 2022b). In most European countries, the growth rate for organic products declined after 2020. While general retail remained at 2019 levels in several countries, many specialised traditional organic/natural food stores, as well as direct farm marketing, had a decline in sales from 2021 on (Soil Association, 2022; Agence Bio France, 2022; BOELW, 2023). However, in some countries, such as Germany, discount retailers had more sales of organic products but also of non-organic products (BOELW, 2023).

Different marketing systems were compared in a study conducted by AMI & Ecozept (Rampold et al., 2022) on alternatives in milk marketing in the German state of Saxony in 2021. Market actors (n = 65) estimated the market potential for the next 3–5 years from 2023 on in a Delphi survey. The results are summarised in Table 2 and were supplemented by the authors with examples from Switzerland and Austria based on expert surveys.

Although the focus in the table was on German-speaking countries, there are comparable developments in other countries. The importance and size of large retail chains and specialised independent retailers might be different in France (Agence Bio, 2022) or in UK (Soil Association, 2022).

Many creative and innovative ways of alternative marketing have been developed by farmers in other countries to gain more independence from supermarkets and to farm more sustainably. These strategies often combine different marketing systems. For example, in Scotland, a farm with a smaller cow herd (50) of a local breed started direct marketing their milk without single-use milk bottles in 2019. The farmer substituted glass bottles and reusable plastic for the single-use bottles and was able to sell the milk in schools via vending machines. The farmer then took a step further and started to replace all plastic components with metal, thus making the system plastic and single-use plastic free. He also tried to get away from plastics for silage. With other outlets, the farmer now supports other local small farms, and other councils are looking to copy the model. This can be seen as an example of 'completely alternative markets' (The Guardian, 2019).

3.4 Comparison of quality labels relevant for organic dairy products in selected countries

Consumers and market actors have been confronted for several years with an increasing number of quality labels/initiatives that do not apply to organic food and farming. Some of these labels compete with organic products, while others can be combined with organic labelling, as shown by a recent study

Table 2 Milk marketing systems for organic and other labelled milk and milk products in Germany (DE), Austria (AT), and Switzerland (CH)

Distribution systems	Description	Characteristics	Coverage relevance	Potential expert-view[a]
Large supermarket chains	Via large processors & wholesalers to food retailers	• Large quantities • Price advantage for consumers • Standardisation	Dominant sales channel also for organic products	Slower growth potential for yogurt, milk, and cheese
Organic supermarkets	Only organic products	• Wide choice • Moderate prices for consumers	Especially in large cities	Slower growth; central locality
Natural food and organic shops	Only organic products	• Customer advice • Cost pressure	Widespread for organic	Declining, with exceptions
Discount retailers	Limited range at low prices	• Lower prices for consumers • Limited offer	DE: Approx. 10-15 % of organic market share	Stronger growth with organic and label products
Specialised shops (often linked with dairies)	Often with artisanal processing	• Authentic products • Specialities • High staff costs	Widespread depending on region	Expandable with regional and/or with organics
Gastronomy and public catering/procurement	Different systems, mainly cities	• Complex distribution	Different share of organic, partly regional	Potential for gastronomy & municipalities
Farm shops, often with own farm processing	Different systems and offers	• Authentic • Lower consumer prices • Personnel-intensive	Market share of 5–10% Increase during pandemic 2020	Decline in sales after lockdown; partly regaining market share
Milk-vending machine	Vending machines on farms or shops	• Labour saving for farmers • Investment costs high	Mostly on farm, with direct marketing	Slight expansion, also with other products
School milk projects	Often local projects	• Subsidies needed • Costly	Mostly with regional milk, partly organic	Limited, good distributor needed
Community supported farms (solidarity farming)	Close farmer–consumer relation (participation)	• Risk shared • Local distribution • Personnel-intensive	Few farms with milk	Expandable but limited
Consumer cooperatives	Non-profit shops	• Lower prices for members • high engagement	Few, especially in larger cities	Expandable; interesting for dairy products
E-commerce with farms or other actors	Varies according to farm	• Customer oriented • Logistics costly	Increasing importance during lockdown	Expandable for organic and other labels

[a]Own expert consultations.
Source: Rampold et al. (2022).

in Saxony (Germany) on alternative milk marketing (Rampold et al., 2022). Several labelling programmes in Germany were described and their potential was evaluated. The following summarises the results, which are generalised for other German-speaking countries (Austria and Switzerland), based on expert knowledge and studies (AMA, 2022; Stolz, 2019; Richter et al., 2004).

1 Different labelling of organic products based on EU regulation alone or with private labels.

In the EU, the use of the organic EU label with the leaf-shaped stars is mandatory on all products. This label is well established and controlled and known to many consumers.

In some countries, private label organisations have developed their own standards, some of which have stricter requirements for dairy products than the EU regulations, especially for feeding. For example, in the Bio Suisse standards, only a maximum of 5% concentrated feed (excluding mill by-products) can be used as a supplement in the daily ratio of ruminants (Bio Suisse, 2023). In the last year, the market for such organic milk products with private labels has grown. However, the extent to which further growth is possible depends on several factors, such as the commitment of actors in the supply chain, cooperation, communication, competition from other labels, consumer perception, and their willingness to pay the often higher prices for organic milk products with private labels, which is also influenced by the general economic situation.

In Switzerland, two small private initiatives have started to market milk from cows that have not been treated with antibiotics and are labelled as 'antibiotic-free organic milk'. This type of initiative provides some limited potential for exporting milk powder to the US, because USDA organic regulations do not allow milk from cows treated with antibiotics to be marketed, whereas the EU allows the marketing of milk from treated cows as organic as long as a double lock-in period (or withdrawal period) after treatment is observed for the sale of such milk.

2 Different non-organic labels for milk production with strong use of hay, pasture, or grass.

There are various labels used to indicate the kind of feeding/pasturing. The term 'hay milk' is protected by the EU for milk from cows not fed silage of fodder. The use of this label is very limited (in Germany, it represented only 1% of all cow milk in 2020). The private label 'pasture milk' is used more often (in Germany, it accounted for ca. 4% of total milk volume in 2020). In Austria, farmer groups also use the label 'pasture milk'. The requirement for this label is a minimum number of days grazing on pasture (e.g. 130 days/year). In Switzerland, the label 'grass milk' is used by the Swiss Integrated farmer association (IP-Suisse). There is potential for further growth of these labels, which are combined with the organic label(s) or with regional labels by some initiatives (Rampold et al., 2022).

3 Regional origin labelling for milk and cheese products (partly with organic).

The market potential of origin-labelled milk and milk products with a regional label has been growing in several regions. This was more widespread in regions where regional labels have been strongly promoted by a regional state programme. There is potential for expansion, with or without organic labels, as consumer studies indicate a growing interest in regional products (Rampold et al., 2022).

In mountain areas, the term 'Mountain milk' is sometimes used, mainly in Austria and Switzerland by small farmer groups. There is a limited potential for further growth.

4 Other indications and labels for animal welfare and fair prices in dairy products.

Animal welfare labels have been launched by some private animal welfare associations in Germany (with several levels/grades) and Switzerland and are used by some retail chains. There might be limited potential for growth for non-organic dairy and meat products.

Several small initiatives in Germany, Austria, and Switzerland use a 'fair milk' label to generate a higher price for milk. Until now, the market potential has been limited.

In Germany, the term 'GMO-free milk', which means no fodder from GMO plants, is a standard for conventional milk, which is widespread (>70% of all cow milk), although it is of little market relevance (Rampold et al., 2022).

Many political and socio-economic factors will play a role in how different market actors can and will exploit these potentials and change their strategy in different regional and national contexts and frameworks.

Additionally, there are significant differences in the way milk and dairy products can be produced and processed. For example, consumers may purchase different kinds of drinking milk such as raw milk, pasteurised milk, UHT (ultra high temperature) milk, ESL (extended shelf life) milk, milk processed under high pressure, non-homogenised milk, lactose-free milk, etc. Packaging can also be very different, with less or more environmental impact. Consumer information and visibility is a key, but that will not be described further in this text.

4 Consumer attitudes

4.1 Attitudes towards organic food

In order to develop the market for organic dairy products, it is important to understand the attitudes and expectations of consumers towards organic food in general. Various consumer studies have been published on this topic, particularly in Europe. A literature review in the ProOrg project showed that the main motivators for consumers to buy organic food were health, taste, environmental protection, and animal well-being (Meier, Stolz and Koch, 2021).

Consumers expect that organic products are healthier than conventional ones because of the reduction or non-use of chemical substances and food additives during production and processing (Zanoli & Naspetti, 2002; Hughner et al., 2007; Naspetti and Zanoli, 2009; Stolz et al., 2009; Schäufele and Janssen, 2021). In addition, consumers assume that organic products contain more nutrients than their conventional counterparts (Hill and Lynchehaun, 2002; Lea and Worsley, 2005; Lee et al., 2013; Jensen et al., 2013; Kahl et al., 2012; von Meyer-Höfer et al., 2015; Agence Bio, 2022).

Many consumers believe that organic food tastes better than conventional food (Stobbelaar et al., 2007; Lee et al., 2013; Agence Bio, 2022). A positive label effect was observed in sensory evaluation studies (Hemmerling et al., 2013) although a better taste is not always achieved by organic production for several reasons, e.g. variety, processing of food, etc. Organic products are also perceived to be more environmentally friendly, mainly because no synthetic pesticides are used on organic farms (Agence Bio, 2022). Consumers also assume that organic farming supports more small-scale agriculture and local rural communities (Hamzaoui Essoussi and Zahaf, 2009). Furthermore, people who are concerned about animal welfare are more likely to consume organic food (Honkanen et al., 2006). This was found to be particularly important for consumers of organic food in Germany and Switzerland (Stolz, 2019; BLE, 2020). Of course, there are some differences regarding consumer attitudes towards organic products both between countries (due to cultural differences) and consumer groups (regular or occasional organic consumers).

Looking at organic milk, in particular, Schiano et al., 2022, revealed that parents consider organic milk to be safer, healthier, and more pure as compared to conventional milk since it is hormone free and more animal friendly in its production.

4.2 Consumer milk-purchasing criteria in Germany and Switzerland

Consumers are buying more processed organic food since the supply has continuously increased during the past decade. This also holds for milk and dairy products in Germany and Switzerland (Kretzschmar and Schmid, 2011; Stolz et al., 2019; Schaack, 2019, 2020; BOELW, 2022, 2023; BLW, 2022).

In order to better understand the milk-purchasing criteria and the consumer attitudes with regard to the processing of milk, a quantitative online survey combined with a consumer choice experiment was conducted in the ProOrg project. In total, data from 600 consumers in Germany and 687 consumers in Switzerland were collected (Meier, Stolz and Koch, 2021). Initially, consumers were asked to rank the six most important purchase criteria for milk out of a total of 12 criteria. In both countries, fat content, taste, freshness, price, shelf life, and region of origin were among the six most selected criteria for milk purchase. Although the production system (organic/conventional) was the first priority for a considerable number of respondents, it was of lower importance in the overall ranking. In contrast, processing was of relatively low importance to respondents although some of the most selected criteria (freshness, shelf-life, taste, and fat content in Germany) are highly influenced by the processing method. See *Fig.* 7.

4.3 Acceptance of new processing technologies by German and Swiss consumers of organic milk

The consumer survey in the ProOrg project (Meier et al., 2021) showed that generally consumers know very little about processing and that they are first

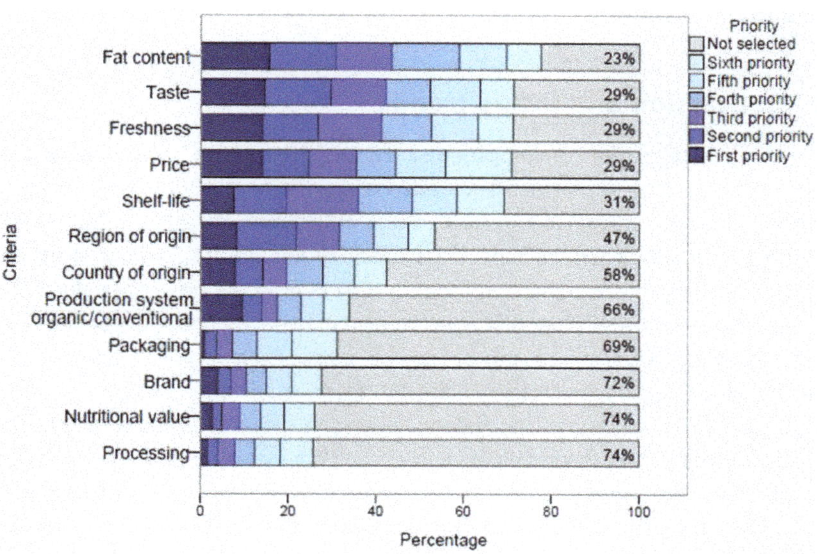

Figure 7 Criteria German and Swiss respondents considered as most important when buying milk. More information on differences between the two countries is provided in the project report (Meier et al., 2021). Source: Adapted from: Meier et al. (2021), ProOrg Report on the outcome of the quantitative consumer survey (D.6.2).

sceptical towards new processing methods such as high-pressure processing (HPP). HPP is an increasingly used innovative technology which has the same goal as pasteurisation, which is to remove potentially harmful microorganisms from milk and other products. However, unlike pasteurisation which uses high temperatures to eliminate these organisms and can therefore affect the taste and quality of milk, HPP uses high pressure to do the job. Because the milk is not subjected to heat, the flavour, nutrient profile, colour, and smell remain fully intact. The milk can be kept unopened and cooled for about 21 days. The interest in the applications of HPP on milk and dairy products has increased in the last decades as an alternative to pasteurisation (Sanchez et al., 2020).

The acceptance of different milk processing technologies for organic milk was tested with a consumer choice experiment in Germany and Switzerland. In this experiment, some consumers were provided with additional information on the impact of different processing methods. The choice experiment showed that the additionally informed organic consumers preferred HPP milk over pasteurised milk, while the acceptance was lowest for micro-filtrated milk. In addition, organic consumers significantly preferred regional milk over other consumers, and it was found that organic consumers in Germany consume significantly less UHT milk than other types (Meier et al., 2021).

The survey also showed that the concepts of careful processing and organic processing seem to match very well with each other. For both concepts 'the maintenance of nutrients', 'the maintenance of naturalness', and 'no/low-product exposure' are important components/aspects.

5 Conclusion and future trends

The analysis of market developments, market shares, and sales channels for organic milk and dairy products showed steady growth until 2020, but at different levels in different countries. Since 2021, the market structure has changed. The main sales channels are still the large retail chains. While traditional organic stores are losing market share, discount retailers are gaining. Other alternative market channels are still insignificant although they have growth potential. There is also growing competition with other labels, especially regional labels, animal welfare labels, and hay milk labels although these are sometimes combined with organic. In order to respond to these challenges, efficient distribution systems, good cooperation between market actors, and convincing consumer information will be very important to maintain or gain market share for organic products.

Understanding consumer attitudes towards organic food is very important to develop the market for organic dairy products. Better and more targeted communication on the added value of organic milk and dairy is important as well as providing a product that fulfils the high expectations of consumers regarding health, nutrient content, and taste (Kozup et al., 2003).

Consumers not only need to know more about organic farming and food processing but also need to be able to experience the organic products themselves (e.g. with tasting, and/or with stories).

When new processing technologies for milk are taken up in the organic EU Regulation and in private organic standards, it is crucial to raise consumers' awareness about the impacts and potential benefits of different processing technologies in order to help them shape their preferences.

The concept of careful processing offers organic producers and processors the possibility to further differentiate their food products from others on offer. However, it is important to communicate to consumers, what 'carefully processed' means practically. The benefits, such as the preservation of a food's natural taste and nutritional content, should be highlighted (Meier et al., 2021, Kilic et al., 2021; Hüppe & Zander, 2021).

In this section the authors look – more from their personal experiences – at trends, challenges, and possible perspectives for the organic dairy sector in production, processing, marketing, and consumption.

As a result of the climate change debate, the role of animals in sustainable agriculture is more and more discussed and questioned also in the organic sector and the organic research community. This has been explored well by Vaarst et al., 2020, in a conference paper for the Organic Animal Husbandry Alliance (IAHA), summarising results from several Core Organic EU research projects on the subject. A key question is if we as society can still justify so much milk and meat consumption from ruminants, largely because of greenhouse gas emissions (in particular methane). This question is raised not only by the vegan movement. Most likely the change to less farm animals is inevitable although it will not be easy.

If a change is to be done, it could be done regionally in different ways. There are large areas in the world with meadows and pasture, where animal husbandry can transform grassland into valuable protein. Therefore, in areas suitable for animal husbandry, it will be more about minimising the negative effects by finding the appropriate and site-specific stocking densities, ensuring carbon sequestration with good pasture/meadow management and promoting highly diverse landscapes. Some negative effects can also be offset with soil improvement with methods such as the integration of grass–clover into rotations, which has several benefits including the fixation of nitrogen. Systems-based methods such as agro-silvo-pastoral systems show interesting perspectives for the environment and the animals as well as for people.

However, farmers need to be supported by incentives and policies in this transformation. For example, there is a need for financial compensation for carbon/climate friendly farming with meadows and leys (e.g. with grass–clover and agroforestry systems, etc.). Market actors and processors have an important role to play, as animal slaughter will always play a role in the beef and dairy industries. The farmers and processors can both take steps to become more respectful and animal friendly with proper support, and more regional and local methods of slaughter can be considered (e.g. on-farm slaughter to avoid the stress and costs of transport).

On the other end of the value chain, consumers need to be motivated to eat a more plant-based diet instead of the meat-heavy diet that has become the norm in many countries.

Another aspect of the argument against a meat-heavy diet is the amount of feed versus food produced. Should farmers produce more food for humans instead of feed for animals? From an ethical perspective, clearly yes. In addition to reducing the number of farm animals raised for milk and meat, the amount of crops grown for feed need to be reduced.

Research has shown that farmers can reduce the use of concentrates step-wise without economic loss, if breeding lines are more appropriate and feeding systems are improved. As an example, Bio Suisse organic standards have allowed a maximum of 5% concentrates in ruminant diets since 2022. This represents a steady process of reduction over 10 years and proves that organic farmers can show the way! In addition to good research, support from political policies can be helpful to promote more roughage-based systems. We expect also a growing importance of special consumer labels such as hay milk or pasture-beef in combination with organic as already shown earlier.

Another remaining challenge in the future will be how to drastically reduce the use of too many antibiotics for farm animals, which remains an issue in organic farms (although they are not allowed to use antibiotics preventively). There will be more label programmes and consumers asking for antibiotic-free milk. However, the farmers are often left without suitable options if they have to treat animals, and proper pathways are not always available to market-treated animals and their products. Solutions have to be found. A much broader preventive approach is needed to ensure the well-being of animals and to reduce the use of antimicrobials.

Also an important challenge is to further improve animal welfare, in particular for calves, e.g. with other stable systems allowing longer cow–calf contact. This will also help to reduce veterinary treatments. Such developments might be promoted by special labels (see https://cowpassion.ch/cow-calf-international) and higher prices for calf–cow friendly systems because so far consumers are not aware of the issue (Herrler et al., 2023). Newer results from research projects and farmer experimentations with enhanced cow–calf contact can be helpful for interested farmers (Vaarst et al., 2020).

With regard to market structures, a current trend that can be observed in many European countries is the displacement of small and local dairies by big enterprises that apply more industrial and standardised processing technologies. This trend, however, is not only observed in dairy but also in other food sectors. Accordingly, the EU has recently started to reverse this process again by supporting research and development to foster shorter supply chains that benefit local, small producers in order to reduce the dependence on multinational food enterprises and to support value creation in rural areas.

Additionally, an increasing dairy product diversification can be observed in the food market that also applies to the organic dairy sector: dairy product ranges are expanded with products whose recipes originate from abroad, also offering new perspectives for small organic dairies to produce for emerging organic niche markets.

Finally, the current vegan trend that promotes the replacement of milk and dairy products with plant-based protein products (e.g. oat-based milk replacement) has an impact on the organic milk and dairy sector. The organic sector could even be more affected by vegan replacements than the non-organic sector since environmental and climate protection, apart from ethical reasons, are the major consumer motives for choosing vegan products, while at the same time being particularly relevant motives among organic consumers and for organic food choices. It is expected that the current market growth of vegan milk and dairy replacement products will continue in the future. The offer of more vegan products in organic quality will be necessary, but with much less additives than conventional ones which are often highly processed. Dairy producers will have to be creative to maintain relevance with those consumers in this sector.

What we need is a transformation of the whole dairy sector to be more climate neutral and animal friendly. Organic farming should maintain the lead in this process and continue to embrace and educate the public about the benefits of an organic dairy sector lead by the basic ethical principles of the organic movement as explained by IFOAM – health, ecology, fairness, and care.

6 Acknowledgements

The consumer study on organic milk processing was financed by the ERA-net CORE Organic-funded project ProOrg ('Code of practice for organic food processing') based on funds from participating countries and the EU. Market data was provided by Dr Helga Willer from FiBL Research Institute of Organic Agriculture, Switzerland, and DG Agriculture calculations based on Eurostat data (EU Brussels).

7 Where to look for further information

- In the EU project ProOrg several publications (scientific papers and reports) have been published on the processing of organic food. Website: https://www.proorgproject.com/ (accessed 16 January 2023).

- More information about the global market development are found in: The World of Organic Agriculture 2023. https://www.organic-world.net/yearbook/yearbook-2023.html (accessed 16 January 2023).
- Global organic market: Interactive online database with global and European data: https://statistics.fibl.org/data.html
- IFOAM Animal Husbandry Alliance: several proceedings from IAHA conferences (2017, 2020, 2021). https://www.ifoam.bio/about-us/our-network/sector-platforms/ifoam-international-animal-husbandry-alliance

8 References

Agence Bio (2022). Baromètre de consommation et perception des produits biologiques en France. Etude n°2100912 du janvier 2022. Available at: https://www.agencebio.org/wp-content/uploads/2022/03/Barometre-de-consommation-et-de-perception-des-produits-bio-Edition-2022_VF.pdf (accessed 27 February 2023).

AMA (2022). *Nachfrage nach Bioprodukten Bleibt Im Jahr 2021 Ungebrochen. AgrarMarkt Austria. Wien*. Press release. Available at: https://www.ama.at/marktinformationen/milch-und-milchprodukte/aktuelle-informationen/2022/nachfrage-nach-bioprodukten-bleibt-im-jahr-2021-ungebrochen (accessed 27 February 2023).

Bio Suisse (2023). *Standards for the Production, Processing, and Trade of "Bud" Products. Bio Suisse Association*. Basel. Available at: https://international.bio-suisse.ch/dam/jcr:47dc9f0b-3dae-4936-aed8-2f844ab88497/Bio%20Suisse%20Standards%202023%20-%20EN.pdf (assessed 27 February 2023)

Bundesamt für Landwirtschaft (BLW) (2022). *Bio-Konsum – Nachfragetrends und Hintergründe. Fachbereich Marktanalysen*. Available at: https://biobarometer.fibl.org/ (accessed 30 January 2023).

Bund Ökologische Lebensmittelwirtschaft e.V. (BÖLW) (2022). Branchen Report. *Ökologische Lebensmittelwirtschaft*. Berlin (vol. 2022). Available at: https://www.boelw.de/fileadmin/user_upload/Dokumente/Zahlen_und_Fakten/Broschuere_2022/BOELW_Branchenreport2022.pdf (accessed 27 February 2023).

Bund Ökologische Lebensmittelwirtschaft e.V. (BÖELW) (2023). Branchen Report. *Ökologische Lebensmittelwirtschaft*. Berlin (vol. 2023). Available at: https://www.boelw.de/fileadmin/user_upload/Dokumente/Zahlen_und_Fakten/Broschuere_2023/BOELW_Branchenreport2023.pdf (accessed 27 February 2023).

Bundesanstalt für Landwirtschaft und Ernährung (BLE), Geschäftsstelle Bundesprogramm Ökologischer Landbau und andere Formen nachhaltiger Landwirtschaft (BÖLN) (Ed.) (2020). Ökobarometer 2019–Umfrage zum Konsum von Biolebensmitteln.

Council Regulation (EU) 2018/848 of the European Parliament and of the Council of 30 May 2018 on organic production and labelling of organic products and repealing Council Regulation (EC) No 834/2007, Pub. L. No. 32018R0848, 150 OJ L 1 (2018). Available at: https://eur-lex.europa.eu/legal-content/EN/ALL/?uri=CELEX%3A32018R0848 (accessed 27 February 2023).

European Commission (2023). Organic farming in the EU. A decade of organic growth. Agricultural market brief. No. 20. Brussels. Available at: https://agriculture.ec.europa.eu/system/files/2023-01/agri-market-brief-20-organic-farming-eu_en_1.pdf (acessed 30 January 2023).

European Commission. Directorate-general for agriculture and Rural Development (2022). *Study on CAP Measures and Instruments Promoting Animal Welfare and Reduction of Antimicrobials Use*. Brussels. Available at: https://data.europa.eu/doi/10.2762/122586.

Hamzaoui Essoussi, L. and Zahaf, M. (2009). Exploring the decision-making process of Canadian organic food consumers: motivations and trust issues. *Qualitative Market Research* 12(4), 443–459. https://doi.org/10.1108/13522750910993347.

Haskell. (2020). What to do with surplus dairy calves? Welfare, economic, and ethical considerations. Available at: https://d-nb.info/121326569X/34 (assessed 6 July 2023).

Hemmerling, S., Asioli, D. and Spiller, A. (2016). Core organic taste: preferences for naturalness-related sensory attributes of organic food among European consumers. *Journal of Food Products Marketing* 22(7), 824–850. https://doi.org/10.1080/10454446.2015.1121428.

Herrler, M., Chagunda, M. G. G. and Stroebele-Benschop, N. (2023). Public awareness, attitude and empathy regarding the management of surplus dairy calves. *Journal of Agricultural and Environmental Ethics* 36(2), 11. https://doi.org/10.1007/s10806-023-09905-x. Available at: https://link.springer.com/article/10.1007/s10806-023-09905-x (assessed 6 July 2023).

Hill, H. and Lynchehaun, F. (2002). Organic milk: attitudes and consumption patterns. *British Food Journal* 104(7), 526–542. https://doi.org/10.1108/00070700210434570.

Honkanen, P., Verplanken, B. and Olsen, S. O. (2006). Ethical values and motives driving organic food choice. *Journal of Consumer Behaviour* 5(5), 420–430. https://doi.org/10.1002/cb.190.

Hughner, R. S., McDonagh, P., Prothero, A., Shultz, C. J. and Stanton, J. (2007). Who are organic food consumers? A compilation and review of why people purchase organic food. *Journal of Consumer Behaviour* 6(2–3), 94–110. https://doi.org/10.1002/cb.210.

Hüppe, R. and Zander, K. (2021). Consumer perspectives on processing technologies for organic food. *Foods* 10(6), 1212. https://doi.org/10.3390/foods10061212.

Jensen, M. M., Jørgensen, H. and Lauridsen, C. (2013). Comparison between conventional and organic agriculture in terms of nutritional quality of food–A critical review. *CABI Reviews* 8(45). https://doi.org/10.1079/PAVSNNR20138045.

Kahl, J., Alborzi, F., Beck, A., Bügel, S., Busscher, N., Geier, U., Matt, D., Meischner, T., Paoletti, F., Pehme, S., Ploeger, A., Rembiałkowska, E., Schmid, O., Strassner, C., Taupier-Letage, B. and Załęcka, A. (2014). Organic food processing: a framework for concept, starting definitions and evaluation: organic food processing. *Journal of the Science of Food and Agriculture* 94(13), 2582–2594. https://doi.org/10.1002/jsfa.6542.

Kahl, J., Baars, T., Bügel, S., Busscher, N., Huber, M., Kusche, D., Rembiałkowska, E., Schmid, O., Seidel, K., Taupier-Letage, B., Velimirov, A. and Załęcka, A. (2012). Organic food quality: a framework for concept, definition and evaluation from the European perspective. *Journal of the Science of Food and Agriculture* 92(14), 2760–2765. https://doi.org/10.1002/jsfa.5640.

Kilic, B., Cubero Dudinskaya, E., Proi, M., Naspetti, S. and Zanoli, R. (2021). Are They Careful enough? Testing consumers' perception of alternative processing methods on the quality of organic food. *Preprints* 2021, 2021060540.

Kozup, J. C., Creyer, E. H. and Burton, S. (2003). Making healthful food choices: the influence of health claims and nutrition information on consumers' evaluations of packaged food products and restaurant menu items. *Journal of Marketing* 67(2), 19–34. https://doi.org/10.1509/jmkg.67.2.19.18608.

Kretzschmar, U. and Schmid, O. (2011). Quality and safety aspects of organic and low-input food processing: results of a Delphi survey from an expert consultation in 13 European countries. *NJAS* 58(3–4), 111–116. https://doi.org/10.1016/j.njas.2011.09.002.

Lea, E. and Worsley, T. (2005). Australians' organic food beliefs, demographics and values. *British Food Journal* 107(11), 855–869. https://doi.org/10.1108/00070700510629797.

Lee, W. J., Shimizu, M., Kniffin, K. M. and Wansink, B. (2013). You taste what you see: Do organic labels bias taste perceptions? *Food Quality and Preference* 29(1), 33–39. https://doi.org/10.1016/j.foodqual.2013.01.010.

Meier, C., Stolz, H. and Koch, K. (2021). *WP6: Consumer Acceptance, Preferences, and Communication Report on the Outcome of the Quantitative Consumer Survey (D.6.2).* Deliverable D 6.2. ProOrg Project. Available at: https://www.proorgproject.com/_files/ugd/88a346_fb91cbe28f5d42db8317cf49d2a75795.pdf (accessed 27 February 2023).

Naspetti, S. and Zanoli, R. (2009). Organic food quality and safety perception throughout Europe. *Journal of Food Products Marketing* 15(3), 249–266. https://doi.org/10.1080/10454440902908019.

Rampold,C., Schaak, D., Böhm, M. and Schaer, B. (2022). Alternativen in der Milchvermarktung. AMI & Ecozept. Schriftenreihe, Heft 5/2022. Landesamt für Umwelt, Landwirtschaft und Geologie. Freistaat Sachsen. Available at: https://publikationen.sachsen.de/bdb/artikel/39292 (assessed 30 January 2023).

Reiber, C., Wollmeister, M. and Chagunda, M. (2020). Status quo und Determinanten der Kälbervermarktung von ökologischen und konventionellen Milchviehbetrieben in Baden-Württemberg. *Züchtungskunde* 92(5), 320-338. Available at: https://www.zuechtungskunde.de/artikel.dll/zueku-2020-05-reiber-et-al_gy3dqmbrhe3q.pdf?UID=0176E6991BADD5D1A6E296894B686BECC16BF8E45581A0 (assessed 6 July 2023).

Richter, T., Schmid, O., Hempfling, G. and Bahrdt, K. (2004). Ermittlung von zusätzlichen Nachfragepotentialen zur Erhöhung des Absatzes von Biomilch in der Schweiz. FiBL. Available at: https://orgprints.org/6275/1/richter-etal-2004-biomilch.pdf (accessed 27 February 2023).

Sanchez, L., Pérez, M. D. and Paron, J. A. (2020). Chapter 11. HPP in dairy products: impact on quality and applications. In: *Present and Future of High Pressure Processing. A Tool for Developing Innovative, Sustainable, Safe and Healthy Foods* (pp. 245–272). https://doi.org/10.1016/B978-0-12-816405-1.00011-X.

Schaack, D. (2019). Zahlen • Daten • Fakten–die Bio-Branche. *BÖLW (Bund Ökologische Lebensmittelwirtschaft).* (pp. 16–17). Available at: https://www.boelw.de/fileadmin/user_upload/Dokumente/Zahlen_und_Fakten/Broschüre_2019/BOELW_Zahlen_Daten_Fakten_2019_web.pdf (assessed 27 February 2023)

Schaack, D. (2020). Branchenreport. *Bund Ökologische Lebensmittelwirtschaft e.V. (BÖLW).* (pp. 14–29). Available at: https://www.boelw.de/fileadmin/user_upload/

Dokumente/Zahlen_und_Fakten/Broschüre_2020/BÖLW_Branchenreport_2020_web.pdf (accessed 27 February 2023).

Schäufele, I. and Janssen, M. (2021). How and why does the attitude-behavior gap differ between product categories of sustainable food? Analysis of organic food purchases based on household panel data. *Frontiers in Psychology* 12, 595636. https://doi.org/10.3389/fpsyg.2021.595636.

Schiano, A. N., Nishku, S., Racette, C. M. and Drake,M. A. (2022). Parents' implicit perceptions of dairy milk and plant-based milk alternative. *Journal of Dairy Science* 105(6), 4946–4960. https://doi.org/10.3168/jds.2021-21626.

Soil Association (2022). UK organic market trends. Presentation at session "The European Market for Organic Food" at Biofach-Kongress 2022. *Nürnberg Messe*, July 26. Nürnberg. Available at: https://orgprints.org/id/eprint/38258/3/HOLDSTOCK-2022-07-26-UK-MARKET.pdf (assessed 27 February 2023).

Stobbelaar, D. J., Casimir, G., Borghuis, J., Marks, I., Meijer, L. and Zebeda, S. (2007). Adolescents? Attitudes towards organic food: a survey of 15- to 16-year old school children. *International Journal of Consumer Studies* 31(4), 349–356. https://doi.org/10.1111/j.1470-6431.2006.00560.x.

Stolz, H. (2019). Biobarometer Schweiz 2018–was bewegt KonsumentInnen dazu, Bio zu Kaufen? Available at: https://orgprints.org/id/eprint/34436/1/Biobarometer%202018_Stolz.pdf (assessed 27 February 2023).

Stolz, H., Bodini, A., Stolze, M., Hamm, U. and Richter, T. (2009). Lebensmittelqualität aus der Verbraucherperspektive – eine Synthese qualitativer Studien zur Wahr- nehmung und Beurteilung verschiedener Qualitätskriterien bei Öko-Produkten. *Berichte Über Landwirtschaft* 87(1), 153–182.

The Guardian (2019). Cut the wrap. UK dairy farm aims to be first to go single-use plastic. Available at: https://www.theguardian.com/environment/2019/nov/28/scottish-dairy-farm-aims-to-become-uks-first-to-go-single-use-plastic-free-mossgiel-ayrshire (assessed 6 July 2023).

von Meyer-Höfer, M., Nitzko, S. and Spiller, A. (2015). Is there an expectation gap? Consumers' expectations towards organic: an exploratory survey in mature and emerging European organic food markets. *British Food Journal* 117(5), 1527–1546. https://doi.org/10.1108/BFJ-07-2014-0252.

Vaarst, M., Roderick, S., Martin, G., Gunnarson, S., Spengler Neff, A., Bieber, A., Kongsted, A.G. (2020). Potentials, challenges, visions for future European organic animal farming. In: Schmid, O., Johnson, M., Vaarst, M. and Früh, B. (Eds). *Organic Animal Husbandry systems – challenges, performance and potentials. Proceedings of the IAHA Video-Conference on Animal Husbandry on 21. and 22. September 2020, linked to the 20th Organic World Congress of IFOAM 2021*. International Federation of Organic Agriculture Movements - Animal Husbandry Alliance (IFOAM-IAHA), Bonn and Research Institute of Organic Agriculture (FiBL), Frick. Available at: https://orgprints.org/id/eprint/38391/ (assessed 5 July 2023).

Willer, H., Trávnícek, J., Meier, C. and Schlatter, B. (2020). *The world of organic agriculture: statistics & emerging trends 2020*. Forschungsinstitut für Biologischen Landbau (FiBL). International Federation of Organic Agriculture Movements - Organics International. Available at: https://www.fibl.org/fileadmin/documents/shop/5011-organic-world-2020.pdf (assessed 27 February 2023).

Willer, H., Trávnícek, J., Meier, C. and Schlatter, B. (2021). *The world of organic agriculture: statistics & emerging trends* (p. 336). Forschungsinstitut für Biologischen Landbau (FiBL). International Federation of Organic Agriculture Movements - Organics International. Available at: https://www.fibl.org/de/shop/1150-organic-world-2021.html (accessed 27 February 2023).

Willer, H., Trávnícek, J., Meier, C. and Schlatter, B. (2022a). *The World of Organic Agriculture: Statistics & Emerging Trends 2022*. Research Institute of Organic Agriculture FiBL, Frick, and IFOAM. Organics International. Bonn. Available at: https://www.fibl.org/fileadmin/documents/shop/1344-organic-world-2022.pdf (assessed 27 February 2023).

Willer, H., Trávnícek, J. and Schlatter, B. (2022b). The European market for organic food 2020. Presentation at session "The European Market for Organic Food" at Biofach-Kongress 2022. *Nürnberg Messe*, July 26, 2022. Nürnberg. Available at: https://orgprints.org/id/eprint/38258/4/willer-2022-european-market.pdf (assessed 27 February 2023).

Willer, H., Schlatter, B. and Trávnícek, J. (Eds.) (2023). *The World of Organic Agriculture. Statistics and Emerging Trends 2023*. Research Institute of Organic Agriculture FiBL, Frick, and IFOAM. Organics international. Bonn. Available at: http://www.organic-world.net/yearbook/yearbook-2023.html (assessed 27 February 2023).

Zanoli, R. and Naspetti, S. (2002). Consumer motivations in the purchase of organic food: a means-end approach. *British Food Journal* 104(8), 643–653. https://doi.org/10.1108/00070700210425930.

Chapter 14

Organic dairy farming as part of sustainable ecological and social food systems

Mette Vaarst, Aarhus University, Denmark; Adrian Muller, Christian Schader, Florian Leiber and Bernadette Oehen, FiBL, Switzerland; Lindsay K. Whistance, Organic Research Centre, UK; and Stephen Roderick, Duchy College Rural Business School, UK

1 Introduction

Dairy farming has played an important role in many countries over many centuries under a wide range of circumstances. In some countries, the availability of grasslands was a key driver for farming with ruminants whereas, in other situations, the shift from cheap grain imports for food to cheap grain imports for animal feed was a major factor in expanding and intensifying animal production. After the Second World War, dairy farming in Europe gradually became part of a rapidly changing agricultural sector with a strong focus on increasing food production to combat hunger caused by wartime conditions.

High yields and high-quality dairy products became the focus of post-war production, which also stimulated an increasing detachment from grasslands as some governments began to actively support more intensive forms of production away from purely pasture-based systems. The main product of dairy farming was milk, often marketed with government support to make it more affordable (as e.g. described in Switzerland by Moser and Brodbeck (2007)).

http://dx.doi.org/10.19103/AS.2024.0138.15

Milk was also actively promoted as a healthy food, especially in public contexts such as milk provided free to pupils in schools. The productive, high milk-yielding dairy cow was the main focus (Steinfeld, 2006).

Biodynamic farming started to emerge as an alternative to an increasingly industrialized agriculture, including cattle farming. As organic farming developed as a movement during the middle of the twentieth century through the codification of organic principles, standards and regulations, dairy cattle continued to play a key role in those countries that already had a strong dairy sector and a tradition of high consumption of milk products. In some countries, organic milk became an important part of the daily diet of many families, especially those with children, and organic milk was quickly developed as a distinct market segment, as outlined by Schmid et al. in Chapter 13. A similar development has been observed in some regions of the United States of America over the past 30 years (Dimitri and Nehring, 2022). From a farm perspective, conversion to organic agriculture has been aided by the inclusion of ruminants in the farming system given their ability to close nutrient cycles. As grass and forage consumers, they are able to transform marginal grazing areas into milk and meat. They can also consume nitrogen-fixing legumes and provide valuable manure, long considered as a gold standard in promoting soil health, and thereby provide a cornerstone for organic crop production.

The chapters in this book show the great advances in many aspects of organic dairy farming. They demonstrate developments which have successfully created a strong market with growing volumes of produce sold, and which has also become increasingly efficient and productive. Technology is playing an ever-increasing role in agriculture, including organic dairy farming, and Rose et al. (Chapter 10) examine the use of technologies in relation to organic principles. Various chapters also pose questions about the development of specialized, monocultural organic dairy farming, where large dairy companies often control local, regional and international milk markets, with little contact between dairy farmers, local communities and consumers. In many countries, breed diversity is currently very low, and organic dairy farming often relies on high-yielding breeds designed for conventional production (Kargo et al. in Chapter 2).

As in most dairy systems, early separation of calves and mothers has also become a common practice in organic dairy farming. This has been increasingly questioned within farming circles and in wider public debate (Barth et al. Chapter 6). In addition, calves not intended as replacement animals often leave the farms at an early age to be used as conventional veal or are sent to beef fattening farms (Rell et al., 2020). Both of these issues are particularly critical in relation to the core organic principles which call for fairness, health and care, and where no animals should be regarded as surplus to the system. This debate reminds us that dairy cattle are also living, sentient beings, and need to be treated in ways that allow them to live lives worth living, and on which Whistance (Chapter 5) provides important perspectives.

These debates also remind us of the way in which we have developed a view of dairy cattle as 'milk cows', where performance often refers only to high milk yield. We need to rethink this within and beyond the organic sector. All animals – dairy cows, bulls and calves – are living sentient beings which, according to core organic principles, means they should all enjoy a good quality of life as seen from their perspectives. Furthermore, although 'high yield' is not necessarily excluded, cattle are much more than milk producers from a systems perspective. Cattle also produce meat, and, if managed well, they play major roles as 'systems contributors' in supporting the surrounding eco-system through improved soil, air and water quality as well as enhanced biodiversity.

2 Organic dairy farming and sustainability

During recent decades it has become increasingly obvious that the way in which we currently produce and consume food needs to be changed if we want to keep within planetary boundaries and meet the enormous global environmental, social and food systems challenges faced by us and future generations (Richardson et al., 2023). As highlighted by Caro (2022), food systems themselves are responsible for a large share of negative impacts, producing approximately one-third of global anthropogenic greenhouse gas (GHG) emissions, over 90% of humanity's water footprint, and accounting for 40% of all habitable land use, causing 80% of deforestation and 70% of biodiversity loss on land. Our food system also has multiple negative health effects leading to disease and death, with high global rates of obesity combined with high global rates of food insecurity and malnutrition, with more than about 3 billion people still unable to afford a healthy diet. A transition is needed towards more sustainable farming and food production systems and, as is highlighted throughout this book, organic dairy farming needs to be part of this transition.

Sustainability is a complex concept, and the debate around what is 'sustainable' has often been partial and confusing (Schader et al., 2014a). The debate on sustainability is often polarized, e.g. in relation to assessments of productivity on a per product or per animal basis versus a farm- or even landscape-level perspective, where 'sustainability' also covers the contribution of farms to areas such as biodiversity, soil quality, rural livelihoods and landscapes (Eyhorn et al., 2019). Organic dairy farming needs to balance its potential positive contribution to our food systems in terms of high-quality milk, with its impacts on the environment, animal and human welfare including, for example, fair working conditions.

Although many examples provide valuable lessons across countries and regions, site-specific solutions need to be highlighted, as discussed by Grassauer et al. (2022a), who conducted an environmental assessment of an Austrian dairy farm in a less favourable production area (with relatively poor pasture). They concluded that production was feasible under these extensive conditions, allowing for closed production cycles within the region, and producing high-quality dairy products, highlighting the site-specific characteristics central to this outcome. To fully evaluate sustainability, holistic and system-wide measures are also required. Schader et al. (2014b) emphasize the importance of a full systems approach, e.g. by combining full grazing systems, composting of manure and the use of dual-purpose cattle breeds in combination with heat recovery from milk cooling devices, to provide options for mitigating the climate change impact of a dairy farm in a Swiss context. Other authors within this book also present alternative visions for sustainable organic dairy farming, such as the contribution of genetic diversity (Kargo et al in Chapter 2), diversifying and integrating farms (Roderick and Whistance in Chapter 11) and increasing the focus on reducing environmental impacts (Simonson in Chapter 12) through a focus on ecosystem biodiversity.

Despite their national and local importance, issues around equity in the farming and food system, and impacts on surrounding social and ecological systems, are often absent in professional dialogue and literature within the dairy sector. Incorporation of these issues within our debates on the future development of organic farming is of great importance, and experiences in different environments that involve cattle provide us with a broad foundation to learn from each other.

In this book, we have mainly focused on Europe and invited European authors to contribute. The organic dairy farming sector is of considerable importance in Europe and the enormous diversity of systems in different European countries can serve as sources of inspiration and learning beyond Europe. However, it is also clear that any future development pathway for organic or any other form of farming will always need full consideration of broader social, cultural, environmental and climatic diversity and influences. This chapter aims at investigating organic dairy farming in a wider systems context, with particular focus on food and social systems, of which dairy farms, the dairy cattle – including calves and young animals – and the whole dairy sector, are a part. We want to explore how organic dairy systems can contribute to the future sustainable development of these food, social and ecological systems. How can organic dairy cattle, especially in contexts relevant in Europe, contribute to public goods, e.g. in relation to climate change mitigation, biodiversity protection, or developing ecological systems where farming landscapes can contribute positively, or at least not harm, surrounding natural systems, and how can they be part of socially more sustainable, inclusive and fair food systems? Changing complex systems, like a food system, does not happen without significant governance changes and efforts, or without the contribution of all involved actors from farming and non-farming citizens, educational and advisory systems, and market structures.

3 Organic dairy farming as part of surrounding ecological systems

The starting point for developing organic farming has always been based on a holistic systems perspective, with emphasis on ecological cycles and diversity, as stated in the organic principle of ecology: *Organic management must be adapted to local conditions, ecology, culture and scale. Inputs should be reduced by reuse, recycling and efficient management of materials and energy in order to maintain and improve environmental quality and conserve resources* (IFOAM, 2005). The harmony between animals and landscapes – seen from the point of view of both – cannot be overemphasized, and is touched upon from an animal welfare point of view by Whistance (Chapter 5), and from a landscape point of view by Roderick and Whistance (Chapter 11), who write about integration of dairy cattle into crop production as well as into agroforestry systems, and Roderick and Jones (Chapter 3), who discuss diverse species pastures to meet animal and ecological needs. Thamsborg and co-authors (Chapter 9) describe the balance between animals and their landscapes with regard to endoparasites.

It is important to consider the interaction between animals and the ecological systems and think of these as system partners. Healthy, robust animals, which are locally adapted and capable of fully utilizing local resources as they roam in landscapes, are important for a sustainable farming system aimed at enabling animals to meet natural needs. Keeping different age groups together helps to imitate natural group living and can have the advantage of older animals being immune to parasite infections and, in this way, protecting the young. Inter-generational learning and early introduction to natural feeding are other potential advantages. The animals themselves must stay resilient and robust and, in that way, contribute to the entire system. Health, fertility and longevity (as a sign that animals live a life which keeps them healthy) are priorities in organic farming. Likewise, it is a priority to allow animals to have freedoms such as those related to social interaction, exercise and feeding. This is why outdoor living, with access to species-relevant feeding and natural elements in the landscapes, is a priority.

The role of diversity in animal feed contributes strongly to their health, and in this respect, ruminants naturally consume a broad array of grasses, herbs, shrubs and trees, comprising leaves, stems, flowers, fruits, bark and even wood (Van Soest, 1994). With regard to fatty acid composition, an effect of forage biodiversity on the increase of healthy omega-3 fatty acids in milk and meat is evident (Leiber, 2021), which both supports animal health and improves the nutritive value of ruminant products. Schaub et al. (2020) have found that botanical diversity increased yields. Botanical diversity also elicits biochemical diversity, and different bioactive molecules (e.g. phenols, essential oils or alkaloids) have different digestive, metabolic, antioxidative or immunological effects that influence the health and performance of ruminants (Walkenhorst et al., 2020; Zeineldin et al., 2020). Groet et al. (Chapter 8) address the topic of ruminant self-medication, which is possible if they are grazing on diverse meadows and have had the opportunity for learning through trying multiple different plants over time (Rutter, 2010; Villaba et al., 2014; Leiber et al., 2020).

The aims of supporting health, robustness and the synergy between animals and landscapes should also be reflected in the breeds chosen for organic farming and inappropriate breeding goals. It is a challenge that breed diversity has decreased and that the breeds that dominate in organic herds are largely the same as those breeds that have been selected for non-organic, conventional dairy farming. The different goals of organic farming require new breeding strategies that prioritise animals that are healthy and efficient in free-range and pasture-based systems. These strategies are referred to as site-adapted breeding strategies, as they are strongly dependent on location, particularly with regard to nutrition and climatic suitability. Aiming at a balance between cows and calves has led to many choosing crossbreeding or looking for dual or multi-purpose breeds as a breeding strategy, paying attention to both milk and meat, in addition to valuing and acknowledging cattle for their role in maintaining and shaping landscapes.

Seen from a systems perspective, the farming system should be formed in such a way that all animals born there can be fed from farm pasture and live lives worth living. This could also be done through collaboration between different farms, forming a larger farming system that emphasizes landscape rather than artificial boundaries. There are ongoing challenges regarding protein feed and, in particular, feed imports to the farm, often from other countries or continents where crops have been grown specifically to feed cattle and which are typically transported over long distances. Whilst there are opportunities to use imported food by-products as animal feed, thereby contributing to an improved circular economy, there is generally a severe over-dependence on imports which do not contribute to a balanced system.

Although ruminants are grazers by nature, a complete reliance on grazing could become challenging in the face of climate change, resulting in extreme weather patterns such as very dry summers and heavy rainfall affecting pasture quality and availability. This could point to strategies of more integration of grazing within crop rotations. This could be achieved through the utilization of temporary grass-clover leys in a rotation to build soil fertility, thus contributing to a more balanced farm and the opportunity for restoration of natural grasslands.

An emphasis on grazing and local feed may also positively contribute to improved water conservation. In 2010, it was estimated that 20% of the world's freshwater use went to animal farming (Molden et al., 2010), of which only 2% was estimated to be drinking water, with the rest used for crop feed production (Peden et al., 2007). Water for milk production, including what is used during milking and when cleaning the milking equipment, is an unavoidable part of dairy production and the evaluation of water use can be complex.

To summarize, farms involving dairy cattle should be organized so as to feed animals mainly from grassland so that human nutrition can be provided from arable land, leaving space for other forms of life than farm animals, such as soil microorganisms, pollinators and wildlife, and to ensure high animal welfare throughout the animal's lifetime. From an ecological systems point of view, there is a need to strengthen the systems approach across the organic dairy farming sector. As the chapters in the book show, there are an encouraging number of initiatives, developments and examples to learn from.

4 Organic dairy farming as part of surrounding social systems

There are many social aspects of organic dairy farming systems to consider, as they are often part of local communities and social systems. Linkages to local systems are discussed later in this chapter. One of the social systems to consider is clearly the relationships and shared living conditions between humans and animals, in this case, the farmers, cows and calves that share their lives on dairy farms. There are many stories about farmers who have a close connection with their animals that is very much in accordance with the IFOAM principles of care and health. Human relationships with calves, and with the wider herd, are discussed in detail by Barth and co-authors (Chapter 6). Here, we consider social sustainability in terms of the humans who work on farms as well as future farmers.

4.1 Satisfaction in being an organic dairy farmer

Schanz et al. (2023) investigated organic farmers' work satisfaction in multispecies farming in seven different European countries using a mixed-methods approach, including dairy farmers. They concluded that farmers were generally very satisfied despite a high and complex workload. They enjoyed working with complexity and finding solutions to challenging issues. Not only were organic systems considered resource efficient, but they were also appreciated by consumers, and added to farmers' feelings both of being appreciated and of autonomy and personal development. This study suggests that organic farmers thrive through continuous learning and management of complex systems.

Bouttes et al. (2020) surveyed organic dairy farmers following different conversion strategies to organic dairy farming. They found that, despite the uncertainties farmers faced, the general feeling of satisfaction was high regarding economic, agronomic, work-related and other criteria. An interview-based study by Bouttes et al. (2019) concluded that conversion to organic production was perceived to increase the autonomy of the farm and to stimulate learning, especially through a collective dynamic and an open exchange of experience between like-minded farmers. A similar conclusion was reached by Vaarst and Christiansen (2023) in a study of organic dairy farmers learning together about cow–calf contact systems in so-called stable schools (Vaarst et al., 2007). This was very much motivated by the perceived value of continuous development and social learning, which enabled the farmers to develop technical skills, practical solutions, a common understanding of issues related to these systems and a broader feeling of belonging to a supportive group sharing experience and insights. Gonzales-Garcia et al. (2012) highlight that the complex nature of some farming systems requires complex knowledge, which can be rewarding for some farmers. At the other end of the scale, Kalloniemi et al. (2016) investigated stress among Finnish dairy farmers (without mentioning whether they were organic or conventional). They found that shifting government and corporate policies, sometimes negative public views on dairy farming, market unpredictability and high workloads were causing significant levels of burn-out and stress.

Other authors have studied conditions for labourers in big industrial dairy herds (though not specifying if they were organic or conventional). Harrison and Getz (2015) investigated work quality among hired dairy farm labourers in the US State of Wisconsin in relation to farm size. They found no significant differences between large vs. smaller-scale farms, e.g. split shifts appeared on both (e.g. 4:00 am–9:00 am and 4:00 pm–9:00 pm). The study found that many migrant workers work solely as 'milkers'. Based on both in-depth interviews and following them in their work, the authors concluded that their work was often dirty, monotonous, risky and physically arduous. Sexsmith (2022) also studied working conditions for migrant workers on dairy farms in New York State (USA) (although again they did not identify whether these farms were organic or conventional). They identified a range of risks such as the risk of being deported (because they were illegal workers) and hazardous work conditions, e.g. in some cases without having received proper training on the use of (and risks associated with) cleaning and disinfection chemicals. Some large organic farms may also use farm employees solely for milking who are not offered opportunities for more varied work, are not involved in planning or decision-making, or are offered opportunities for education or career progression on the farm. On the other hand, large farms may also have a potential advantage in terms of greater flexibility and better shift structures (Harrison & Getz, 2015). These findings point to the importance of considering working conditions for everybody on the farm, from farm owners and managers to employees of all types, in accordance with the IFOAM principle of fairness (IFOAM, 2005). The various studies highlight the importance of varied and stimulating work, of continuous learning and of social interaction as important future developments and elements of social sustainability.

4.2 Who are the organic farmers now and in the future?

Today's farming and rural landscapes have changed dramatically over recent decades. In many European countries, only a very small percentage of the population is involved in primary agricultural production. This means that few young people have role models within farming, or know about farming practices and what it means to be a farmer, including the responsibility of caring for dairy cattle. Furthermore, structural changes in recent decades have made it difficult to get access to and purchase farmland. In general, farms have grown in size and the number of farms has declined (Eurostat Statistics Explained, 2022). Farmland is frequently bought up by investors or operated by corporations. In animal husbandry, farms tend to be part of integrated businesses with critical decisions often made by senior managers responsible for all aspects of meat, milk or egg supply chains. Farming has become more and more of an industrialized 'production operation' with rigid structures, increasingly operated using digital and other technologies, and with increased bureaucracy (whether from the government or from multinational food processors and large retailers that control much of the food supply chain). In addition, farming and rural areas are considered by some (including rural youth) to be unattractive, 'boring' and with limited social opportunities.

Within the 'family' of farming approaches using organic methods, there are many alternative initiatives, often driven by young people, such as permaculture, agroecology and regenerative farming. Urban farms and an increased focus on rural–urban food systems are gaining increasing interest in many countries and are becoming institutionalized (e.g. https://www.fao.org/in-action/food-for-cities-programme/overview/what-is-the-crfs-programme/en/). There are also initiatives in some countries such as Community Supported Agriculture, share farms, community gardens and farms that offer new and innovative opportunities for land tenure and engagement from those who have not traditionally been involved in agriculture.

In organic dairy farming, a critical question arises as to who takes responsibility for animals all year round, the daily milking and feeding routines and the combination of the many other tasks, such as rotational planning and management, including breeding goals. The question becomes more complex when other system components of resilient farming are considered, such as agroforestry, multifunctional and multispecies farming and interactions with surrounding communities, schools, natural areas, and local food systems.

There are other key questions that go beyond the scope of this book but remain critical to the future shape of organic dairy farming. How will the next generation of organic dairy farmers work, be motivated to enter the sector or stay in it, shape futures, connect to surrounding landscapes and learn to handle all the changes and challenges of climate change, biodiversity loss, digitization, commercialization, etc.? What roles do the current generation of farmers, mentors, advisors, consumers, educators, and researchers play in the development of organic dairy farming? In part, the answers to these questions lie within the current generation of farmers, non-farming citizens, and policymakers, both from within and from outside the dairy sector and within other realms of society. In this respect, communication and a shared sense of direction based on organic principles would appear to be critical elements.

5 Organic dairy farming as part of food systems

In this chapter, we refer to the FAO definition of food systems as the activities around the value-adding chain including production, aggregation, processing, distribution, consumption and disposal of food products that originate from agricultural systems (FAO, 2018), and include all types of actors (Garnett, 2014). Food systems of many different types co-exist in parallel and interact at local, national, regional and global levels (FAO, 2017), with surrounding ecosystems (FAO, 2017) and within social–ecological systems (Colding and Barthel, 2019). The concept links closely with the idea of food security (FAO, 2020), which was explored in depth in 2007 in relation to organic agriculture (FAO, 2016). In this analysis, the four dimensions of food security (availability, accessibility, stability and utilization) were analyzed with the conclusion that organic farming and food systems contributed positively to all of them. Organic agriculture is built on a set of organic principles (IFOAM, 2005), and food sovereignty should be considered as a key part of the fairness principle. Food systems are highly complex, depending on multiple structures related to the social, political, cultural, technological, economic and bio-physical environment, and any transformation needs to involve all relevant actors and systems players (Ingram and Thornton, 2022).

5.1 Dairy cattle in multi-species and multifunctional farms

A transformation of food systems towards more resilient and fair systems also requires a transformation of farming structures. Organic dairy farming tends to be specialized in milk production, occasionally with beef as a by-product. Keeping dairy cattle alongside other animal species (discussed by Roderick and Whistance in Chapter 11) breaks with the dominant trend towards specialization and intensification in the dairy industry.

Benoit et al. (2023) summarized a study of multispecies farming involving 95 farms from six countries and identified four types of farms among study farms. In the one type where dairy cattle were involved, an important feature was on-farm processing and direct sale of milk products which added value to products. This allowed some of the bigger farms with dairy cattle to also have a high number of monogastric animals. In these cases, there was some reliance on purchasing feed with a focus on autonomy in ruminant feeding wherever possible The other three groups did not keep dairy cattle as a main enterprise. Findings were, of course, very context-specific, e.g. the Swedish farms in the study generally did not process milk on the farm. However, the authors found consistent patterns of interactions between farm structure, the way in which animals were kept and managed, how sales and labour were organized as well as the combining of ecological, economic, social and food systems priorities. Schanz et al. (2023) highlight how this could be used in designing policy interventions with the aim to scale up organic farming through diversified, smaller farms.

The option of developing organic dairy farms into more multifunctional farms can also be considered in terms of multiple outputs (OECD, 2001). Grassauer et al. (2022b) assessed the eco-efficiency of Austrian multifunctional dairy farms and found a wide range of outputs and that organic multifunctional dairy farms had a higher mean eco-efficiency score than multifunctional conventional dairy farms. O'Brien et al. (2023) analyzed three different nature-based diversification strategies on dairy farms, including one conversion to organic production. They concluded that this approach resulted in a lower environmental impact, although lower productivity led to higher land use with a negative potential environmental impact (in putting pressure on uncultivated land). The authors acknowledged the huge range of different systems with different possible consequences but overall concluded that 'taking the best elements' could optimize environmental sustainability.

Schader et al. (2008) compared four different European examples of moving to non-commodity multifunctional farms and highlighted the diverse needs and options in different regions. Ulukan et al. (2022) investigated management principles implemented on organic multi-species livestock farms in Europe. Interestingly, farms outperforming their peers do not maximize their diversification at different levels but ensure a balance which ensures complexity is manageable by the farmers.

5.2 The demand for organic dairy products

Organic milk products have been one of the key drivers of organic sector development in many countries. Schmid et al. in Chapter 13 point to consumer motivations such as taste, concern to improve health, environmental protection and animal well-being. Parents consider organic milk as safer, healthier and purer compared to conventional milk, being hormone free and more animal friendly. When consumers purchase milk, the fat content, taste, freshness, price, shelf life and region of origin were found to be the six most relevant criteria. However, Schmid et al. (Chapter 13) also indicate that market structures have recently changed in many places, with increased competition from other labels, especially regional labels, animal welfare labels and special labels such as 'hay-milk' as well as those emphasizing product carbon footprint.

Local food is important and many connect it to particular cultural traditions. Local food is considered authentic, emphasizing the direct connection between producers and consumers (Home et al., 2020). Jia et al. (2024) analyzed Short Food Supply Chains (SFSC) based on reviewing 92 articles on different SFSCs from around the world. Although not focusing particularly on dairy products, they concluded that there was increasing consumer demand for local food products. Ruiz et al. (2024) analyzed issues related to SFSC in Andalusia in Spain (also not specifically addressing dairy products) based on a participative approach. They concluded that SFSC were seen as an alternative to the global food system, but concluded that there was a need for upscaling if they were to play a significant role.

In the case of dairy farming, the need for upscaling is probably greater, considering the current organization of the dairy industry to focus on national, export and global markets. The potential of local food production in regional development was recognized at least 25 years ago as a response to increasingly globalized markets (Marshden et al., 2000). Prache et al. (2022) reviewed organic animal-derived food products in relation to quality and authenticity and highlighted contradictory findings on quality. They concluded that farms and farming practices varied hugely both within the organic and the conventional sectors and that it is hard to generalize comparisons between organic and conventional systems. For example, even though organic farming may reduce the risk of drug residues in food products, outdoor conditions may expose animals to higher environmental contamination, depending on region and practices. The authors highlighted the issue of whether, in the future, potentially increasing demand for organic products would result in more intensification, and whether this would have consequences for product quality or result in consumers seeking plant-based foods as an alternative. In general, dairy production systems have intensified over the past 50 years with resulting environmental and social costs. In many regions, dairy farms have larger herd sizes, and are concentrated in fewer hands with fewer farms overall (Clay et al., 2020). This has driven concerns about the conventionalization of organic agriculture, as discussed by Guptill (2009). In this respect, Kujala et al. (2022) investigated the uneven regional distribution of organic dairy farming in Finland and found that the three regions with the highest organic dairy production generally had a combination of a long organic heritage, a focus on dairy farming, and a significant reliance on subsidies.

5.3 Societal debate about protein from animal sources

Frehner et al. (2022a) investigated national diet recommendations for animal products across five countries (Bulgaria, Malta, the Netherlands, Sweden and Switzerland), and whether they could be produced primarily with domestic feed within a circular food system. Through modelling, they concluded that between 22% (Netherlands) and 47% (Switzerland) of recommended protein contributions could be met by local animal protein production, although consumption of plant-based food should also be increased. They recommended that national dietary recommendations for animal products should be revised and recommended quantities be lowered.

Looking at the Swiss food system, Frehner et al. (2022b) concluded that a shift to more plant-based diets, in combination with a higher share of organic produce, an emphasis on circularity principles, increased domestic production and reduced food waste, would increase the overall sustainability of the Swiss food system. Frehner et al. (2021) concluded that reducing the consumption of different types of meat would contribute to improving different aspects of sustainability. However, based on interviews with Swiss consumers, Richter et al. (2023) concluded that direct measures to reduce meat intake were largely rejected, but other indirect measures such as food system education were more positively received. Frehner et al. (2021) showed that any intervention to move consumption from certain types of food to others (to improve different aspects of sustainability) affects societal groups differently. Interestingly, they showed that nationalities, age groups, and smoking status seemed to have more influence on acceptance of change than income or educational groups, which may contradict concerns about the price of organic food as a factor limiting demand.

Muller and Schader (2017) argue for using GHG emissions as a key sustainability indicator to assess animal farming as a protein source in human nutrition. However, they have also emphasized that the important role of ruminants in farming (as well as ecological) systems should not be forgotten, and a much more nuanced and multidimensional picture is needed. They also point out that alternatives such as a shift towards more monogastric animal production (e.g. poultry) may have even more serious environmental impacts.

5.4 Organic dairy farming in future sustainable rural-urban food systems

In relation to food systems, organic dairy farming needs to consider the increasing urban population (in towns, suburbs and cities), which has grown globally from 746 million in 1950 to 3.9 billion in 2014 and 4.52 billion in 2022 (https://ourworldindata.org/urbanization), a billion more than the rural population. This has stimulated a significant and rapidly increasing interest in urban and peri-urban areas to develop more local food systems. Jennings et al. (2015) analyzed how well-planned and governed city-region food systems could contribute to better food security for different groups of citizens as well as stable incomes for producers, as well as other goals such as circular economies and greater resilience.

There are many different types of urban agricultural systems, from small or medium-sized food communities (with local producers selling to local consumers) to community-supported agriculture, farmers' markets, box schemes (where farms supply directly to consumers) and urban farms. Many such initiatives are based on organic or agro-ecological farming, mostly focusing on vegetable and fruit production, as well as smaller animals (poultry and small ruminants such as sheep). However, there are visions for developing larger city-region food systems which combine urban, peri-urban and rural food production to create comprehensive rural–urban linkages. These could involve organic dairy production e.g via including herbal leys in rotation with the production of crops such as fresh vegetables and fruits within peri-urban areas, though integrating and managing larger animals well can be challenging in smaller and more urban systems where significant investment in infrastructure and knowledgeable animal management would be required.

'Becoming more sustainable' does not only happen at the farm level, and certainly not just at the level of a single commodity, but must involve the surrounding systems in which products are processed, traded and consumed. In line with the principles of agroecology, Vaarst et al. (2018) suggested that food systems could be characterized according to the following criteria: Minimizing the use of external inputs, Extent of internal resource recycling, Resilience, Multifunctionality, Building on complexity and incorporating greater systems integration, Contextuality, Equity and Nourishment.

These criteria should allow for diverse and context-relevant food systems which can be realized through multi-actor planning. This clearly would call for diversification and clever zoning of rural–urban landscapes, which might require a complete re-think of the way in which we keep dairy cattle. This might then create synergies but also challenges in addition to many other changes covered in this book e.g. cow–calf systems and integrated crop–animal farming. There might also be obvious system benefits through stronger links between food and farming systems, for example with regard to circular economies utilizing food by-products as protein feeds for animals.

5.5 Organic dairy farming and the 'feed-the-world' question

The current food system, as a whole, is obviously not functioning effectively and needs systemic transformation. Organic agriculture can contribute to this transformation. It can help to provide sufficient nourishing food for a future growing population while simultaneously reducing environmental impacts from agriculture, or even contributing positively by nourishing the surrounding ecosystems, as well as improving social systems around agriculture. It is a precondition for this that systems are well-designed and implemented to provide effective solutions to current challenges (Muller et al., 2019). The amount of animal feed and the number of animals need to be reduced, and there must be a better balance between animals and land. This needs corresponding changes in diets in many situations, as well as a reduction in food wastage. In addition, there remain major challenges, such as the very large urban populations in megacities, which currently are dependent on the constant transport of huge amounts of food, and ongoing climate changes and the collapse of some ecosystems.

Schader et al. (2015) have modelled feeding animals from grassland and reserving arable land for the production of crops for human consumption. Their model assumed a reduced per capita intake of animal products. On this basis, they showed that sufficient food (in terms of human-digestible energy and protein/calorie ratio) can be produced despite the fact that the efficiency of animal production being reduced as a consequence of relying on a purely grassland-based diet. In other words: cattle can contribute to the food supply without exacerbating 'food-feed' competition. Muller et al. (2019) outlined future scenarios on feeding the world from organic agriculture. They showed that organic farming can contribute only if adequately high amounts of legumes are produced and with significant reductions in food-competing feed use, fewer animal-based products and less food wastage. They emphasized that a shift is needed away from focusing on food availability and production to a focus on accessibility and other aspects of food security and the food system. Based on such studies, our conclusion is that organic dairy cattle and dairy farming have roles to play in the future food systems, in balanced and primarily grazing and roughage-based systems.

6 Organic dairy farming as part of future sustainable development

As highlighted above, the use of permanent and temporary grasslands of high diversity is fundamental for organic dairy cattle (see Roderick and Jones, Chapter 3). Grasslands play a potentially significant role in the mitigation of climate change (Leiber et al., 2020). The debate on GHG emissions and dairy farming is complex and controversial. One of the controversies concerns the differences between organic and non-organic farming rather than dairy farming per se. This was illustrated by Muller et al. (2016), in response to the so-called McGee study (2015), where the omission of synthetic mineral fertilizer production in the comparison skewed the results.

Landert et al. (2020) looked at the environmental impacts of agroecological practices (which can be considered comparable to organic farming), on different types of farms, including dairy farms. They tested the results using three different sustainability tools and concluded that, while biodiversity and water quality were consistently better in agroecological farms, GHG emissions varied and were case-dependent. Many studies point to organic dairy production as less efficient and therefore more environmentally damaging. However, a modelling study on data from 3,074 French dairy farms by Lambotte et al. (2023) concluded that organic farms had a lower carbon footprint than conventional farms.

There is no doubt that protecting the large carbon stocks in grasslands is highlighted as essential to avoid contributing to climate change due to potential CO_2 release from these sources if they are not appropriately managed. Many studies have shown that grasslands and agroforestry systems have the potential to sequester carbon, despite the context dependency, time limitations and potential reversibility of soil carbon sequestration processes. Godde et al. (2020) have pointed to the importance of promoting sequestration, especially on degraded lands. However, Godde et al. (2020) have also warned about potentially unrealistic expectations of soil carbon management in grazing systems, since mitigation estimates are associated with large uncertainties and will ultimately depend on many factors in larger systems, not least time frames. In conclusion, much scientific evidence across a range of different contexts seems to echo that 'It is not about the cow, but about the how'.

More broadly, concepts like resilience, diversification and learning in communities of practice are seen as key to the future development of food and farming systems. These concepts are deeply embedded in organic principles which highlight diversity, integration, resilience and fair social systems for humans and animals. Diversity and diversification at different levels have been highlighted throughout this and other chapters.

It is abundantly clear that there are no current farming systems that can be considered sustainable without the adoption of relevant supporting policies and broader social change. Eyhorn et al. (2019) suggest four policy interventions which could help drive the transformation of agriculture through the development of organic agriculture

1 supporting transformative systems while improving their performance;
2 stimulating demand for more sustainable products;
3 incentivizing incremental improvements in all systems; and
4 improving legal requirements and industry norms.

These interventions play out at different levels of the food system, but generally suggest an important role for organic farming, including organic dairy farming, in driving this transformation. Clay et al. (2020) call for policy approaches that consider dairy system transitions within broader processes of social–environmental change taking environmental impact, animal welfare, rural development (social, cultural and economic dimensions) and human health into account.

7 Conclusion

Organic agriculture should provide everyone involved with a good quality of life, ... This principle insists that animals should be provided with the conditions and opportunities of life that accord with their physiology, natural behaviour and well-being. (From the IFOAM principle of Fairness)

Dairy cattle have been ecologically important animals for farms and rural landscapes in many places and over long periods of time. They can provide us with food and other products as meaningful and valuable parts of food systems. All animals born on the farm should matter and be treated fairly according to organic principles. Dairy cattle can thrive in local, circular ecological systems (e.g. with added features such as agroforestry) and in balance with the landscapes to which they contribute. Circular systems require feeding minimal amounts of human edible concentrates to cows (as argued by e.g. Steinfeld, 2006 and discussed by Faux et al., 2022). Organic dairy systems have multifunctional roles, particularly in marginal areas with limited amounts of arable land, and there is good potential for integration within rotational arable systems with grass leys.

In recent decades, organic dairy systems increasingly followed the productivist narratives of conventional agriculture that focus on maximizing milk yield. However, based on studies and arguments as outlined above, we find it relevant to consider how future development could go in another direction: organic circular, grassland-based dairy systems could instead be a role model for many conventional farms. In conclusion, in exploring the potential future development pathways for organic farming, we want to highlight the following:

Firstly, in Europe in particular, dairy farming has long been a central production system in many regions, partly based on the ample availability of grasslands. The intensification and industrialization that have occurred over the last 50 years have created dependencies on concentrate feed use and housing systems which have had negative environmental, health and welfare impacts. However, these developments should not detract from the role dairy cattle can play in more sustainable forms of production.

Secondly, organic dairy cattle can play a central role in future circular, organic food systems, based on locally available grassland resources and aiming at utilizing those efficiently in organic systems which can simultaneously provide healthy and high-quality food, reduce environmental impacts, increase animal welfare and maintain landscapes. Stocking rates should be locally adapted to the ecosystem contexts in terms of the availability of permanent and temporary grasslands. In many areas, organic animal production needs to reduce its reliance on importing feed and using cropland-based concentrate feed. This requires site-adapted breeding strategies, adjusted yield expectations, changes in consumption patterns of animal products, and enhanced grassland management. Such changes also call for appropriate governance strategies and consumer awareness.

Thirdly, current organic dairy systems often rely on many animals leaving the system to be fattened and slaughtered conventionally, with a corresponding welfare and environmental impact. There needs to be a focus on ways to reduce this, including options such as improving animal longevity, herds combining calves and beef cattle, extended lactations and later first-calving ages, but also policy- and consumer-focused measures, aiming at more balanced meat–milk consumption consistent with the production characteristics of dairy production systems.

Fourthly, particular focus should be put on changing potentially adverse incentive schemes that stem from product-based environmental footprints, in particular those related to GHG emissions. Organic dairy farming should be organized around the overall contribution of animals to food systems. The performance of dairy cattle should not be assessed only on the basis of one product (milk), but rather with respect to the totality of nutrients produced from a landscape over several years. This more holistic perspective will inform the optimal design of organic dairy production as an integral part of agricultural landscapes.

Looking more generally, the worldwide organic movement has committed to the four core principles of health, ecology, fairness and care (IFOAM, 2005). These continue to serve as a robust guide to future development, calling for consideration, care and commitment at all levels of farming and food systems across time and space.

8 Where to look for further information

FiBL Info centre: FiBL provides a wide range of information covering various aspects of organic agriculture. The Info centre consolidates FiBL's publications, both digital and print, multimedia resources, news, press releases, event schedules and more. Stay informed about FiBL's activities and products by accessing the Info centre. https://www.fibl.org/en/info-centre#c73696

Organic Research Centre: Find the results and publications from the research conducted at the Organic Research Centre here: https://www.organicresearchcentre.com/research/

ICROFS is based on an organic philosophy and the established organic principles. The centre is established by the Ministry of Environment and Food of Denmark. The research both in Denmark and internationally is described here: https://icrofs.dk/en/research

9 References

Benoit, M., et al. (2023). Interactions between animal enterprises and marketing strategies shape organic multispecies farming systems [Article]. *Agronomy for Sustainable Development*, 43(6), 15, Article 77. https://doi.org/10.1007/s13593-023-00930-8

Bouttes, M., et al. (2020). *Conversion to organic farming increases dairy farmers' satisfactionindependently of the strategies implemented. Agronomy for Sustainable Development*, 40, 1–12. https://doi.org/10.1007/s13593-20-00616-5

Bouttes, M., et al. (2019). Converting to organic farming as a way to enhance adaptive capacity. *Organic Agriculture*, 9, 235–247. https://doi.org/10.1007/s13165-018-0225-y

Caro, D. (2022). Sustainability of food systems and reinforcement of the science-policy interface: Re-focusing on priorities. *Resources, Environment and Sustainability*, 11, 100100. https://doi.org/10.1016/j.resenv.2022.100100

Colding, J. & Barthel, S. (2019). Exploring the social-ecological systems discourse 20 years later. *Ecology and Society*, 24(1), 2. https://doi.org/10.5751/ES-10598-240102

Clay, N., et al. (2020). Dairy intensification: Drivers, impacts and alternatives. *Ambio* 49, 35–48. https://doi.org/10.1007/s13280-019-01177-y

Dimitri, C. & Nehring, R. (2022). Thirty years of organic dairy in the United States: the influences of farms, the market and the organic regulation [Article]. *Renewable Agriculture and Food Systems*, 37(6), 588–602. https://doi.org/10.1017/s1742170522000254

Eurostat Statistics Explained. (2022). Farms and farmland in the European Union – statistics. https://ec.europa.eu/eurostat/statistics-explained/index.php?title=Farms_and_farmland_in_the_European_Union_-_statistics#The_evolution_of_farms_and_farmland_between_2005_and_2020 (accessed 18th July 2024; planned update 24th November 2025)

Eyhorn, F., et al. (2019). Sustainability in global agriculture driven by organic farming [Editorial Material]. *Nature Sustainability*, 2(4), 253–255. https://doi.org/10.1038/s41893-019-0266-6

FAO. (2016). Extract from the report Sustainable agricultural development for food security and nutrition: what roles for livestock? *Summary and Recommendations*, 12. https://www.fao.org/fileadmin/user_upload/hlpe/hlpe_documents/HLPE_S_and_R/HLPE_2016_Sust-Agr-Dev-FSN-Livestock_S-R_EN.pdf (accessed 31st March 2024)

FAO. (2017). HLPE report 12. Nutrition and food systems. In *A report by the High Level Panel of Experts on Food Security and Nutrition of the Committee on World Food Security*. FAO, Rome, p. 152. https://www.fao.org/3/I7846E/i7846e.pdf (accessed 31st March 2024)

FAO. (2018). Sustainable food systems Concept and framework, p. 8. https://www.fao.org/3/ca2079en/CA2079EN.pdf (accessed 31st March 2024)

FAO. (2020). HLPE report 15. Food security and nutrition: building a global narrative towards 2030 (A report by the High Level Panel of Experts on Food Security and Nutrition of the Committee on World Food Security, Issue. https://www.fao.org/3/ca9731en/ca9731en.pdf (accessed 31st March 2024)

Faux, A. M., et al. (2022). Feed autonomy in organic cattle farming systems: a necessary but not sufficient lever to be activated for economic efficiency. *Organic Agriculture*, 12, 335–352. https://doi.org/10.1007/s13165-021-00372-0

Frehner, A., et al. (2022a). The compatibility of circularity and national dietary recommendations for animal products in five European countries: a modelling analysis on nutritional feasibility, climate impact, and land use [Article]. *Lancet Planetary Health*, 6(6), E475–E483. https://www.zora.uzh.ch/id/eprint/227980/1/PIIS254251962200119X_compatibility_frehner.pdf

Frehner, A., et al. (2022b). Consumer strategies towards a more sustainable food system: insights from Switzerland [Article]. *American Journal of Clinical Nutrition*, 115(4), 1039–1047. https://doi.org/10.1093/ajcn/nqab401

Frehner, A., et al. (2021). How food choices link sociodemographic and lifestyle factors with sustainability impacts [Article]. *Journal of Cleaner Production*, 300(14), Article 126896. https://doi.org/10.1016/j.jclepro.2021.126896

Garnett, T. (2014). Three perspectives on sustainable food security: efficiency, demand restraint, food system transformation. What role for life cycle assessment? *Journal of Cleaner Production*, 73, 10–18. https://doi.org/10.1016/j.jclepro.2013.07.045

Godde, C. M., et al. (2020). Soil carbon sequestration in grazing systems: managing expectations [Article]. *Climatic Change*, 161(3), 385–391. https://doi.org/10.1007/s10584-020-02673-x

González-García, E., et al. (2012). The complex nature of mixed farming systems requires multidimensional actions supported by integrative research and development efforts [Article]. *Animal*, 6(5), 763–777. https://doi.org/10.1017/s1751731111001923

Grassauer, F., et al. (2022a). Environmental assessment of austrian organic dairy farms with closed regional production cycles in a less favorable production area [Article]. *Frontiers in Sustainable Food Systems*, 6(12), Article 817671. https://doi.org/10.3389/fsufs.2022.817671

Grassauer, F., et al. (2022b). Assessing and improving eco-efficiency of multifunctional dairy farming: the need to address farms' diversity [Article]. *Journal of Cleaner Production*, 338(13), Article 130627. https://doi.org/10.1016/j.jclepro.2022.130627

Guptill, A. (2009). Exploring the conventionalization of organic dairy: trends and counter-trends in upstate New York [Article]. *Agriculture and Human Values*, 26(1–2), 29–42. https://doi.org/10.1007/s10460-008-9179-0

Harrison, J. L. & Getz, C. (2015). Farm size and job quality: mixed-methods studies of hired farm work in California and Wisconsin [Article]. *Agriculture and Human Values*, 32(4), 617–634. https://doi.org/10.1007/s10460-014-9575-6

Home, R., et al. (2020). The importance of being local: the role of authenticity in the concepts offered by non-themed domestic restaurants in Switzerland [Article]. *Sustainability*, 12(9), 16, Article 3907. https://doi.org/10.3390/su12093907

IFOAM. (2005). The organic principles. https://www.ifoam.bio/why-organic/shaping-agriculture/four-principles-organic (accessed 31st March 2024)

Ingram, J. & Thornton, P. (2022). What does transforming food systems actually mean? *Nature Food*, 3(11), 881–882. https://doi.org/10.1038/s43016-022-00620-w

Jennings, S., et al. (2015). Food in an Urbanized World: The Role of City Region Food Systems in Resilience and Sustainable Development. FAO and ISU. http://www.fao.org/fileadmin/templates/FCIT/documents/Food_in_an_Urbanised_World_Report_DRAFT_February_2015.pdf

Jia, F., et al. (2024). Promoting resilient and sustainable food systems: a systematic literature review on short food supply chains [Article]. *Journal of Cleaner Production*, 435(15), Article 140364. https://doi.org/10.1016/j.jclepro.2023.140364

Kallioniemi, M. K., et al. (2016). Stress and burnout among Finnish dairy farmers [Article]. *Journal of Agromedicine*, 21(3), 259–268. https://doi.org/10.1080/1059924x.2016.1178611

Kujala, S., et al. (2022). Factors affecting the regional distribution of organic farming [Article]. *Journal of Rural Studies*, 92, 226–236. https://doi.org/10.1016/j.jrurstud.2022.04.001

Lambotte, M., et al. (2023). Organic farming offers promising mitigation potential in dairy systems without compromising economic performances [Article]. *Journal of Environmental Management*, 334(15), Article 117405. https://doi.org/10.1016/j.jenvman.2023.117405

Landert, J., et al. (2020). Assessing agro-ecological practices using a combination of three sustainability assessment tools [Article]. *Landbauforschung-Journal of Sustainable and Organic Agricultural Systems*, 70(2), 129–144. https://doi.org/10.3220/lbf1612794225000

Leiber, F., et al. (2020). The relevance of feed diversity and choice in nutrition of ruminant livestock [Article]. *Landbauforschung-Journal of Sustainable and Organic Agricultural Systems*, 70(1), 35–38. https://doi.org/10.3220/lbf1592393539000

Leiber, F. (2021). Let them graze! Potentials of ruminant production outside the feed-food competition. In E.-H. Scialabba (Ed.), *Managing Healthy Livestock Production and Consumption*. Elsevier: Academic Press, pp. 137–148.

Marsden T K, et al. (2000) "Food supply chain approaches: exploring their role in rural development" Sociologia Ruralis 40 424-438

McGee, J. (2015). Does certified organic farming reduce greenhouse gas emissions from agricultural production? *Agriculture and Human Values*, 32(2), 255–263. https://doi.org/10.1007/s10460-014-9543-1

Molden, D., et al. (2010). Improving agricultural water productivity between optimism and caution. Agricultural Water Management, 97, 528-535. http://dx.doi.org/10.1016/j.agwat.2009.03.023

Moser, P. & Brodbeck, B. (2007). *Du lait pour tous: portrait en images, documents et analyses de l'économie et de la politique laitières en Suisse au 20e siècle*. Hier + Jetzt, Verlag. für Kultur und Geschichte, Baden, Switzerland.

Muller, A., et al. (2016). Does certified organic farming reduce greenhouse gas emissions from agricultural production? Comment on the McGee study [Article]. *Agriculture and Human Values*, 33(4), 943–947. https://doi.org/10.1007/s10460-016-9706-3

Muller, A. & Schader, C. (2017). Efficiency, sufficiency, and consistency for sustainable healthy food [Editorial Material]. *Lancet Planetary Health*, 1(1), E13–E14. https://doi.org/10.1016/s2542-5196(17)30012-8

Muller, A., et al. (2019). Strategies for feeding the world more sustainably with organic agriculture [Article]. *Nature Communications*, 8(13), Article 1290. https://doi.org/10.1038/s41467-017-01410-w

O'Brien, D., et al. (2023). Environmental impact of grass-based cattle farms: a life cycle assessment of nature-based diversification scenarios [Article]. *Resources Environment and Sustainability*, 14(14), Article 100126. https://doi.org/10.1016/j.resenv.2023.100126

OECD. (2001). Multifunctionality. Towards an analytical framework. Report, p. 157. https://www.oecd-ilibrary.org/docserver/9789264192171-en.pdf?expires=1711886231&id=id&accname=ocid177114&checksum=3191284486B474A6ACC56B99A7859944 (accessed 31st March 2024)

Peden, D., et al. (2007). Water and livestock for human development. In D. Molden (Ed.) Water for food, water for life: a comprehensive assessment of water management in agriculture (pp. 485-515). Interantional Water Management Institution, Colombo, Sri Lanka. Earthscan, London.

Prache, S., et al. (2022). Review: quality and authentication of organic animal products in Europe [Review]. *Animal*, 16(13), Article 100405. https://doi.org/10.1016/j.animal.2021.100405

Rell, J., et al. (2020). Stakeholders' perceptions of the challenges to improving calf health and reducing antimicrobial use in Swiss veal production. Preventive Veterinary Medicine, 179, 11, Article 104970. https://doi.org/10.1016/j.prevetmed.2020.104970

Richardson, K., et al. (2023). Earth beyond six of nine planetary boundaries. *Science Advances*, 9, eadh2458. https://doi.org/10.1126/sciadv.adh2458

Richter, S., et al. (2023). Acceptance of meat reduction policies in Switzerland [Article]. *Iscience*, 26(3), 21, Article 106129. https://doi.org/10.1016/j.isci.2023.106129

Ruiz, A. M., et al. (2024). Participative assessment of agroecological short food supply chains and proposals for improvement: a case study of Andalusia (Spain) [Article; Early Access]. *Agroecology and Sustainable Food Systems*, 28. https://doi.org/10.1080/21683565.2024.2311124

Rutter, S. M. (2010). Review: grazing preferences in sheep and cattle: implications for production, the environment and animal welfare. *Canadian Journal of Animal Science*, 90, 285–293.

Schader, C., et al. (2014a). Scope and precision of sustainability assessment approaches to food systems [Article]. *Ecology and Society*, 19(3), 15, Article 42. https://doi.org/10.5751/es-06866-190342

Schader, C., et al. (2014b). Quantification of the effectiveness of greenhouse gas mitigation measures in Swiss organic milk production using a life cycle assessment approach [Article]. *Journal of Cleaner Production*, 73, 227–235. https://doi.org/10.1016/j.jclepro.2013.11.077

Schader, C., et al. (2007, September 17–20). *Societal demand for commodity and noncommodity outputs: a regional perspective* [Rural landscapes and agricultural policies in Europe]. MEA-Scope Workshop, Florence, Italy.

Schader, C., et al. (2015). Impacts of feeding less food-competing feedstuffs to livestock on global food system sustainability [Article]. *Journal of the Royal Society Interface*, 12(113), 12, Article 20150891. https://doi.org/10.1098/rsif.2015.0891

Schader, C., et al. (2008). Uptake of agri-environmental measures on organic and conventional farms in Switzerland [Article]. *Agrarforschung*, 15(10), 506–511.

Schanz, L., et al. (2023). High work satisfaction despite high workload among European organic mixed livestock farmers: a mixed-method approach [Article]. *Agronomy for Sustainable Development*, 43(1), 15, Article 4. https://doi.org/10.1007/s13593-022-00852-x

Schaub, S., et al. (2020). Plant diversity effects on forage quality, yield and revenues of semi-natural grasslands. *Nature Communications*, 11, 768.

Sexsmith, K. (2022). The embodied precarity of year-round agricultural work: health and safety risks among Latino/a immigrant dairy farmworkers in New York [Article]. *Agriculture and Human Values*, 39(1), 357–370. https://doi.org/10.1007/s10460-021-10252-8

Steinfeld, H. (2006). *Livestock's long shadow: environmental issues and options*. Food and Agriculture Organization of the United Nations, Rome.

Ulukan, D., et al. (2022). Positive deviant strategies implemented by organic multi-species livestock farms in Europe. *Agricultural Systems*, 201, 103453.

Vaarst, M. & Christiansen, I. A. (2023). Three years of situated social learning and development of diverse cow-calf contact systems in Danish organic dairy farms [Article]. *Journal of Dairy Science*, 106(10), 7020–7032. https://doi.org/10.3168/jds.2022-22755

Vaarst, M., et al. (2018). Exploring the concept of agroecological food systems in a city-region context [Article]. *Agroecology and Sustainable Food Systems*, 42(6), 686–711. https://doi.org/10.1080/21683565.2017.1365321

Vaarst, M., et al. (2007). Danish stable schools for experiential common learning in groups of organic dairy farmers [Article]. *Journal of Dairy Science*, 90(5), 2543–2554. https://doi.org/10.3168/jds.2006-607

Van Soest, P. J. (1994). *Nutritional ecology of the ruminant*. Cornell University Press, Ithaca. 488 pp.

Villaba, J. J., et al. (2014). Ruminant self-medication against gastrointestinal nematodes: evidence, mechanism, and origins. *Parasite*, 21, 31.

Walkenhorst, M., et al. (2020). A multicomponent herbal feed additive improves somatic cell counts in dairy cows - a two stage, multi center, placebo controlled long term on-farm trial. *Journal of Animal Physiology and Animal Nutrition*, 104, 439–452.

Zeineldin, M. M., et al. (2020). Potential contribution of plants bioactive in ruminant productive performance and their impact on gastrointestinal parasites elimination. *Agroforestry Systems*, 94, 1415–1432.

Index

www.ingramcontent.com/pod-product-compliance
Ingram Content Group UK Ltd.
Pitfield, Milton Keynes, MK11 3LW, UK
UKHW021826100926
8556IPUK00003B/173